T0335780

Space-Time Symmetry and Quantum Yang–Mills Gravity

How space-time translational gauge symmetry enables
the unification of gravity with other forces

ADVANCED SERIES ON THEORETICAL PHYSICAL SCIENCE
A Collaboration between World Scientific and Institute of Theoretical Physics

ISSN: 1793-1495

Series Editors: Dai Yuan-Ben, Hao Bai-Lin, Su Zhao-Bin
(Institute of Theoretical Physics Academia Sinica)

Published

Advanced Series on Theoretical Physical Science

Volume **11**

Space-Time Symmetry and Quantum Yang–Mills Gravity

How space-time translational gauge symmetry enables the unification of gravity with other forces

Jong-Ping Hsu
University of Massachusetts Dartmouth, USA

Leonardo Hsu
University of Minnesota, USA

 World Scientific

NEW JERSEY · LONDON · SINGAPORE · BEIJING · SHANGHAI · HONG KONG · TAIPEI · CHENNAI

Published by

World Scientific Publishing Co. Pte. Ltd.

5 Toh Tuck Link, Singapore 596224

USA office: 27 Warren Street, Suite 401-402, Hackensack, NJ 07601

UK office: 57 Shelton Street, Covent Garden, London WC2H 9HE

British Library Cataloguing-in-Publication Data
A catalogue record for this book is available from the British Library.

Advanced Series on Theoretical Physical Science — Vol. 11
SPACE-TIME SYMMETRY AND QUANTUM YANG–MILLS GRAVITY
How Space-Time Translational Gauge Symmetry Enables the Unification
of Gravity with Other Forces

Copyright © 2013 by World Scientific Publishing Co. Pte. Ltd.

All rights reserved. This book, or parts thereof, may not be reproduced in any form or by any means, electronic or mechanical, including photocopying, recording or any information storage and retrieval system now known or to be invented, without written permission from the Publisher.

For photocopying of material in this volume, please pay a copying fee through the Copyright Clearance Center, Inc., 222 Rosewood Drive, Danvers, MA 01923, USA. In this case permission to photocopy is not required from the publisher.

ISBN 978-981-4436-18-2

Typeset by Stallion Press
Email: enquiries@stallionpress.com

Printed in Singapore

Taiji Symmetry

Taiji Symmetry

Taiji Symmetry

Taiji Symmetry

太极对称

To

The Physics Alumni, George Leung and the University of Massachusetts Dartmouth. Their support made possible the research on space-time symmetry and quantum Yang-Mills gravity, as well as its unification with other forces.

太極對稱

Quantum Yang-Mills Gravity

Quantum Yang-Mills Gravity

Quantum Yang-Mills Gravity

One new result of quantum Yang–Mills gravity is that the apparent curvature of space-time appears to be simply a manifestation of the flat space-time translational gauge symmetry for the motion of quantum particles in the classical limit.

Jong-Ping Hsu and Leonardo Hsu, 'Overview', p. xxi, (Jan. 2013).

The most glaring incompatibility of concepts in contemporary physics is that between Einstein's principle of general coordinate invariance and all the modern schemes for a quantum-mechanical description of nature.

Freeman J. Dyson, "Missed Opportunities", Bull. of Amer. Math. Soc., 78, Sept. 1972.

Evidently, the usual statements about future positions of particles, as specified by their coordinates, are not meaningful statements in general relativity. This is a point which cannot be emphasized strongly enough. . . . Expressing our results in terms of the values of coordinates became a habit with us to such a degree that we adhere to this habit also in general relativity, where values of coordinates are not per se meaningful.

Eugene P. Wigner, *Symmetries and Reflections* (The MIT Press, 1967), pp. 52–53.

This talk has, I think, something to do with general relativity, and I don't understand general relativity. You should therefore take these speculations with some caution.

Victor F. Weisskopf, on "Mach's Principle and the Existence of Gravity Waves", Lec. in Theor. Phys., (Ed. W. E. Brittin, B. W. Downs and J. Downs, Interscience, 1961), p. 60.

The important thing is to be able to make predictions about images on the astronomers' photographic plates, . . . and it simply doesn't matter whether we ascribe these predictions to the physical effect of gravitational fields on the motion of planets and photons or to a curvature of space and time. (The reader should be warned that these views are heterodox and would meet with objections from many general relativists.)

Steven Weinberg, *Gravitation and Cosmology* (John Wiley & Sons, 1972) p. 147.

Preface

It began with the idea that the four-dimensional space-time for inertial frames could be formulated solely on the basis of a single postulate, i.e., the principle of relativity for physical laws. This idea, named 'taiji relativity,' was stimulated by a university term paper by one of the authors (LH) and discussed in the book *A Broader View of Relativity*. Although this foundational idea in and of itself was not new, it turned out that the framework of taiji relativity could also be generalized to derive explicit space-time transformations for non-inertial frames that simplified to the Lorentz transformations in the limit of zero acceleration, something never before accomplished. The taiji framework also supported the quantization of the physical fields of the strong and electroweak interactions, as well as all established conservation laws. Thus, with faith in the Yang–Mills idea and taiji relativity, it was not unreasonable to think that such a framework might also accommodate the gravitational interaction. This idea was additionally bolstered by the fact that conservation of energy and momentum is a consequence of the space-time translational symmetry of the action for a physical system, and that the energy–momentum tensor might be related to the source of the gravitational field.

The idea that translational symmetry in flat space-time could be the gauge symmetry for gravitational potential fields has emerged slowly. At present, such an idea is unorthodox and is definitely a minority viewpoint, but it is an extremely intriguing one! Why should gauge symmetry in flat space-time be so successful for modeling all known interactions except gravity?

At first, deriving field equations and an equation of motion for classical objects that were logically and experimentally consistent seemed an insurmountable difficulty. Enlightenment came from the observation that, in the classical limit, translational gauge symmetry requires that all wave equations (with the exception of the gravitational wave equation) reduce to Hamilton–Jacobi type equations with an effective Riemannian metric tensor. In other words, even though the underlying physical space-time is flat, all objects behave as if they were in a curved space-time.

Many more years of search and research finally resulted in the discovery of a gauge invariant action within a generalized Yang–Mills framework that was consistent with all classical tests of gravity. Almost all the advantages of the Yang–Mills idea were then available to tackle such a quantum gravity, except that the gravitational coupling constant was not dimensionless.

Einstein's theory of gravity is formulated in the more complicated framework of curved space-time and as such appears to be too general to be compatible with the quantization of fields and the conservation of energy–momentum. Yang–Mills gravity represents an alternative approach that brings gravity back into the arena of gauge fields in flat space-time, in which the quantization of fields and the conservation of energy–momentum are already well-established.

The aim of this book is to provide a treatment of quantum Yang–Mills gravity with an emphasis on its ideas and the experimental evidence that might support it. The two big ideas of Yang–Mills gravity are:

(1) There exists a fundamental space-time symmetry framework that can encompass all interactions in physics, including gravity, and is valid in both inertial and non-inertial frames of reference. This framework (the 'taiji symmetry' framework) is based on a flat space-time with arbitrary coordinates for all reference frames and can accommodate a generalized Yang–Mills idea for external gauge groups and a totally unified model of all interactions, along the lines of the ideas of Glashow, Salam, Ward, and Weinberg.

(2) The gravitational field is the manifestation of flat space-time translational gauge symmetry, which is the only exact gauge symmetry. All internal gauge symmetries are violated subtly by gravity, so that classical objects move in gravitational fields as if they were in a curved space-time. Furthermore, the space-time translational gauge symmetry enables the unification of Yang–Mills gravity with the electroweak theory and quantum chromodynamics.

This monograph summarizes our collaborative research and striving to bring forth new viewpoints and ideas over the past two decades since that university term paper. We hope the book will be useful for students and researchers interested in a unified picture of nature. The present work would not have been possible were it not for the publication of our earlier volumes, *A Broader View of Relativity, General Implications of Lorentz and Poincaré Invariance* (World Scientific, 2006) and *100 Years of Gravity and Accelerated Frames, The Deepest Insight of Einstein and Yang–Mills* (with D. Fine, World Scientific, 2005).

We would like to thank Dana Fine, Howard Georgi and colleagues for their interest and discussions, and we are indebted to Kazuo Cottrell and Leslie Hsu for the reading and proofing of many chapters. One of us (JP) greatly benefitted from discussions with C. C. Chen, J. M. Wang, T. S. Cheng and W. T. Yang (Taida), T. Y. Wu (Tsing Hua and Academia Sinica), S. Okubo and R. E. Marshak (University of Rochester), H. W. Huang (Cornell), and E. C. G. Sudarshan, C. B. Chiu, D. A. Dicus and T. N. Sherry (Center for Particle Theory, Austin) in earlier years. Special acknowledgment is made to Bonnie Chiu Hsu for her tireless support throughout our lives.

Jong-Ping Hsu (University of Massachusetts Dartmouth),
Leonardo Hsu (University of Minnesota, Twin Cities)

Contents

Overview

0-1. Motivation: 'The most glaring incompatibility of concepts in contemporary physics'

Despite the success of Einstein's theory of gravity,[1] there are difficult and long-standing problems with how general relativity fits into the big picture of physics. In the January 1972 Josiah Willard Gibbs Lecture, given under the auspices of the American Mathematical Society, Freeman Dyson pointed out one of the problems with the principle of general coordinate invariance,[2] saying "The most glaring incompatibility of concepts in contemporary physics is that between Einstein's principle of general coordinate invariance and all the modern schemes for a quantum-mechanical description of nature."

General coordinate invariance implies that space-time points can be given arbitrary labels.[3] Because the presence of gravity implies a change in the space-time metric, the physical separation of two bodies cannot be fixed in the sense that one can no longer talk about two particles having a fixed distance between each other. Similarly, the relative rates of two clocks at two different locations are related in an arbitrary way and hence, time has no physical meaning. The metric tensor $g_{\mu\nu}$ is needed so that physical laws described by space-time can be independent of the coordinate systems.

In a lucid discussion of coordinates and momenta in general relativity, Wigner wrote 'The basic premise of this theory [the general theory of relativity] is that coordinates are only auxiliary quantities which can be given arbitrary values for every event. Hence, the measurement of position, that is, of the space coordinates, is certainly not a significant measurement if the postulates of the general theory are adopted: the coordinates can be given any value one wants. The same holds for momenta. Most of us have struggled with the problem of how, under these premises, the general theory of relativity can make meaningful statements and predictions at all.'[3]

This property of coordinates is at odds with the physically meaningful and operationally defined space-time of quantum mechanics and gauge field theories with Poincaré invariance. Coordinate transformations in flat space-time form a much smaller group than that in curved space-time, and space and time coordinates in inertial frames can be operationally defined. Physics based on

Poincaré invariance is endowed with great mathematical simplicity and elegance and supported by uncountable experiments. In particular, the concept of local commutativity, which is essential for quantum field theories, holds only if the space-time framework includes the Poincaré group.

A second problem with Einstein's principle of general coordinate invariance is that it does not include a conservation law for energy–momentum. From the group-theoretic viewpoint in mathematics, Einstein's gravity is based on the group formed by general coordinate transformations in curved space-time. This is a Lie group with a continuously infinite number of infinitesimal generators. In 1918, Noether's second theorem[4] suggested that energy could be conserved if and only if a physical theory is invariant under a finite or countably infinite number of infinitesimal generators, a result consistent with Hilbert's claim (1918) that the failure of the law of conservation for energy was characteristic of Einstein's theory of gravity.[4]

These serious problems have inspired physicists to search for a consistent and complete theory of classical and quantum gravity over the past century, a search that still goes on today.

0-2. The Taiji symmetry framework

The taiji symmetry framework, or 'taiji space-time,' is the maximally symmetric space-time with zero curvature.[a] It is capable of accommodating both inertial and non-inertial frames of reference and all known conservation laws, as well as all known interactions, including gravity, and the quantization of all fields.

Traditionally, discussions in particle physics, field theory and other branches of physics have all been based on inertial frames with the Lorentz and Poincaré transformations. However, to obtain a broader and more complete understanding of physics, it is desirable to generalize the physical framework to include non-inertial frames. Inertial frames represent only limiting and idealistic cases and moreover, there is now strong evidence that the observable universe is expanding with a non-zero acceleration.

In the first six chapters, we derive explicit coordinate transformations between inertial frames and non-inertial frames (including rotating frames

---

[a] In ancient Chinese thought, the word 'taiji' or 'taichi' denotes The Absolute, i.e., the ultimate principle or the condition that existed before the creation of the world.

and frames with a co-linear acceleration and velocity where the acceleration is an arbitrary function of time) within the taiji symmetry framework. These transformations are derived on the basis of the 'principle of limiting continuation,' i.e., in the limit of zero acceleration, the transformations must simplify to the Lorentz transformations and the Poincaré metric tensor $P_{\mu\nu}$ of non-inertial frames must reduce to the Minkowski metric $\eta_{\mu\nu} = (1, -1, -1, -1)$. The existence of finite space-time transformations between inertial frames and such accelerated frames implies that the space-time associated with the non-inertial frames is flat,[5,6] i.e., the Riemann–Christoffel curvature tensor $R_{\lambda\mu\nu\alpha}$ is zero. The geometry of flat space-time with arbitrary coordinates is much simpler than the curved space-time of general relativity.

We discuss testable experimental predictions resulting from these coordinate transformations, some of which have already been performed, such as the lifetime dilation of unstable particles traveling in a circular storage ring, as well as new group properties of these transformations. Because inertial and non-inertial frames are not equivalent and no objects may travel faster than a light signal, transformations between inertial and non-inertial frames form only pseudogroups,[7] rather than the usual Lie groups, and map only a portion of the space-time in an accelerated frame to the entire space-time of inertial frames.

Finally, we note that as stated by Noether's first theorem, the physical manifestations of symmetry are the conservation laws. Ideally, well-established laws such as the conservation of energy–momentum and angular momentum should be natural consequences of the symmetries of a physical framework. The taiji symmetry framework, which is based mathematically on a flat space-time with arbitrary coordinates and the Poincaré metric tensor, and accommodates both inertial and non-inertial frames, should be the most general framework for physical theories for which this is true. In contrast, a theory based on a curved space-time such as general relativity cannot automatically conserve the energy–momentum tensor, but only the 'energy–momentum pseudotensor' $t^{\mu\nu}$ as expressed by $\partial_\mu t^{\mu\nu} = 0$. Physically, this is because in a curved space-time with arbitrary coordinates, one can remove the gravitational field from any given volume element so that unlike the energy associated with electromagnetic fields, it is meaningless to define an energy associated with a gravitational field at a given point, which is not completely satisfactory.

0-3.   Quantum Yang–Mills gravity

In the second part of this book (chapters 7 through 13), we construct Yang–Mills gravity, which is a classical and quantum theory of gravity consistent with all known experimental results. It is based on a generalized Yang–Mills framework and translational gauge symmetry in the flat space-time of the taiji symmetry framework. As might be anticipated by the reader, perhaps the biggest challenge of the theory is to reproduce all the known experimental effects that are consistent with a curved space-time, within a framework based on a flat space-time.

Around 1920, Weyl realized that electromagnetic fields and the conservation of charge could be understood in terms of the symmetry of the U_1 gauge transformation.[8] Weyl's idea was subsequently generalized to the SU_2 gauge symmetry in 1954 by Yang and Mills, in order to understand isospin conservation in strong interactions.[9] This highly non-trivial generalization was Yang and Mills' deepest insight and was crucial to further progress in physics, especially in the unification of the electromagnetic and weak forces.

In 1955, Utiyama proposed a gauge-invariant interpretation of interactions and discussed a general gauge theory involving a Lie group of N generators, based on the work of Yang and Mills. Utiyama further formulated a gauge theory of gravity based on the Lorentz group and identified Einstein's field equations as the gauge field equations, which paved the way for future research on gauge theories of gravity.[10] Since then, quantum field theories with gauge symmetries of internal groups (which are independent of the external space-time symmetries) have been successfully applied to modeling all fundamental interactions except for the gravitational interaction. Decades later, gravity still stands alone as the last challenge to Yang, Mills, Utiyama, and Weyl's idea of gauge symmetry.

In tackling the challenge of creating a quantum theory of gravity, however, it appears that the original Yang–Mills framework for internal gauge groups is inadequate. In the approach we describe here, we generalize that framework to include external space-time gauge groups,[11] and we call this generalized Yang–Mills framework the Yang–Mills–Utiyama–Weyl (YMUW) framework.

The guiding principle in constructing Yang–Mills gravity is that there must be a gauge symmetry that is compatible with both the conservation law of energy–momentum and the quantization of gauge fields. One promising candidate in flat space-time is the global translational symmetry. Suppose a Lagrangian is invariant under a global space-time translation,

$$x^\mu \to x'^\mu = x^\mu + \Lambda_o^\mu, \tag{0.1}$$

where the shifts of space-time coordinates Λ_o^μ are infinitesimal constants. This invariance implies the conservation of energy–momentum through Noether's theorem. The set of these transformations in flat 4-dimensional space-time forms the T_4 group, which is an Abelian subgroup of the Poincaré group. However, the global transformations (0.1) are inadequate for the formulation of a gauge theory. We must enlarge the global group to include local translations in flat space-time,

$$x^\mu \to x'^\mu = x^\mu + \Lambda^\mu(x). \tag{0.2}$$

It suffices to consider the infinitesimal and arbitrary vector function $\Lambda^\mu(x)$. It is important to realize that the infinitesimal transformations (0.2) have two conceptual interpretations[11]: (i) They are local translations in flat space-time, and (ii) they are general coordinate transformations. This dual role of (0.2) implies that local translational gauge invariance is inseparable from general coordinate invariance in flat space-time.[b]

The invariance under the coordinate translations (0.2) is the mathematical basis of Yang–Mills gravity. In inertial frames, the fundamental space-time symmetry group is the Poincaré group, which contains two sub-groups: the global translational group T_4 and the Lorentz group (which contains rotations as a subgroup). The Abelian sub-group of space-time translations T_4 is our focus because it leads to the conservation of energy–momentum and it is the external T_4 gauge symmetry that enables us to construct a quantum theory of gravity that is consistent with experiments.

Although a full derivation and justification for Yang–Mills gravity must wait until later, we here give just enough detail to indicate how a theory of gravity based on a flat space-time can replicate the effects of a curved space-time. Consider a particle-physics approach to gravity based on the YMUW framework with external space-time translational gauge symmetry and for now, consider only inertial frames

---

[b] In this connection, it is interesting to note that Einstein followed the general coordinate invariance in *curved* space-time, and hence employed space-time curvature and a covariant derivative to formulate his theory. In contrast, Yang–Mills gravity follows the local translational invariance in flat space-time, and hence uses T_4 gauge curvature and a gauge covariant derivative to formulate a theory of the gravitational interaction. The difficulties of Einstein gravity discussed in section 0-1 are thus avoided.

with the Minkowski metric tensor $\eta_{\mu\nu} = (1, -1, -1, -1)$. The formulations of electromagnetic and Yang–Mills-type gauge theories associated with internal gauge groups are all based on the replacement,

$$\partial_\mu \rightarrow \partial_\mu + ifB_\mu(x), \tag{0.3}$$

in a Lagrangian. The field $B_\mu = B_\mu^a(x)t_a$ involves constant matrix representations of the generators t_a of internal gauge groups. However, in the YMUW framework, the generators of the external space-time translation group T_4 are the momentum operators $p_\mu = i\partial_\mu$ or the displacement operator $T_\mu = \partial_\mu$, where $p_\mu = iT_\mu$ ($c = \hbar = 1$). Hence, in Yang–Mills gravity, the replacement is assumed to take the form

$$\partial_\mu \rightarrow \partial_\mu - ig\phi_\mu^\nu p_\nu = \partial_\mu + g\phi_\mu^\nu \partial_\nu, \quad J_\mu^\nu = \delta_\mu^\nu + g\phi_\mu^\nu, \tag{0.4}$$

similar to (0.3) in non-Abelian gauge theories.

This replacement is the basic postulate for the local formulation of Yang–Mills gravity. Since the generators of the T_4 group are $T_\mu = \partial_\mu$, we have a tensor gauge field $\phi^{\mu\nu} = \phi^{\nu\mu} = \eta^{\lambda\nu}\phi_\lambda^\mu$ (i.e., a spin-2 field) rather than a 4-vector field (i.e., a spin-1 field) in the T_4 gauge covariant derivative,

$$\Delta_\mu \equiv \partial_\mu + g\phi_\mu^\nu \partial_\nu \equiv J_\mu^\nu \partial_\nu. \tag{0.5}$$

Because the tensor field ϕ_μ^ν is associated with the most general flat space-time symmetry, as shown in the transformations (0.2), the gravitational tensor field $\phi_{\mu\nu}$ may be called the space-time translational gauge field or the space-time gauge field.

Two consequences of the replacements (0.4) and (0.5) can now be seen. First, we note that the replacement (0.4) in the Lagrangian differs from the U_1 replacement $\partial_\mu\psi \rightarrow (\partial_\mu + ieA_\mu(x))\psi$ in the free Lagrangian of the charged fermion field, ψ. Because the term $g\phi_\mu^\nu \partial_\nu$ in the T_4 replacement (0.4) has the same sign as its complex conjugate, it implies that the gravitational interaction is purely attractive for all matter and anti-matter, unlike the electromagnetic interaction, where the difference in sign of the complex conjugate implies the existence of both attractive and repulsive forces.

Second, the T_4 gauge covariant derivative Δ_μ in (0.5) has remarkable power in that it dictates that in flat space-time, all particle wave equations, including the electromagnetic wave equations, reduce to an equation of the Hamilton–Jacobi type with basically the same[c] 'effective Riemannian metric tensor' $G_{\mu\nu}(x)$,

$$G^{\mu\nu}(\partial_\mu S)(\partial_\nu S) - m^2 = 0, \quad m^2 \geq 0, \quad c = \hbar = 1, \tag{0.6}$$

$$G^{\mu\nu}(x) = \eta_{\alpha\beta} J^{\alpha\mu}(x) J^{\beta\nu}(x), \quad J^{\mu\nu}(x) = \eta^{\mu\nu} + g\phi^{\mu\nu}(x), \tag{0.7}$$

in the classical (short-wavelength) limit.[12] Equation (0.6) shows the effects of a universal gravitational force acting on all classical objects. Because formally, it is the same as the corresponding equation in Einstein's gravity based on curved space-time, (0.6) implies that classical objects, such as planets and light rays, behave as if they were in a 'curved space-time,' even though the underlying physical space-time is actually flat.

Therefore, one new result of quantum Yang–Mills gravity is that the apparent curvature of space-time appears to be simply a manifestation of the flat space-time translational gauge symmetry for the motion of quantum particles in the classical limit. In the first order approximation, Yang–Mills gravity is consistent with the experimentally measured gravitational red shift. In the second order approximation, it correctly predicts the perihelion shift of Mercury and the gravitational quadrupole radiation of binary pulsars.[13] Furthermore, Yang–Mills gravity allows a quantum theory of gravity to be formulated and Feynman–Dyson rules for the gravitational tensor fields and the associated 'ghost particles'[13] to be derived, as we will show in chapters 10 and 11. Thus, Yang–Mills gravity brings the gravitational interaction back into the experimentally established arena of gauge fields based on flat space-time.

In contrast, Einstein's approach involving the geometrization of physical fields in curved space-time appears to be successful only in the formulation of classical gravity, with the conservation of the 'energy–momentum pseudo-tensor' rather than the true energy–momentum tensor. If one were to adopt a similar approach to, say, the electromagnetic interaction, which is velocity-dependent, it would be natural to employ the Finsler geometry, where the fundamental metric tensors depend on both positions and velocities (i.e., the differentials of coordinates), rather than the Riemannian geometry.[14] However, all attempts thus far to geometrize classical electrodynamics have not even begun to approach the usefulness of quantum electrodynamics.

---

[c]There is a small difference between the effective Riemannian metric tensor of the fermions and that of the vector gauge bosons. See eqs. (8.36) and (8.40) in ch. 8.

The classical theory of Einstein gravity is usually considered to be most coherent mathematically and satisfying aesthetically.[2] However, there is little chance of making progress simply by going over its successes. Because of the success of Einstein gravity, many have made arguments that attempted to make it unique or necessary.[15] There is an old argument due to Feynmann (and perhaps others) that the only consistent theory (with no more than two derivatives) of a spin-2 field coupled to itself is, in fact, general relativity. It is not necessary to analyze the detailed arguments one by one, all that is necessary is to present a counterexample. Yang–Mills gravity, with its translational gauge symmetry in flat space-time and consistency with all known experiments, as well as other desirable features such as the preservation of the conservation law of energy–momentum, operational definitions of space-time coordinates, and quantizability of gravitational fields, is just such a counterexample from outside the curved space-time framework of general relativity.[d]

0-4.   Unification of all interactions based on the Yang–Mills–Utiyama–Weyl framework

In chapter 14 through 16, we consider a model for unifying all interactions in nature (including gravity) within the taiji symmetry and the generalized Yang–Mills frameworks, following the ideas of Glashow, Salam, Ward, and Weinberg for the unified electroweak interaction.[16] We here present a brief preview of the unification of the gravitational interaction with just the electroweak interaction.

The T_4 gauge covariant derivative (0.5) paves the way for a unification of the gravitational and electroweak ('gravielecweak') interactions based on $T_4 \times SU_2 \times U_1$. Consider a unified 'gravielecweak' model based on a generalized covariant derivative d_μ,[17]

$$d_\mu \equiv J_\mu^\nu \partial_\nu - i f W_\mu^a t^a - i f' U_\mu t_o, \quad J_\mu^\nu = \delta_\mu^\nu + g \phi_\mu^\nu. \tag{0.8}$$

This is a linear combination of the usual gauge covariant derivative in the electroweak theory and the gravitational T_4 gauge covariant derivative (0.5). The gauge curvatures, $C_{\mu\nu\alpha}$, $W_{\mu\nu}^a$, and $U_{\mu\nu}$, associated with each group in the 'gravielecweak' covariant derivative d_μ are given by

$$[d_\mu, d_\nu] = C_{\mu\nu\alpha} T^\alpha - i f W_{\mu\nu}^a t^a - i f' U_{\mu\nu} t_o, \quad [t^a, t^b] = i \epsilon^{abc} t^c, \tag{0.9}$$

---

[d] For key features of Yang–Mills gravity and Einstein gravity, see a comparison at the end of this overview.

and the $T(4)$ gauge curvature $C_{\mu\nu\alpha}$ in flat space-time is[17]

$$C_{\mu\nu\alpha} = J_{\mu\sigma}\partial^\sigma J_{\nu\alpha} - J_{\nu\sigma}\partial^\sigma J_{\mu\alpha} = \Delta_\mu J_{\nu\alpha} - \Delta_\nu J_{\mu\alpha} = -C_{\nu\mu\alpha}. \qquad (0.10)$$

The new SU_2 and U_1 gauge curvatures in the presence of the gravitational gauge potential are then given by

$$W^a_{\mu\nu} = J^\sigma_\mu \partial_\sigma W^a_\nu - J^\sigma_\nu \partial_\sigma W^a_\mu + f\epsilon^{abc} W^b_\mu W^c_\nu \qquad (0.11)$$

and

$$U_{\mu\nu} = J^\sigma_\mu \partial_\sigma U_\nu - J^\sigma_\nu \partial_\sigma U_\mu = \Delta_\mu U_\nu - \Delta_\nu U_\mu, \qquad (0.12)$$

respectively. We observe that the T_4 gauge potential appears in SU_2 and U_1 gauge curvatures in (0.11) and (0.12), but the SU_2 and U_1 gauge potentials do not appear in the T_4 gauge curvature in (0.10). This reflects the universal nature of the gravitational interaction, in which all things participate. In the absence of gravity, i.e., $g = 0$, the gauge curvatures in (0.11) and (0.12) have the usual $SU_2 \times U_1$ gauge transformation properties in the Weinberg–Salam theory.[16]

One new implication of the preceding unification is that there is an extremely small violation of the local $SU_2 \times U_1$ gauge invariance due to gravity, i.e., the presence of J^σ_μ in the gauge curvatures (0.11) and (0.12). Such a violation of internal gauge symmetries can be traced to the presence of J^ν_μ in the unified covariant derivative d_μ in (0.8) and, in the case of the U_1 gauge symmetry for example, appears to be just right for the emergence of the effective Riemannian metric tensors in the eikonal equations in the geometric-optics limit of the electromagnetic wave equations. Thus, in the gravielecweak model, the effective curved space-time for the motion of classical objects that enables its compatibility with classical tests of general relativity[17] and the violations of gauge symmetries of internal groups ($SU_2 \times U_1$) due to gravity are related.

It appears that all observed fundamental interactions in the physical world are manifestations of U_1, SU_2, SU_3, and T_4 gauge symmetries. The intimate relationship between interactions and gauge symmetries was first advocated by Utiyama, Yang, Mills and others.[18] The unification of all forces through our understanding of the profound and intertwined internal and external gauge symmetries based on the taiji symmetry framework may be the key to uncovering nature's deepest secrets.

A Comparison of Key Features of Yang–Mills Gravity, Einstein Gravity and Electromagnetic Theory

<u>Yang–Mills Gravity</u> *(Flat spacetime)*[$]

$\phi_{\mu\nu}$: *gauge potential fields,*

Physical frames of reference : defined,

Coordinates x^μ and momenta p_μ have operational meaning,

Inv. under local spacetime translations in flat spacetime:

$$x'^\mu = x^\mu + \Lambda^\mu(x), \quad \Lambda^\mu(x) = \text{infinitesimal local translation,}$$

$$\phi'_{\mu\nu}(x) = \phi_{\mu\nu}(x) - \Lambda^\lambda(x)\partial_\lambda\phi_{\mu\nu}(x) - \phi_{\mu\alpha}(x)\partial_\nu\Lambda^\alpha(x) - \phi_{\alpha\nu}(x)\partial_\mu\Lambda^\alpha(x);$$

Space − time translation (T_4) group generators: iD_μ,[$$]

T_4 gauge covariant derivative:

$$(D_\mu + g\phi_{\mu\nu}D^\nu) = J_{\mu\alpha}D^\alpha, \qquad J_{\mu\alpha} = P_{\mu\alpha} + g\phi_{\mu\alpha};$$

Gauge field strength (T_4 gauge curvature): $C_{\mu\nu\lambda}$,

$$[J_{\mu\alpha}D^\alpha, J_{\nu\beta}D^\beta] = C_{\mu\nu\lambda}D^\lambda, \qquad C_{\mu\nu\lambda} = J_{\mu\alpha}D^\alpha J_{\nu\lambda} - J_{\nu\beta}D^\beta J_{\mu\lambda};$$

Lagrangian: $L = \left[\dfrac{1}{4g^2}\left(C_{\mu\nu\lambda}C^{\mu\nu\lambda} - 2C_{\mu\alpha}{}^\alpha C^{\mu\beta}{}_\beta\right)\right]\sqrt{-\det P_{\mu\nu}}$;

Maximum graviton self-coupling: $4 - vertex.$

[$] Inertial and non-inertial frames with the Poincaré metric tensors $P_{\mu\nu}$.

[$$]D_μ is the covariant derivative associated with the Poincaré metric tensors in a general frame of reference in flat space-time. It satisfies the relations $[D_\mu, D_\nu] = 0$.

Einstein Gravity *(Curved spacetime)*

$\Gamma^{\lambda}_{\mu\nu}$: *gauge potential fields,*

Physical frames of reference:undefined,[$]

Coordinates x^{μ} and momenta p_{μ} have no operational meaning,[$]

Inv. under general coord. transformations in curved spacetime:

$x'^{\mu} = x^{\mu} + \xi^{\mu}(x), \quad \xi^{\mu}(x) = $ *arbitrary infinitesimal vector functions;*

$$g'_{\mu\nu}(x) = g_{\mu\nu}(x) + g_{\alpha\beta}(x)[E_{\mu\nu}{}^{\alpha\beta}{}_{\lambda}]\xi^{\lambda}(x); \quad [\$\$]$$

'Infinite continuous group generators': $E_{\mu\nu}{}^{\alpha\beta}{}_{\lambda} = -(\delta^{\alpha\beta}_{\mu\nu}\overleftarrow{\partial}_{\lambda} + \delta^{\alpha\beta}_{\mu\lambda}\partial_{\nu} + \delta^{\alpha\beta}_{\lambda\mu}\partial_{\mu}),$

Covariant derivative: $\quad \Delta_{\mu}V_{\nu} = \partial_{\mu}V_{\nu} + \Gamma^{\lambda}_{\mu\nu}V_{\lambda},$

Riemann $-$ Christoffel (spacetime) curvature: $\quad R^{\alpha}_{\lambda\mu\nu},$

$$[\Delta_{\mu}, \Delta_{\nu}]V_{\lambda} = R^{\alpha}_{\lambda\mu\nu}V_{\alpha}, \quad R^{\alpha}_{\lambda\mu\nu} = \partial_{\nu}\Gamma^{\alpha}_{\lambda\mu} - \partial_{\mu}\Gamma^{\alpha}_{\lambda\nu} - \Gamma^{\alpha}_{\mu\beta}\Gamma^{\beta}_{\nu\lambda} + \Gamma^{\alpha}_{\nu\beta}\Gamma^{\beta}_{\mu\lambda};$$

Lagrangian: $\quad L = \left\{\dfrac{R}{16\pi G}\right\}\sqrt{-\det g_{\mu\nu}}, \quad R = g^{\mu\nu}R^{\beta}_{\mu\beta\nu};$

Maximum graviton self-coupling: $\infty - $ *vertex.*

[$] See Wigner's comments in ref. 3.

[$$] The expression for $E_{\mu\nu}{}^{\alpha\beta}{}_{\lambda}$ should be used with some caution. It is merely to show the difficulty of interpreting Einstein's gravitational fields as gauge fields in the Yang–Mills approach.[19] For rigorous discussions of the infinite continuous group and its generators, see Noether's paper in ref. 4. The transformation $\delta g_{\mu\nu} = g'_{\mu\nu} - g_{\mu\nu}$ can also be expressed as:

$$\delta g_{\mu\nu} = E_{\mu\nu}{}^{\alpha'\beta'}{}_{\lambda''}\, g_{\alpha'\beta'}\xi^{\lambda''}$$

$$\equiv -\int \left[(\delta^{\alpha\beta}_{\mu\nu}\partial_{\lambda}\delta(x-x')\delta(x-x'') + \delta(x-x')\{\delta^{\alpha\beta}_{\mu\lambda}\partial_{\nu}\delta(x-x'')\right.$$

$$\left. + \delta^{\alpha\beta}_{\lambda\nu}\partial_{\mu}\delta(x-x'')\}\right]g_{\alpha\beta}(x')\xi^{\lambda}(x'')dx'\,dx'',$$

$$\delta(x-x') \equiv \delta^4(x-x'), \quad dx' \equiv d^4x', \text{ etc.}$$

Electromagnetic Theory *(Minkowski spacetime)* [$]

A^μ :   *gauge potential fields,*

Physical frames of reference : *defined,*

Coordinates x^μ and momenta p_μ have operational meaning,

Inv. under U_1 gauge transformations,

$$A'_\mu = A_\mu + \frac{1}{e}\partial_\mu \Lambda(x), \quad \Lambda(x) = \text{arbitrary infinitesimal function,}$$

U_1 group generator :   1,

U_1 gauge covariant derivative : $\Delta_\mu = \partial_\mu + ieA_\mu,$

Gauge field strength (U_1 gauge curvature) : $F_{\mu\nu},$

$$[\Delta_\mu, \Delta_\nu] = ieF_{\mu\nu}, \quad F_{\mu\nu} = \partial_\mu A_\nu - \partial_\nu A_\mu;$$

$$\text{Lagrangian}: \quad L = \left[-\frac{1}{4}F_{\mu\nu}F^{\mu\nu}\right];$$

Photon self-coupling : *none.*

[$] *inertial frames.*

References

1. A. Einstein and M. Grossman, in *100 Years of Gravity and Accelerated Frames, The Deepest Insights of Einstein and Yang–Mills* (Ed. J. P. Hsu and D. Fine, World Scientific, 2005) p. 48; A. Einstein, *ibid.*, p. 65. D. Hilbert, *ibid.*, p. 120. Hilbert elegantly embedded Einstein's idea of general coordinate invariance in an action to derive the gravitational field equation, independent of Einstein in 1915.
2. F. J. Dyson, Bulletin of the American Math. Soc., 78, Sept. 1972.
3. E. P. Wigner, *Symmetries and Reflections, Scientific Essays* (The MIT Press, 1967), pp. 52–53. Wigner wrote " Evidently, the usual statements about future positions of particles, as specified by their coordinates, are not meaningful statements in general relativity. This is a point which cannot be emphasized strongly enough and is the

basis of a much deeper dilemma than the more technical question of the Lorentz invariance of the quantum field equations. It pervades all the general theory, and to some degree we mislead both our students and ourselves when we calculate, for instance, the mercury [sic] perihelion motion without explaining how our coordinate system is fixed in space.... Expressing our results in terms of the values of coordinates became a habit with us to such a degree that we adhere to this habit also in general relativity, where values of coordinates are not per se meaningful."

4. E. Noether, Goett. Nachr., 235 (1918). English translation of Noether's paper by M.A. Tavel is online. Google search: M.A. Tavel, Noether's paper.

5. Ta-You Wu and Y. C. Lee, Intern. J. Theoretical Phys. **5**, 307 (1972); Ta-You Wu, *Theoretical Physics, vol.4, Theory of Relativity* (Lian Jing Publishing Co., Taipei, 1978) pp. 172–175.

6. J. P. Hsu and L. Hsu, *A Broader View of Relativity, General Implications of Lorentz and Poincaré Invariance* (World Scientific, 2006), Chapters 18, 19, 24, 25 and 26. Online: 'google books, a broader view of relativity, Hsu.'

7. O. Veblen and J. H. C. Whitehead in *The Foundations of Differential Geometry* (Cambridge Univ. Press, 1954), pp. 37–38.

8. H. Weyl, *Space-time-matter*, 4th ed. (1922), Trans. H. L. Brose (Dover, 1930). See also F. London, Z. Phys. **37**, 375 (1927) and V. Fock, Z. Phys. **39**, 226 (1927). The physical meaning of the U_1 gauge invariance was analyzed by V. I. Oglevetski and I. V. Polubarinov, Nuovo Cimento, XXIII, 173 (1962). They first decomposed the field A_μ as

$$A_\mu = (A_\mu - \partial^{-2}\partial_\mu\partial^\nu A_\nu) + \partial^{-2}\partial_\mu\partial^\nu A_\nu \equiv A_\mu^1 + A_\mu^0,$$

showing that the physical spin 1 components, A_μ^1, of the massless vector field were unchanged under the U_1 gauge transformation, $A_\mu' = A_\mu + \partial_\mu \Lambda(x)$, and that only the unphysical spin 0 component, A_μ^0, was changed. The observable field strength $F_{\mu\nu}$ does not involve the unphysical spin 0 component. Thus, the physical predictions of the electromagnetic theory are gauge invariant.

9. C N. Yang and R. L. Mills, Phys. Rev. **96**, 191 (1954). See also T. D. Lee and C. N. Yang, Phys. Rev. **98**, 155 (1955) and R. Shaw, Cambridge PhD thesis (1955).

10. R. Utiyama, Phys. Rev. **101**, 1597 (1956); T. W. B. Kibble, J. Math. Physics, **2**, 212 (1961). See also *100 Years of Gravity and Accelerated Frames, The Deepest Insights of Einstein and Yang–Mills* (Eds. J. P. Hsu and D. Fine, World Scientific, 2005), Chapters 7 and 8, (available online: 'google books, 100 years of gravity, J. P. Hsu.'); T. W. Hehl, P. von der Heyde, G. D. Kerlick and J. M. Nester, Rev. Mod. Phys. **48**, 393 (1976); G. Grignani and G. Nardelli, Phys. Rev. D **45**, 2719 (1992); D. Cangemi and R. Jackiw, Phys. Rev. Lett. **69**,

233 (1992); T. Strobl, Phys. Rev. D **48**, 5029 (1993); G. Grignani and G. Nardelli, Nucl. Phys. B **412**, 320 (1994); N. A. Phys. Letts. B **391**, 59 (1997); Y. Ne'eman, Acta Physica Polonica, B **29**, 827 (1998); V. J. L. Jaramillo and J. Guerrero, J. of Math. Phys., **44**, 5166 (2003); T. Kawai, Phys. Rev. D **62**, 104014/1 (2000).

11. J. P. Hsu, Intl. J. Mod Phys. A **21** 5119 (2006).

12. L. Landau and E. Lifshitz, *The Classical Theory of Fields* (Trans. by M. Hamermesh, Addison-Wesley, Cambridge, MA. 1951), pp. 312–323, pp. 136–137 and p. 319. See also J. P. Hsu, Phys. Rev. Lett. **42**, 934 (1979) for an early discussion of the geometric-optics limit of the Dirac equation with maximum four-dimensional gauge symmetry.

13. J. P. Hsu, Intl. J. Mod Phys. A **24**, 5217 (2009); Eur. Phys. J. Plus **126**, 24 (2011).

14. J. P. Hsu, Nuovo Cimento, B **108**, 183 (1993); B **108**, 949 (1993); B **109**, 645 (1994).

15. A century ago, Einstein observed "Concepts which have proved useful for ordering things easily assume so great an authority over us, that we forget their terrestrial origin and accept them as unalterable facts. They then become labeled as 'conceptual necessities,' '*a priori situations*,' etc. The road to scientific progress is frequently blocked for long periods by such errors. It is therefore not just an idle game to exercise our ability to analyze familiar concepts, and to demonstrate the conditions on which their justification and usefulness depend, and the way in which these developed, little by little, from the data of experience. In this way they are deprived of their excessive authority." A. Einstein, Phys Zeitschr. **17** 101 (1916).

16. Kerson Huang, *Quarks, Leptons and Gauge Field* (World Scientific, 1982) pp. 108–117 and pp. 152–156.

17. J. P. Hsu, Mod. Phys. Letts. A **26**, 1707 (2011); Chin. Phys. C **36**(5), 403 (2012).

18. See *100 Years of Gravity and Accelerated Frames, The Deepest Insights of Einstein and Yang–Mills* (Eds. J. P. Hsu and D. Fine, World Scientific, 2005), Chapters 4(C) and 10(C).

19. C. N. Yang, ref. 18, pp. 387–389. Utiyama (ref. 10) concluded that Einstein's equations are gauge field equations. But Yang believed that was an unnatural interpretation of gauge fields. He obtained gravitational equations, which are third-order differential equations for $g_{\mu\nu}$.

About the Authors

Jong-Ping Hsu received his B.S. from the National Taiwan University and M.S. from the National Tsing Hua University. He earned his Ph.D. degree in 1969 studying particle physics at the University of Rochester (New York) with Professor S. Okubo. He has done research at McGill University, Rutgers, the University of Texas at Austin, the Marshall Space Flight Center, NASA, and the University of Massachusetts Dartmouth. He has been a visiting scientist at Brown, MIT, Taiwan University, Beijing Normal University, and the Academy of Science, China. His research is concentrated in the areas of gauge field theories, Yang–Mills gravity, total-unified model of interactions and broad views of 4-dimensional symmetry. He has published more than 150 papers and articles, and two books, "A Broader View of Relativity, General Implications of Lorentz and Poincaré Invariance" (with L. Hsu, World Scientific, 2006) and "Lorentz and Poincaré Invariance" (with Y. Z. Zhang, World Scientific, 2001). He has also co-edited four conference proceedings and the book "100 Years of Gravity and Accelerated Frames — The Deepest Insights of Einstein and Yang–Mills" (with D. Fine, World Scientific, 2005). He is currently a chancellor professor, the chair of physics department and the Director of the Jing Shin Research Fund at the University of Massachusetts Dartmouth.

Leonardo Hsu earned his A.B. at Harvard University and his Ph.D. studying semiconductor physics at the University of California, Berkeley. After doing a postdoc at the Center for Innovation in Learning at Carnegie Mellon University, he joined the faculty of the University of Minnesota, Twin Cities in 2000. His current research includes both experimental and theoretical work. His primary interest is in studying how students learn to solve problems in introductory physics courses and the attitudes and beliefs of non-science majors about physics and the learning of physics. He also studies transport processes in III-Nitride semiconductors and their alloys, as well as the implications of the principle of relativity and different systems of units for different relativity theories. He was a co-editor of the "JingShin Theoretical Physics Symposium in Honor of Professor Ta-You Wu" (World Scientific, 1998). His work has been supported by the Potz Science Fund and by the National Science Foundation.

Part I

The Taiji Symmetry Framework

Leonardo Hsu and Jong-Ping Hsu

1.

Space-Time Symmetry, Natural Units and Fundamental Constants

1-1.   Underpinnings

Before discussing the development of the taiji relativity framework, it is worth taking some time to examine some of the supporting structures of the theory that are often taken for granted, namely the units used for specifying physical quantities and the fundamental constants that do or do not appear in the theory. In developing taiji relativity, we hope to create the simplest space-time framework that can accommodate all the laws of physics in a unified manner, where by 'simplest' we mean that it is based on the smallest number of postulates. Correspondingly, in our discussions, we choose to use a system of units that is also, in a way, the 'simplest,' in the sense that it has only one unit. This is the system of 'natural units,' in which all physical quantities are specified in terms of a power of the same single unit.[1-3] One desirable feature of natural units is that the equations expressing physical laws are simpler to write down and have fewer constants that can obscure the essential physics they embody. For example, both the speed of light c and Planck's constant $\hbar$ have unit magnitude with no dimensions and thus do not appear in any of the equations. The system of natural units is widely used by particle physicists for just this reason, although it is far from being the most convenient or useful choice for many everyday applications.

One might ask however if the natural unit system is simply a calculational convenience or whether there is a physical justification for, in effect, setting c and $\hbar$ equal to one. If there is to be a physical justification, one must be able to operationally define each of the base units in a well-established unit system, say, the SI system of units, in terms of one single unit in a physically meaningful way. For example, one could arbitrarily decide that 1 meter is equivalent to 5 kelvin, but such an equivalence would not be physically meaningful. On the other hand, saying that 1 second is equivalent to 299 792 458 meters is physically meaningful, as we shall see.

Below, we argue that there is indeed a physical basis for the natural unit system and thus its use in constructing a framework for physical laws, by presenting definitions for the most relevant base SI units in terms of a single unit. We also discuss the implications of these definitions for the nature of physical constants such as the speed of light c, Planck's constant $\hbar$ and Boltzmann's constant k_B.

1-2.   Physical basis for the system of natural units

Equivalence of length and time

Prior to the twentieth century, space and time seemed to be completely different entities and thus two different units, the meter and second, were invented to quantify them. From a modern viewpoint however, the space-time symmetry of special relativity implies that space and time are not independent and separate, but parts of a four-dimensional space-time. This view is reflected in the fact that in the SI system, the unit of length (the meter), is defined in terms of the unit of time (the second).[3]

Such a definition is possible because in our physical laws, four coordinates, three spatial and one temporal, are necessary to specify the space-time location of an event. Although there is nothing logically wrong with using different units to express these four quantities, it is mathematically simpler to use the same single unit, either meter or second. As an analogy, there is nothing wrong with expressing north–south distances in meters and east–west distances in miles. Doing so, however, introduces an extra conversion constant into physical equations that is clearly artificial. The mathematical form of physical laws is much simpler when distances in both directions are expressed using the same unit.[4] The same is true for intervals of both space and time.

In order to establish a physically meaningful equivalence between the meter and the second, we must choose a physical phenomenon that involves the dimensions of both length and time. The constant speed c of a particle with a zero rest mass provides a convenient conversion factor between the two units. For practical reasons of precision and reproducibility, the unit of time, the second is chosen to be the base unit in the SI system. The modern (1983) definition of the meter is then 'The meter is the length of the path travelled by light in vacuum during a time interval of 1/299 792 458 of a second.'[3]

One could imagine that had the four-dimensional symmetry of our physical universe been understood from the very beginning, we might now only talk about the single dimension of length (or time), instead of the two dimensions of length and time, and only a single unit for both spatial and temporal intervals might have arisen, rather than the two separate units of second and meter. In fact, some textbooks on special relativity[4,5] intentionally write the Lorentz transformation equations with both spatial and temporal coordinates expressed using the same unit to emphasize this point. Thus, we see that the constant $c = 299\,792\,458$ m/s originates from an incomplete understanding of physics and is merely a conversion factor between the human-defined units of meter and second.[4,6]

Equivalence of mass and length

Similarly, before the development of quantum mechanics and Einstein's understanding of the equivalence of mass and energy, mass seemed to be a completely separate quantity from time and space. As such, the present definition of the kilogram, the SI unit of mass, is based on a physical artifact, the international prototype of the kilogram. However, just as the four-dimensional symmetry of our universe indicates that a single unit is sufficient to express both spatial and time intervals, the dual wave-particle nature of physical entities provides a physical basis for establishing a physical equivalence between the unit of mass with that of time. Using the relationships between energy and frequency $E = \hbar\omega$ (one could also use de Broglie relation $p = \hbar k$) and Einstein's relationship $E = mc^2$ (which becomes $E = m$ when expressing both length and time using the same units), we can then write $E = m = \hbar\omega$ to establish a relationship between a particle's mass and the angular frequency of its corresponding wave, where Planck's constant $\hbar$ acts as a the conversion factor between mass and time. Using the fact that 1 meter is equivalent to $1/299\,792\,458$ of a second, we have

$$\hbar = 1.054 \times 10^{-34} J \cdot s = 1.054 \times 10^{-34} \left(\frac{1}{299\,792\,458} \right)^2 kg \cdot s = 1, \quad (1.1)$$

so that

$$1\,kg = 8.511... \times 10^{50} s^{-1} = 2.839... \times 10^{42} m^{-1}. \quad (1.2)$$

Following this line of thought, one serious proposal for a future definition of the kilogram that is being considered by the General Conference on Weights and Measures (CGPM) is 'The kilogram is the mass of a body whose equivalent energy is equal to that of a number of photons whose frequencies sum to exactly $\left[(299\,792\,458)^2/66\,260\,693\right] \times 10^{41}$ hertz.'[7]

In the system of natural units used by particle physicists, it is more usual to set $\hbar = h/2\pi = 1$ rather than $h = 1$, so that 1 kg is equivalent to $\left[2\pi(299\,792\,458)^2/66\,260\,693\right] \times 10^{41}\,s^{-1}$ since $E = \hbar\omega$. In this case, one could define the kilogram as "The kilogram is the mass of a body whose equivalent energy is equal to that of a number of photons whose **angular** frequencies sum to exactly $\left[2\pi(299\,792\,458)^2/66\,260\,693\right] \times 10^{41}\,s^{-1}$." We will use this alternative definition. It is important to note that although the numerical value of the equivalence between the kilogram and inverse second is a matter of choice and not unique, the main result, that there is a physical basis for expressing masses and time intervals using the same unit (albeit different powers of that single unit), still stands.

At present, the proposed definition given above for the SI unit kilogram has not yet been adopted because the stability and reproducibility of the mass of the international prototype of the kilogram is still better than that of any atomic standard that we can yet achieve.[8,9] However, if the kilogram were to be redefined in terms of the frequency of a collection of photons, one consequence would be the fixing of Planck's constant $\hbar$ to an exact value, analogous to the case where the redefinition of the meter in 1983 had the effect of fixing the speed of light c to an exact value, with no uncertainty. Planck's constant would then become analogous to the speed of light, a conversion factor between the human defined units of kilogram and second.

Although the ability to express lengths, masses, and time intervals using the same single unit implies that all physical quantities can be expressed in terms of powers of that unit,[10] one can also develop corresponding operational equivalences for the other base SI units. Let us consider current and temperature as examples.

Equivalence of current and length

The present definition of the SI unit of current ampere is based on the force between two parallel current carrying wires

$$\frac{F}{l} = \frac{2I_1 I_2}{c^2 d} \tag{1.3}$$

written in Gaussian units where F/l is the force per unit length between two infinite current-carrying wires, I_1 and I_2 are the currents in the two wires, c is the speed of light, and d is the perpendicular distance between the wires. Using the equivalences previously developed between the units of second, meter, and kilogram, we can insert 2×10^{-7} newton per meter for F/l, $c = 1$, 1 ampere for both I_1 and I_2, and 1 meter for d, to obtain the conversion 1 ampere $= 5.331... \times 10^{17} \, second^{-1} = 1.778... \times 10^9 \, meter^{-1}$. Thus the definition of the ampere can be rewritten in terms of the unit meter as follows: 'The ampere is that constant current which, if maintained in two straight parallel conductors of infinite length, of negligible circular cross-section, and placed 1 meter apart in vacuum, would produce between these conductors a force equal to $2.109... \times 10^{10} \, meter^{-2}$ per meter of length (or $5.683... \times 10^{35} \, second^{-2}$ per second of length)'.

Equivalence of temperature and length

The present definition of the SI unit of temperature is 'The kelvin, unit of thermodynamic temperature, is the fraction 1/273.16 of the thermodynamic temperature of the triple point of water.'[3] However, just as special relativity shows that space and time are both parts of a four-dimensional space-time and quantum mechanics shows that mass and frequency (or wavelength) are simply different aspects of the same quantity, the kinetic theory of gases implies that temperature is not an independent characteristic of a system, but simply one type of energy scale. For example, the temperature of a monatomic ideal gas is related to the average kinetic energy of its individual gas molecules through $E = (3/2)k_B T$. In natural units, both energy and temperature are dimensionally equivalent to the kilogram, the inverse meter, or the inverse second.

To establish a physically meaningful conversion factor between the kelvin and these other units, we note that the Boltzmann constant k_B already provides a conversion between the energy and the temperature of a system. In analogy with using c and $\hbar$ as conversion factors between the meter, second, and kilogram, a conversion which fixes the value of the Boltzmann constant k_B to unity ($k_B = 1$) is convenient and results in the equivalence 1 kelvin $= 1.309... \times 10^{11} \, second^{-1} = 436.7... \, meter^{-1}$. A new definition of the unit kelvin might then be 'The kelvin, unit of thermodynamic temperature, is the thermodynamic temperature of a system whose energy is equal to the energy of a collection of photons whose angular frequencies sum to $1.309... \times 10^{11} \, second^{-1}$ (or, equivalently, $4.367... \times 10^2 \, meter^{-1}$)'.

Were this definition to be adopted, then, as was the case for the speed of light c and as may become the case for Planck's constant $\hbar$, the Boltzmann constant

k_B would take on the status of a conversion constant between Kelvin and second, a historical artifact resulting from an incomplete understanding of the atomic basis of temperature and energy. In this sense, the Boltzmann constant would become similar to Joule's constant, which was used to relate mechanical energy (in joules) and heat (in calories). When it was realized that both were the same type of quantity, the unit joule was used to express both and Joule's constant of 4.184 J/cal became merely another conversion constant with an equivalent value of one.

Similarly, other SI base units such as mole and luminous intensity can also be written in terms of the unit second.[2]

1-3.   Nature of the fundamental constants

In the equations of physics embodying our theories that allow us to predict and explain the diverse phenomena we observe in the universe, certain quantities appear repeatedly. These quantities, conventionally called fundamental constants or fundamental physical constants, typically either have their origins in the mathematical formulation of physical theories (such as $\hbar$ and Newton's gravitational constant G) or are properties of particles that are currently viewed as the basic building blocks of matter (such as the charge and mass of the electron). Levy-Leblond[11] grouped the physical constants into two categories, (1) constants characterizing whole classes of physical phenomena (such as the electric charge e and the universal gravitational constant G), and (2) universal constants (such as c and h) which act as concept or theory synthesizers (for example, Planck's constant h synthesizes the concepts of momentum and wavelength through the relation $p = h/\lambda$).

In this section, we use the natural system of units to divide the fundamental constants into two categories of our own, (A) those constants that are inherent descriptors of our universe, and (B) those constants that are historical products of an incomplete understanding of our universe and thus have values that are determined by human convention (or more precisely, by the initial definitions of the diverse units that were developed to describe our world, such as meter, second, and kilogram). Our reason for doing so is that in our formulation of what we hope to be the simplest space-time framework and theory that can accommodate all known interactions, the only constants we would like to appear are those from category (A), whose values are inherent properties of the universe and cannot be explained by any known theory. This is not to say that constants from category (B) are not useful. Indeed, the values of those constants are important for interpreting the results of experiments that can check the consistency and explanatory power of our theories and for making connections between physics and other disciplines. However, it should not be necessary for them to appear in a fundamental theory of physics.[12] Thus, papers in metrology that discuss the measurement of various physical constants[13,14] actually describe two conceptually different types of experiments. One is an experiment which seeks to determine more precisely one of the fundamental numbers that characterizes our universe. The second is an experiment that seeks to determine more precisely the conversion factor between two historically independent definitions of two units.

One criterion that can be used to determine the category to which a particular constant belongs is whether or not the value of that constant can be made exact or unity by a suitable physically grounded (re-)definition of units. For example, quantities such as c, $\hbar$, and k_B are category (B) constants because they can be made to have the value unity by redefining the second, meter, kilogram, and kelvin, as previously discussed.[a] Likewise, the vacuum permittivity ϵ_0 and permeability μ_0 are category (B) constants. Although they are not equal to unity under present definitions of the units, their values are exact, and they could be made unity by a suitable redefinition of the ampere. As a final example, the Stefan–Boltzmann constant σ is related to other constants by $\sigma = 2\pi^5 k_B^4 / (15 h^3 c^2)$ and in natural units (where $c = \hbar = k_B = 1$), has the exact value $\pi^2/60$. By a suitable redefinition of the kelvin (or meter, second, or kilogram), its value can be made unity.

Constants whose values cannot be made equal to unity by any physically grounded choice of units, and are therefore category (A) constants or inherent characteristics of our universe, include the electron charge e, the weak mixing (or the Weinberg) angle θ_W (found in the unified electroweak theory), and the strong coupling constant α_s (in quantum chromodynamics). Constants such as the Josephson constant $K_J = 2e/h$ and the von Klitzing constant $R_K = h/e^2$, by

---

[a] It is interesting to compare this result with Dirac's outlook on the physics of the future: 'The physics of the future, of course, cannot have the three quantities $\hbar$, e and c all as fundamental quantities. Only two of them can be fundamental, and the third must be derived from those two. It is almost certain that c will be one of the two fundamental ones. The velocity of light, c, is so important in the four-dimensional picture, and it plays such a fundamental role in the special theory of relativity, correlating our units of space and time, that it has to be fundamental... I think one is on safe ground if one makes the guess that in the physical picture we shall have at some future stage e and c will be fundamental quantities and $\hbar$ will be derived.'[15]

virtue of their relationship to category (A) fundamental constants, are themselves also fundamental constants. However, e, α_e (the fine structure constant), K_J, and R_K are not four independent fundamental constants, since they are all powers or multiples of each other. At present, the only quantities that qualify as category (A) fundamental constants under this criterion are the coupling constants for each of the four fundamental forces G_N (gravitational), $\alpha_e{}^b$ and θ_W (electroweak), and α_s (strong).[16]

In summary, natural units are not merely a calculational convenience, but have a conceptual basis rooted in the nature of our physical universe. Furthermore, the system of natural units provides a means of separating out the constants that should or should not appear in a fundamental physical theory. We shall use natural units throughout this book.

References

1. W. Pauli, Phys. Rev. **58**, 716 (1940). Pauli began his paper on "The Connection Between Spin and Statistics" with a discussion of 'Units and Notations': 'Since the requirements of the relativity theory and the quantum theory are fundamental for every theory, it is natural to use as units the vacuum velocity of light c, and Planck's constant divided by 2π which we shall simply denote by $\hbar$. This convention means that all quantities are brought to the dimension of the power of a length by multiplication with the power of $\hbar$ and c'.

2. L. Hsu and J. P. Hsu, Eur. Phys. J. Plus (2012) **127**: 11. (DOI 10.1140/ epjp/i2012-12011-5). Available online: arXiv, Natural units, L. Hsu.

3. B. N. Taylor (ed.), NIST Special Publication 330: The International System of Units (SI) (National Institute of Standards and Technology, Gaithersburg, MD, 2001), pp. 5–9; J. D. Jackson, *Classical Electrodynamics* (John Wiley and Sons, New York, 1999), 3rd ed., pp. 775–776.

4. E. F. Taylor and J. A. Wheeler, *Spacetime Physics* (2nd. ed., W. H. Freeman, New York, 1992), pp. 1–4.

5. T. A. Moore, *A Traveler's Guide to Spacetime: An Introduction to the Special Theory of Relativity* (McGraw-Hill, New York, 1995).

6. Reference 4, p. 5; J. P. Hsu and L. Hsu, *A Broader View of Relativity, General Implications of Lorentz and Poincaré Invariance* (2nd ed., World Scientific, 2006) pp. 434–436. Available online: 'google books, a broader view of relativity, Hsu.'

7. I. M. Mills, P. J. Mohr, T. J. Quinn, B. N. Taylor, and E. R. Williams, Metrologia **43**, 227 (2006).

---

[b]We shall use Heaviside–Lorentz rationalized units with $\hbar = c = 1$, $\alpha_e = e^2/(4\pi) = 1/137.036$, where $-e = -0.3028$ is the charge of the electron in natural units. ($\alpha_e = e^2/(\hbar c)$ in cgs [Gaussian] units and $\alpha_e = e^2/(4\pi \hbar c \epsilon_0)$ in mks [SI] units.)

8.   B. N. Taylor and P. J. Mohr, Metrologia **36**, 63 (1999).
9.   E. R. Williams, R. L. Steiner, D. B. Newell, and P. T. Olson, Phys. Rev. Lett. **81**, 2404 (1998).
10. The reason is that all units in SI can be expressed in terms of the units meter, kilogram and second. For example, the unit Newton for force can be expressed as $kg \cdot m \cdot s^{-2}$.
11. J.-M. Levy-Leblond, Rivista del Nuovo Cimento 7, 1897 (1977).
12. In a column in Physics Today (F. Wilczek, Physics Today **56**(5), 10 (2003)), Wilczek states that parameters such as e, $\hbar$, and the mass of the electron m_e can be eliminated by a suitable choice of a system of units for length ($\hbar^2/m_e e^2$), time ($\hbar^3/m_e e^4$), and mass (m_e). This is much the same as choosing a system of natural units, in which parameters such as c, $\hbar$ and k_B are eliminated. As Wilczek mentions later in the article, for a more complete theory, additional parameters such as the fine structure constant α inevitably come in.
13. P. J. Mohr and B. N. Taylor, Rev. Mod. Phys. **72**, 351 (2000).
14. V. Kose and W. Woeger, Metrologia **22**, 177 (1986).
15. That the gravitational coupling constant G is not dimensionless is intimately related to the fact that the gravitational interaction is related to the external space-time translational gauge symmetry. See J. P. Hsu and L. Hsu, *A Broader View of Relativity, General Implications of Lorentz and Poincaré Invariance* (2nd ed., World Scientific, 2006) pp. 434–436. Available online: 'google books, a broader view of relativity, Hsu.'; J. P. Hsu, 'Quantum Yang-Mills gravityin flat space-time and effective curved space-time for motions of classical objects' (Euro Phys. J. Plus, **126**:24 (2011)), [available online: arXiv 1102.2253']. In the conventional theory of gravity, it may be attributed to the property that energy–momentum tensor is the source of the gravitational field.
16. P. A. M. Dirac, Sci. Am. **208**, 48 (1963).

2.

The Taiji Relativity Framework

2-1. A new space-time framework

The primary goal of physics is to devise the most logically simple and parsimonious theories that can predict and explain the wide range of observed phenomena in the universe. A fundamental physical theory should be based on the smallest and simplest set of postulates. Einstein formulated special relativity on the basis of two postulates:

(I) the principle of relativity, which states that the laws of physics have the same form in every inertial frame, and

(II) the universality of the speed of light, which states that the speed of light is independent of the motion of the source and of the observer.

In this and the next few chapters, we shall show that from a modern viewpoint, one can formulate a relativity theory that is consistent with the results of all known experiments, solely on the basis of the principle of relativity (postulate I). The idea of formulating a relativity theory using only the principle of relativity is not new. In the literature, one can find statements to the effect that the universality of the speed of light is a direct consequence of the principle of relativity and that the second postulate is not needed.[1,2] However, such discussions typically end there, without further elaboration. In our case, we shall show that a relativity theory based solely on the first postulate, while calculationally equivalent to special relativity, has a very different conceptual foundation, particularly when viewed in light of the physical basis of natural units.

In a previous book, we showed how this different conceptual foundation led to a fresh view of old physics and new results, including making the solution of the relativistic many-body problem more tractable and helping to develop more convenient definitions for thermodynamics quantities in a relativistic framework.[3] In this volume, we demonstrate how this conceptual difference allows us to construct a flat four-dimensional space-time with arbitrary coordinates that can serve as a framework for both a classical and quantum theory of gravity. Unlike the arbitrary coordinates of the curved space-time in general relativity, those in this flat space-time retain their physical significance and can be operationally defined. Furthermore, because the resulting theory of gravity is a gauge theory in a flat space-time, it can be unified with the other known interactions in a way not allowed by the curved space-time of Einstein gravity.

We call this flat four-dimensional space-time framework 'taiji relativity' because it is based on what we believe to be the smallest and simplest set of postulates and system of units (natural units), and 'taiji' in ancient Chinese thought, symbolizes 'The Absolute,' or the ultimate principle or condition that existed before the creation of the world.[4]

2-2.   Taiji relativity

We begin by deriving the taiji coordinate transformations between two inertial frames using natural units. Although these equations will look identical to those in special relativity (the Lorentz transformations), there is an important conceptual difference between them that we will discuss at the end of this chapter. Moreover, in chapters 3–6 we will show how these transformations can be generalized to develop explicit and physically meaningful coordinate transformations between non-inertial frames with arbitrary linear accelerations or rotational motion, a feat that general relativity cannot accomplish, despite its ability to predict physical phenomena such as the perihelion shift of planets and the change in the rate of ticking of clocks in different gravitational fields.

As usual, we describe an event using the four coordinates

$$(w, x, y, z) = x^\mu, \quad \text{and} \quad (w', x', y', z') = x'^\mu, \quad \mu = 0, 1, 2, 3, \tag{2.1}$$

in inertial frames $F(w, x, y, z)$ and $F'(w', x', y', z')$, respectively. The quantities w and w' are the evolution variables and have the dimension of length. We use the letter w to distinguish it from t, which we will take to be the evolution variable expressed using the conventional unit second and also refer to w and w' as the taiji time. In special relativity, the quantities w and t are related by $w = (299\,792\,458\,m/s)t$ in every inertial frame. Such is not necessarily the case in taiji relativity.

Using the usual simplifying conventions that the relative motion between F and F' frames is along the parallel x and x' axes, that the origins of F and F'

coincide at $w = w' = 0$, and that the transformation equations between the two inertial frames is linear,[5] the transformations between the primed and unprimed coordinates can be written as

$$w' = a_1 w + a_2 x, \quad x' = b_1 w + b_2 x, \quad y' = y, \quad z' = z, \tag{2.2}$$

where a_1, a_2, b_1, and b_2 are constants to be determined. The principle of relativity implies that physical laws must have the same form in any inertial frame. This statement is equivalent to demanding that the four-dimensional interval, $s^2 = \eta_{\mu\nu} x^\mu x^\nu = x_\mu x^\mu = w^2 - r^2$, be invariant so that[6]

$$w'^2 - r'^2 = w^2 - r^2. \tag{2.3}$$

Substituting the expressions for w and x from (2.2) into (2.3) and equating the coefficients of w^2, wx, and x^2 in addition to using the condition that when $dx'/dw' = 0$, $dx/dw = \beta$, we can then determine all four unknown constants in terms of β, where β is a dimensionless constant that characterizes the magnitude of the relative motion between F and F'. The four-dimensional coordinate transformations can then be written as

$$w' = \gamma(w - \beta x), \quad x' = \gamma(x - \beta w), \quad y' = y, \quad z' = z; \tag{2.4}$$

$$\gamma = \frac{1}{\sqrt{1 - \beta^2}}, \tag{2.5}$$

with a corresponding inverse transformation, derived from (2.4) and (2.5)

$$w = \gamma(w' + \beta x'), \quad x = \gamma(x' + \beta w'), \quad y = y', \quad z = z'. \tag{2.6}$$

From the solution of the Maxwell equations for a plane wave, it follows that the propagation of a light signal emitted from any source is described by $s^2 = 0$ or equivalently $ds^2 = dw^2 - dr^2 = 0$.[7] In that case, the speed of light β_L implied by $ds = 0$ is thus

$$\beta_L = |dr/dw| = 1. \tag{2.7}$$

Since ds^2 is invariant, this result, which is a consequence of the principle of relativity, holds in all inertial frames: the speed of a light signal is isotropic and independent of the motion of the source and of the observer.

From equations (2.4)–(2.5), we can also derive the velocity transformation,

$$u'_x = \frac{u_x - \beta}{1 - u_x\beta}, \quad u'_y = \frac{u_y}{\gamma(1 - u_x\beta)}, \quad u'_z = \frac{u_z}{\gamma(1 - u_x\beta)}; \tag{2.8}$$

$$\mathbf{u}' = \left(\frac{dx'}{dw'}, \frac{dy'}{dw'}, \frac{dz'}{dw'} \right), \quad \mathbf{u} = \left(\frac{dx}{dw}, \frac{dy}{dw}, \frac{dz}{dw} \right),$$

where $\mathbf{u}$ is the velocity of a particle as measured by an F frame observer and $\mathbf{u}'$ is the velocity of the same particle as measured by an F' observer. Since we use natural units, these velocities are dimensionless quantities, the components of which range in value from -1 to 1.

2-3.   Operationalization of taiji time

If taiji relativity is to be a physical theory, then the taiji time w must have an operational definition. As we are accustomed to the usual kinds of clocks that measure time in seconds, one might ask how the measurement of time as expressed by w can be physically realized. The answer is straightforward. Because taiji relativity and special relativity are mathematically identical if one sets $c = 1$ in special relativity, one can use the usual procedure for synchronizing clocks in special relativity to set up a clock system in taiji relativity. In order to obtain taiji time measurements in units of meters, we merely re-label the clocks from seconds to meters after synchronization and scale the readings so that a time interval Δt of 1 second corresponds to a time interval Δw of 299 792 458 m.

A taiji-time Δw can also be understood in terms of an optical path length. For a finite interval, the propagation of a light signal starting from the origin $\vec{r} = 0$ at $w = 0$ is described by

$$w^2 - r^2 = 0, \tag{2.9}$$

where r is the distance traveled by the light signal during the taiji-time interval $\Delta w = w$. Since Equation (2.9) implies that $r = w$ and the speed of light $\beta_L \equiv r/w = 1$ is invariant, w is also the distance traveled by the light signal. Thus, the time interval Δw between two events can be interpreted as the distance traveled by a light signal between the occurrence of the two events.

This interpretation of w suggests another way of operationalizing the taiji time, which is to use clocks whose rates of ticking are based on the motion of a light signal. Such clocks, which have been discussed in many books on special relativity, consist of a light bulb and light detector mounted next to each other facing in the same direction, and a mirror placed half a meter away facing the two devices.[8] The light bulb flashes, sending out a light signal that reflects off the mirror and returns to the detector. When the detector receives the signal, it triggers the light bulb to flash again. Because a light signal travels a total distance of 1 meter between flashes, the time interval Δw marked by the clock between flashes is $\Delta w = 1$ meter.

2-4. Conceptual difference between taiji relativity and special relativity

As noted previously, the taiji coordinate transformation equations (2.4) and (2.5) are identical to the conventional Lorentz transformation equations if one sets $c = 1$ so that $t = w$. Therefore, taiji relativity shares all of the mathematical and operational characteristics of special relativity. For example, clocks for measuring the taiji-time w can be synchronized in the same way as clock systems for relativistic time, the taiji coordinate transformations share all of the Lorentz and Poincaré group properties of special relativity, and all of the experimental predictions of special relativity are duplicated by taiji relativity. However, there is an important conceptual difference between the two theories.

In special relativity (and as defined by the 1983 meeting of the General Conference on Weights and Measures), the two quantities, which are directly proportional and related by $w = (299\ 792\ 458\ m/s)t$, are merely two different ways of expressing the same idea, namely a quantification of the evolution of a system. As Taylor and Wheeler pointed out,[9] the number 299 792 458 is simply a conversion factor whose value is the consequence of the choices made by humans in defining the units of meter and second.

However, in taiji relativity, we propose a new way of thinking about the relationship between w and t. If the current relationship between them is merely a product of the human perceptual system and the choices made in selecting standards for the dimensions of time and length, why not explore the implications of a different choice in the relationship between w and t? Just because $w = ct$ leads to a relativity theory that is consistent with all experiments does not imply that it is the only relationship that is consistent with all experiments. Because changing the relationship between w and t is equivalent to changing the definition of the meter in terms of the second, a human convention, none of the experimental predictions of the theory verified over the last century would change. Some types of new relationships, such as setting $w = (9.83571 \times 10^8 ft/s)t$ are trivial. However, what if we were to adopt a relationship that is more radically different, such as

$$w = bt \qquad (2.10)$$

where b is not a constant but a function itself of space and time?[10,11]

At first glance, such a relationship may seem impractical. Not only would the speed of light measured in units of meter per second no longer be a universal constant (although the dimensionless speed as measured in natural units would still be a universal constant as demonstrated in (2.7)), clocks set up to display the time t in seconds would appear to run in some crazy non-uniform way.

However, as we have shown in a previous book,[12] such a relationship can, in fact, result in a relativity theory that

(a) keeps all inertial frames equivalent,
(b) still correctly predicts experimental results,
(c) still has all of the Lorentz and Poincaré group properties of special relativity, and most importantly
(d) can be more convenient for certain types of calculations than special relativity (though consequently, it is less convenient than special relativity for other types of calculations).

A useful analogy for thinking about such alternative relationships between w and t is the different choices one can make for coordinate systems: Cartesian, spherical, cylindrical, hyperbolic, etc. One can think of special relativity, with its particularly simple relationship between w and t, as analogous to the Cartesian coordinate system. Changing the scaling factor between w and t while keeping it constant is analogous to adopting a different Cartesian coordinate system with a different grid spacing. Moving to a relationship $w = bt$ where b is a function of space and time is analogous to changing to a coordinate system such as spherical coordinates, with its complex relationships between (x, y, z) and (r, θ, ϕ). Neither system is more correct than any other and all predict the same physics. However, depending on the problem to be analyzed, certain coordinate systems are more convenient and are better suited to revealing symmetries and invariants of the physical systems under consideration (See Fig. 2.1).

2-5. A short digression: The role of a second postulate

Although some authors claimed that the universality of the speed of light is a direct consequence of the principle of relativity and thus that a separate

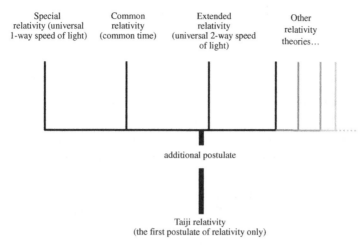

Fig. 2.1　"Tree" of relativity theories showing the logical connection between taiji relativity, based solely on the Principle of Relativity, and other relativity theories for which an additional postulate has been made.

second postulate is unnecessary, Einstein evidently thought that a second postulate WAS necessary to establish the universality of the speed of light.[1,2] In light of Equation (2.7) and our discussion of the relationship between w and t, we can resolve these two points of view.

As we saw in the derivation of Equation (2.7), the principle of relativity by itself implies that the speed of light, when measured in terms of a dimensionless velocity within the framework of natural units, is a universal constant. However, the principle of relativity does not, by itself, specify any particular relationship between the evolution variable w, with the dimension of length, and the evolution variable t, measured in a unit such as the second. The use of two different dimensions to measure space and time is a choice and there is no *a priori* relationship between the meter and the second that is determined by any property of our universe. Thus, we are free to make that choice ourselves, according to whatever is most convenient for us.

Choosing

$$w = ct$$

in every inertial frame, as Einstein did, corresponds to making a second postulate that "The speed of light, measured in units of meter per second, is independent of the motion of the source and of the observer" and is equivalent to defining the unit second as "The interval of time that passes when a light signal travels a distance 299 792 458 m in a vacuum". On the other hand, choosing

$$w = bt,$$

where b is a function, results in a different second postulate and a different definition of the second. Although the speed of light measured in units of meter per second is no longer a universal constant under this choice, this is purely because we have chosen a different definition of the 'second'. *The dimensionless speed of light expressed in natural units is still a universal constant.*

Thus, whether or not a second postulate is necessary for the formulation of special relativity can be seen as a difference in one's viewpoint of space and time. In Einstein's era, space and time were thought to be two very different entities as evidenced by the use of two different dimensions and two independently defined units to quantify them. Einstein (and to a lesser extent, Poincaré) was able to free himself sufficiently from that mode of thought to develop special relativity, in which space and time are unified in a four-dimensional space-time. However, Einstein also did not have the decades of experience working with special relativity as a fundamental part of physics that modern physicists have, and thus was not able to see all of the implications of his theory, such as that spatial and temporal quantities could then be expressed using the same units. For him, it was necessary to introduce a second postulate in order to obtain the universality of the speed of light in the system of units in use at the time. From our modern point of view in which both spatial and temporal intervals can be expressed using the same units, it is revealed that the universality of the speed of light (when expressed as a dimensionless velocity in natural units) is indeed a direct consequence of the principle of relativity.

References

1. L. Landau and E. Lifshitz, *The Classical Theory of Fields* (Transl. M. Hamermesh, Addison-Wesley, 1957) p. 2; E. F. Taylor and J. A. Wheeler, *Spacetime Physics* (Second Ed., W. H. Freeman and Company, 1992) p. 60.
2. A. A. Tyapkin, Sov. Phys. Usp. **15**, 205 (1972).
3. J. P. Hsu and L. Hsu, *A Broader View of Relativity, General Implications of Lorentz and Poincaré Invariance* (Second Ed. World Scientific. 2006), chapters 13–15. Available online: google books, a broader view of relativity, Hsu. See also Jong-Ping Hsu and

Leonardo Hsu, Phys. Letters A **196**, 1 (1994); Leonardo Hsu and Jong-Ping Hsu, *Nuovo Cimento B*, **111**, 1283 (1996).

4. See, for example, *The New Lin Yutang Chinese–English Dictionary* (Ed. Lai Ming and Lin Tai-Yi, Panorama Press, Hong Kong, 1987) p. 90.

5. We can demonstrate that the principle of relativity demands that the transformation be linear in the following way: Let us consider the relations between (w, x) and (w', x'). Suppose they are related by non-linear relations: $w' = aw + bx + cwx + ew^2 + fx^2 + \dots$, and $x' = Ax + Bw + Cwx + \dots$, where $a, b, c, \dots A, B, C, \dots$ are constant. After taking differentiations, one has $dw' = a^* dw + b^* dx$ and $dx' = A^* dw + B^* dx$, where a^*, b^*, A^* and B^* are, in general, functions of w and x. For an object at rest in the inertial frame $F'(w', x')$, i.e., $dx' = 0$, its velocity as measured in the inertial frame $F(w, x)$ is $dx/dw = \beta = $ constant. This relation, together with the infinitesimal 4-dimensional interval, leads to the result $a^* = A^* = \gamma$ and $b^* = B^* = -\gamma\beta$, where γ is given by (2.5). This implies that a^*, b^*, A^* and B^* are constants. It follows that the relations between (w', x') and (w, x) are linear.

6. For a free particle, $r^2/w^2 = \beta^2 = $ constant. Eq. (2.3) can be written as $m^2 w^2/s^2 - m^2 r^2/s^2 = m^2$, which is exactly the same as the 'energy–momentum' relation $(p_0)^2 - p^2 = m^2$ because $p_0 = m/\sqrt{1 - \beta^2}$ and $\mathbf{p} = m\beta/\sqrt{1 - \beta^2}$.

7. The speed of a plane wave in vacuum can be determined by the motion of a particular constant value of its phase, $k_0 w - \vec{k} \cdot \vec{r} = $ *constant*. Since the wave 4-vector of light satisfies $(k_0)^2 - (\vec{k})^2 = 0$, the speed is $dr/dw = \pm k_0/|\vec{k}| = \pm 1$, which implies $dw^2 - dr^2 = 0$ or $ds^2 = 0$. The finite transformation (2.4) is equivalent to the transformation of the differentials, $dw' = \gamma(dw - \beta dx), dx' = \gamma(dx - \beta dw), dy' = dy, dz' = dz$. In the same sense, the finite 4-dimensional interval (2.3) is equivalent to the differential interval $ds^2 = dw'^2 - dr'^2 = dw^2 - dr^2$.

8. See, for example, J. P. Hsu and Y. Z. Zhang, *Lorentz and Poincaré Invariance, 100 Years of Relativity* (World Scientific, New Jersey, 2001) p. xxvi.

9. E. F. Taylor and J. A. Wheeler, *Spacetime Physics* (Second Ed., W. H. Freeman and Company, 1992); T. Moore, *Six Ideas that Shaped Physics* (McGraw-Hill Science, 2002). For an early discussion of using only one unit to express time and length, see H. Poincaré, *Compt. Rend. Acad. Sci. Paris* **140**, 1504 (1905) and *Rend. Circ. Mat. Palermo* **21**, 129 (1906). For an English translation of the latter, see H. M. Schwartz, *Am. J. Phys.* **39**, 1287 (1971), **40**, 862 (1972), **40**, 1282 (1972).

10. J. P. Hsu, *Nuovo Cimento B* **74**, 67 (1983); *Phys. Letters A* **97**, 137 (1983); Editorial, *Nature* **303**, 129 (1983); J. P. Hsu, *Found. Phys.* **8**, 371 (1978); **6**, 317 (1976); J. P. Hsu and T. N. Sherry, *ibid* **10**, 57 (1980); J. P. Hsu and C. Whan, *Phys. Rev. A* **38**, 2248 (1988),

Appendix. These papers discuss a four-dimensional symmetry framework based on the usual first postulate and a different second postulate, namely, a common time $t' = t$ (or $w = bt$ and $w' = b't$) for all observers in different inertial frames.

11. Such a radically different relationship, $w = bt$, also implies a change in the concept of simultaneity in terms of the time t within the 4-dimensional symmetry framework. For a comprehensive discussion of the arbitrariness of the concept of simultaneity, see A. A. Tyapkin, ref. 2.

12. J. P. Hsu and L. Hsu, *A Broader View of Relativity, General Implications of Lorentz and Poincaré Invariance* (Second Ed. World Scientific. 2006), chapters 7–16. Available online: google books, a broader view of relativity; see also Hsu. Editorial, Nature **303**, 129 (1983).

3.

The Principle of Limiting Continuation of Physical Laws and Coordinate Transformations for Frames with Constant Accelerations

3-1. The principle of limiting continuation

One of the most powerful methodologies in physics for understanding a complex system or situation is that of stripping out its complications and reducing it to its bare essentials in order to create a tractable problem that can be modeled using known principles. As the system becomes better understood, the various complications can be restored, perhaps using perturbation theory or the like, to create progressively more sophisticated representations until the predictive and explanatory power of the model satisfies some criteria of precision or adequacy relative to the real-world situation. In this tradition, physical theories have all been formulated for the case of inertial frames even though, strictly speaking, all physically realizable frames are non-inertial due to the gravitational interaction and the accelerated expansion of the universe. Although the inertial frame approximation is adequate for a great number of situations, ideally, our understanding of physical principles should not be restricted to inertial frames.

Thus far, general relativity represents our best theory for understanding physics in non-inertial frames. Nevertheless, it suffers from a number of drawbacks that limit its predictive and explanatory power. One of the most prominent of these drawbacks is that the mathematical framework of general relativity appears to be too general to develop an explicit metric tensor for accelerated frames.[a] Thus, in order to write down an explicit coordinate transformation between a non-inertial and an inertial frame, or between two non-inertial frames, one must make additional assumptions or impose additional conditions that are not part of the framework of general relativity.

---

[a] Space-time coordinates in general relativity have no operational or physical meaning, E. P. Wigner, *Symmetries and Reflections, Scientific Essays* (The MIT Press, 1967), pp. 52–53; see also section 0-1 in this volume.

Even then, one must be careful to impose the 'right' additional conditions (in effect, to choose a 'correct' coordinate system) or else non-physical terms may appear in the calculations.[1] This is fundamentally different from the case in classical physics, where an injudicious choice of coordinates can lead to a mathematically intractable, but still physically correct, set of equations. This demand for the selection of a 'correct' coordinate system appears to be inconsistent with the principle of general coordinate invariance and, although this problem has been recognized for decades, no completely satisfactory resolution has yet been found.

In the meantime, there have been several attempts to develop explicit coordinate transformations involving non-inertial frames,[2] none of which have been completely satisfactory. For example, Møller's transformation, sometimes referred to as Rindler coordinates, reduces to the identity transformation in the limit of zero acceleration, rather than the Lorentz transformation.

In this and the next two chapters, we describe our approach to generalizing the Lorentz transformation to both inertial and non-inertial frames. Unlike many of the previous attempts, which begin with general relativity and choose a particular gauge (such as the harmonic gauge), we base our derivation on what we call the principle of limiting continuation of physical laws. Simply put:

The laws of physics in a reference frame F_1 with an acceleration a_1 must reduce to those in a reference frame F_2 with an acceleration a_2 in the limit where a_1 approaches a_2.[b]

As a practical matter, the space of possibilities of explicit metric tensors for frames with arbitrary accelerations is too large for the principle of limiting continuation to be of much use. However, we can accomplish this task by making the generalization of the metric tensors from inertial frame to accelerated frames in a series of small steps. In this chapter, we first make the minimum generalization by using the principle of limiting continuation to develop an explicit coordinate transformation between an inertial frame and another frame with a constant acceleration point along the same line as its relative velocity. We will call this motion constant-linear-acceleration (CLA) and will discuss later in this chapter an operational definition for "constant acceleration." We then derive the metric tensor

---

[b] Since the accelerations a_1 and a_2 are not specified, in the special case that $a_1 \to a_2 = 0$, this principle of limiting continuation of physical laws reduces to the principle of relativity in the zero acceleration limit, provided the limit exists. In this sense, the principle of limiting continuation of physical laws includes the 'limiting Poincaré and Lorentz invariance' as a special case, and appears to be a simple generalization of the principle of relativity from inertial frames to non-inertial frames.

for CLA frames, explore the group properties of the coordinate transformation and use it to make a number of specific experimental predictions. In chapter 4, we use our results for CLA frames to develop an explicit coordinate transformation between an inertial frame and another frame with an arbitrary-linear-acceleration parallel pointing along the same line as its relative velocity (ALA). Finally, in chapter 5, we use our results to develop an explicit coordinate transformation between an inertial frame and another frame that rotates about a fixed point. In all cases, the space-time coordinates retain their traditional operational definitions, unlike the arbitrary coordinates in general relativity.

All inertial and non-inertial frames will be treated within the taiji symmetry framework, i.e., using a flat space-time with arbitrary coordinates and the Poincaré metric tensor, which is the Minkowski metric tensor generalized from inertial to non-inertial frames. As mentioned previously, the taiji symmetry framework has the advantages that it has the maximum space-time symmetry, guaranteeing the law of conservation of the energy–momentum tensor and enabling the quantization of all physical fields (including gravity) when formulated as a (generalized) Yang–Mills gauge field. These subjects will be discussed later in the book.

3-2. Constant linear acceleration: The Wu transformations

Our first step in constructing a coordinate transformation for accelerated frames is to make a minimal generalization from the Poincaré transformations[3] to a set of coordinate transformations between an inertial frame and a frame with a constant-linear-acceleration (CLA), where both the acceleration and velocity of the non-inertial frame point along the same line (they may be either parallel or anti-parallel). Postulating a minimum departure[4] of the CLA transformations from the Poincaré transformations, we write the relationship between the infinitesimal space-time intervals of an inertial frames $F_I(x_I) = F_I(w_I, x_I, y_I, z_I)$ and a constant-linear-acceleration (CLA) frame $F(x) = F(w, x, y, z)$ as

$$ds^2 = W^2 dw^2 - dx^2 - dy^2 - dz^2 = dw_I^2 - dx_I^2 - dy_I^2 - dz_I^2. \tag{3.1}$$

In general, W is a function of w and x as long as the motion of the CLA frame $F(x)$ is along the parallel x and x_I axes. We note that if one were to postulate the form $ds^2 = W^2 dw^2 - X^2 dx^2 - Y^2 dy^2 - Z^2 dz^2$ or $ds^2 = W^2 dw^2 - R^2(dx^2 + dy^2 + dz^2)$, then there are too many unknown functions to be determined by the principle of limiting continuation. In addition, if one were to postulate

$ds^2 = dw^2 - R^2(dx^2 + dy^2 + dz^2)$, then there is no physical solution for the accelerated transformation. Equation (3.1) leads to the following transformations for the coordinate differentials:

$$dw_I = \gamma(Wdw + \beta dx), \quad dx_I = \gamma(dx + \beta Wdw),$$

$$dy_I = dy, \quad dz_I = dz; \quad \gamma = \sqrt{1 - \beta^2}, \tag{3.2}$$

where $W = W(x, w)$ and $\beta = \beta(w)$ are unknown functions to be determined. In order for a finite space-time transformation between F_I and F to exist, these unknown functions must satisfy the integrability conditions[5]

$$\frac{\partial(\gamma W)}{\partial x} = \frac{\partial(\gamma \beta)}{\partial w}, \quad \frac{\partial \gamma}{\partial w} = \frac{\partial(\gamma \beta W)}{\partial x}. \tag{3.3}$$

Assuming the usual linear relationship between the velocity β and the evolution variable w, i.e., $\beta = \alpha_o w + \beta_o$, leads to the finite transformation[5,6]

$$w_I = \gamma\beta\left(x + \frac{1}{\alpha_o \gamma_o^2}\right) - \frac{\beta_o}{\alpha_o \gamma_o} + w_o,$$

$$x_I = \gamma\left(x + \frac{1}{\alpha_o \gamma_o^2}\right) - \frac{1}{\alpha_o \gamma_o} + x_o, \quad y_I = y + y_o, \quad z_I = z + z_o;$$

$$\tag{3.4}$$

$$\gamma = \frac{1}{\sqrt{1 - \beta^2}}, \quad \gamma_o = \frac{1}{\sqrt{1 - \beta_o^2}}, \quad \beta = \alpha_o w + \beta_o,$$

which satisfies the initial conditions $x_I = x_o$ and $w_I = w_o$ when $w = x = 0$. We call (3.4) the Wu transformations in honor of T. Y. Wu's idea of a kinematical approach to deriving transformations between accelerated frames in a space-time with a vanishing Riemann–Christoffel curvature.[5] As expected, in the special case $\beta_o = 0$ and $w_o = x_o = y_o = z_o = 0$, the Wu transformation (3.4) reduces to Møller's accelerated transformation,[2,7,8]

$$w_I = \left(x + \frac{1}{\alpha_o}\right)\sinh(\alpha_o w^*), \quad x_I = \left(x + \frac{1}{\alpha_o}\right)\cosh(\alpha_o w^*) - \frac{1}{\alpha_o},$$

$$\tag{3.5}$$

$$y_I = y, \quad z_I = z,$$

with a change of the time variable, $w = (1/\alpha_o)\tanh(\alpha_o w^*)$. It further-more reduces smoothly to the Poincaré transformation in the limit of zero acceleration, $\alpha_o \to 0$,

$$w_I = \gamma_0(w + \beta_o x) + w_o, \quad x_I = \gamma_0(x + \beta_o w) + x_o,$$

$$y_I = y + y_o, \quad z_I = z + z_o, \tag{3.6}$$

in accordance with the principle of limiting continuation.[c]

This smooth connection with the Lorentz transformation in the limit of zero acceleration when $w_o = x_o = y_o = z_o = 0$ in (3.6) is a crucial property of the accelerated transformations not found in other proposed transformations[2] between non-inertial frames. The inverse Wu transformation derived from (3.4) is

$$w = \frac{w_I - w_o + \beta_o/(\alpha_o\gamma_o)}{\alpha_o(x_I - x_o) + 1/\gamma_o} - \frac{\beta_o}{\alpha_o},$$

$$x = \sqrt{(x_I - x_o + 1/\gamma_o\alpha_o)^2 - (w_I - w_o + \beta_o/\alpha_o\gamma_o)^2} - \frac{1}{\alpha_o\gamma_o^2}, \tag{3.7}$$

$$y = y_I + y_o, \quad z = z_I + z_o,$$

and differentiation of (3.4) leads to (3.2) with

$$W = \gamma^2(\gamma_o^{-2} + \alpha_o x), \quad \gamma = \frac{1}{\sqrt{1 - \beta^2}} \quad \beta = \alpha_o w + \beta_o. \tag{3.8}$$

We can now write the invariant infinitesimal interval (3.1) in terms of the Minkowski metric tensor $\eta_{\mu\nu}$ for F_I and $P_{\mu\nu}$ for CLA frame F,

$$ds^2 = \eta_{\mu\nu}dx_I^\mu dx_I^\nu = P_{\mu\nu}dx^\mu dx^\nu, \tag{3.9}$$

$$\eta_{\mu\nu} = (1, -1, -1, -1), \quad P_{\mu\nu} = (W^2, -1, -1, -1), \quad W = \gamma^2(\gamma_o^{-2} + \alpha_o x).$$

Thus, the physical space-time of a CLA frame is characterized by $P_{\mu\nu}$, which we call the Poincaré metric tensor, which reduces to the Minkowski metric $\eta_{\mu\nu}$ in the limit of zero acceleration. We will refer to W in the metric tensors $P_{\mu\nu}$ and the differential form of the Wu transformation (3.2) as the Wu factor, whose physical implications will be discussed later.

One can also generalize the Wu transformation to the case where the time-dependent velocity β and the constant-linear-acceleration are in the same

---

[c]In (3.4), we use the approximation $\gamma \approx \gamma_0(1 + \beta_0\alpha_0 w\gamma_0^2)$, which is obtained by expanding γ to first order in α_o (which is assumed to be small), while keeping $\beta_o < 1$ to all orders.

arbitrary direction $\boldsymbol{\beta}/\beta$ not necessarily along the x/x_I-axis. In this case, the finite Wu transformation (3.4) becomes

$$w_I = \gamma\beta\left[\frac{\boldsymbol{\beta}\cdot\mathbf{r}}{\beta} + \frac{1}{\alpha_o\gamma_o^2}\right] - \frac{\beta_o}{\alpha_o\gamma_o} + w_o,$$

$$\mathbf{r}_I = \mathbf{r} + \left[(\gamma-1)\frac{\boldsymbol{\beta}\cdot\mathbf{r}}{\beta} + \left(\frac{\gamma}{\gamma_o}-1\right)\frac{1}{\alpha_o\gamma_o}\right]\frac{\boldsymbol{\beta}}{\beta} + \mathbf{r}_o, \qquad (3.10)$$

$$\gamma = \frac{1}{\sqrt{1-\beta^2}}, \quad \gamma_o = \frac{1}{\sqrt{1-\beta_o^2}}, \quad \beta = \alpha_o w + \beta_o, \quad \alpha_o\|\boldsymbol{\beta}_o.$$

3-3.   Operational meaning of the space-time coordinates and 'constant-linear-acceleration'

It is important to define what "physical space-time coordinates" are in accelerated frames and how one can realize them operationally. In inertial frames, the space-time coordinates (w_I, x_I, y_I, z_I) can be pictured as a rigid and uniform grid of meter sticks and identical clocks that have been synchronized using isotropic light signals.[d] Let us use the term "space-time clock" to denote a clock (on the grid) that shows both the time w_I and the position r_I of the clock, i.e., (w_I, x_I, y_I, z_I). One can always use the space-time clock closest to an event to record the space-time coordinates of that event.

In CLA frames, since the Wu transformation between an inertial and a CLA frame is non-linear, one can no longer picture the space-time coordinates of a CLA frame as a uniform grid of space-time clocks. However, consider the inverse Wu transformation (3.7). The space-time clocks in the CLA frame can be synchronized using only this transformation. A "clock" is simply a device that shows a time and position and, if based on say a computer chip, it can run at an arbitrary rate and show an arbitrary time and position. In order to synchronize clocks in the CLA frame, we can imagine that the computer clocks associated with that frame have the ability to measure their position r_I relative to the F_I frame, to obtain w_I from the nearest F_I clock, and then to compute and display w and $r = (x, y, z)$ using (3.7) (with known velocity β_o and acceleration α_o) on a readout. If these clocks in the CLA frame are uniformly spaced on a regular grid, the differences in their

---

[d]Cf. Ref. 5, sections 7d and 18c.

time readings will not be uniform. However, this grid of computer clocks will automatically become the more familiar Einstein clocks of inertial frames as the acceleration α_o approaches zero, provided that $w = ct$ and $w' = ct'$. This indicates that the coordinates (w, x, y, z) in (3.7) or (3.4) for a CLA frame play the same role and have an analogous physical meaning to the space-time coordinates in the transformations for inertial frames.

We now examine the operational definition of "constant-linear-acceleration." Consider the invariant action S_f for a "free particle" in a CLA frame, which can be associated with the Wu transformations with time variable w

$$S_f = -\int_a^b mds = \int_{w_a}^{w_b} L_w dw, \tag{3.11}$$

$$ds^2 = W^2 dw^2 - dx^2 - dy^2 - dz^2, \quad P_{\mu\nu} = (W^2, -1, -1, -1),$$

$$L_w = -m\sqrt{P_{\mu\nu}u^\mu u^\nu} = -m\sqrt{W^2 - (\beta^i)^2}, \quad u^\mu = \frac{dx^\mu}{dw} = (1, \beta^i),$$

where $W(w, x) = \gamma^2(g_0^{-2} + \alpha_o x)$.

As usual, the covariant momentum p_i and the corresponding energy p_0 (or the Hamiltonian H) with dimensions of mass are given by

$$p_i = -\frac{\partial L_w}{\partial \beta^i} = \left(\frac{-m\Gamma\beta_x}{W}, \frac{-m\Gamma\beta_y}{W}, \frac{-m\Gamma\beta_z}{W}\right), \quad p_i = P_{ik}p^k = -p^i, \tag{3.12}$$

$$p_0 = \left(\frac{\partial L_w}{\partial \beta^i}\beta^i - L_w\right) = m\Gamma W = P_{00}p^0, \quad p^0 = m\frac{dx^0}{ds} = m\frac{dw}{ds}; \tag{3.13}$$

$$\Gamma = \frac{1}{\sqrt{1 - \beta^2/W^2}}, \quad \beta^2 = \beta_x^2 + \beta_y^2 + \beta_z^2 = -\beta_i\beta^i, \quad \beta^i = \frac{dx^i}{dw} = u^i,$$

where $i = 1, 2, 3$. Thus, the covariant momentum $p_\mu = (p_0, p_1, p_2, p_3) = (p_0, -\mathbf{p})$ transforms like the covariant coordinate $dx_\mu = P_{\mu\nu}dx^\nu$. We have

$$p_{I0} = \gamma\left(\frac{p_0}{W} - \beta p_1\right), \quad p_{I1} = \gamma\left(p_1 - \beta\frac{p_0}{W}\right), \quad p_{I2} = p_2, \quad p_{I3} = p_3; \tag{3.14}$$

$$p_{I0}^2 - \mathbf{p}_I^2 = (p_0/W)^2 - \mathbf{p}^2 = m^2. \tag{3.15}$$

Now consider a particle at rest in the CLA frame ($\mathbf{p} = 0$). From (3.14) and (3.15), we have $p_0 = mW$ and $p_{I0} = m\gamma$. Thus, when $\mathbf{r}$ is constant,

$$\left(\frac{dp_{I0}}{dx_I}\right)_x = m\gamma^3 \frac{d\beta}{dx_I} = \frac{m\alpha_0}{(\gamma_0^{-2} + \alpha_0 x)}; \quad \beta = \alpha_0 w + \beta_0. \tag{3.16}$$

Thus, we see that an object moving with a constant-linear-acceleration undergoes a uniform change per unit length of its energy p_{I0}, as measured in an inertial frame F_I. Similarly, a CLA frame is a reference frame that is co-moving with such an object. An object at rest in F and located at, say, $\mathbf{r} = (x, y, z)$ will have a constant change of energy p_{I0} per unit distance traveled as measured from an inertial frame F_I. The magnitude of this constant change depends on the initial velocity β_0 of the CLA frame and the location of the object in a highly non-trivial manner, as shown in (3.16).

The principle of limiting continuation of physical laws (where we treat space-time coordinate transformations as physical laws) is critical in obtaining the result (3.16). Serendipitously, what we mean by constant-linear-acceleration is precisely the type of acceleration that is realized in high energy particle accelerators, facilitating comparison between the theory and experiment, as will be discussed later in this chapter.

3-4.   Singular walls and horizons in accelerated frames

Let us now examine the mapping of physical coordinates between an inertial frame and a CLA frame. First, we consider the inverse Wu transformations (3.7) with $x_o^\mu = 0$, the case in which the origins of the inertial and CLA frame coincide at $w = w_I = 0$. From the transformation equation for w, we see that lines of constant w correspond to straight lines in F_I, and that all such lines pass through the point $(w_I, x_I) = (-\beta_0/(\alpha_0\gamma_0), -1/(\alpha_0\gamma_0))$. In contrast, from the transformation equation for x, lines of constant x correspond to hyperbolic lines that satisfy the equation

$$\left(x + \frac{1}{\alpha_0\gamma_0^2}\right)^2 = \left(x_I + \frac{1}{\gamma_0\alpha_0}\right)^2 - \left(w_I + \frac{\beta_0}{\alpha_0\gamma_0}\right)^2. \tag{3.17}$$

The graphs in Fig. 3.1 show lines of constant x (hyperbolic lines) and of constant w (straight lines) in the (w_I, x_I) plane for the Wu transformations with non-zero inertial velocity.

Fig. 3.1   Lines of constant w and x in the F_I (inertial) frame. The CLA frame has velocity $\beta_o = 0.5$ and accelerations $\alpha_o = -0.001$ (top graph) and $\alpha_o = +0.001$ (bottom graph).

Second, from an examination of the Wu factor W in (3.9), one can see that P_{00} vanishes when $x = -1/(\alpha_o \gamma_o^2)$ (with arbitrary y and z coordinates). This corresponds to a wall singularity in the CLA frame. All points in the inertial frame F_I are mapped into only a portion of the CLA frame F. The location of the wall depends on the magnitude of the acceleration and as would be expected, as the acceleration α_0 approaches zero, the distance between the wall and the origin of the CLA frame approaches infinity.

In the portion of space beyond the wall in the CLA frame ($x < -1/(\alpha_o \gamma_o^2)$), the differentials dw_I and Wdw have different signs, which is unphysical. This situation resembles the negative energy solutions for a classical particle in some ways. This portion of space is inaccessible to observers in inertial frames. The inverse Wu transformation (3.7) implies that as the coordinate x approaches $-1/(\alpha_o \gamma_o^2)$:

(a) Clocks slow down and stop at the singular wall

$$\left(\frac{dw}{dw_I}\right)_{dx=0} = \frac{1}{\gamma W} = \frac{(x_I + 1/\gamma_o \alpha_o)^2 - (w_I + \beta_o/\alpha_o \gamma_o)^2}{\alpha_o (x_I + 1/\alpha_o \gamma_o)^3} \to 0. \quad (3.18)$$

(b) The speed of light as measured by observers in the CLA frame F increases without bound

$$\left(\frac{dx}{dw}\right)_{ds=0} = \pm|W| = \pm \frac{\alpha_o (x_I + 1/\gamma_o \alpha_o)^2}{\sqrt{(x_I + 1/\gamma_o \alpha_o)^2 - (w_I + \beta_o/\alpha_o \gamma_o)^2}} \to \pm\infty. \quad (3.19)$$

Result (a) shows that the rate of ticking of a clock at rest relative to F at $(x, 0, 0)$ is slower than that of a clock at rest relative to F_I at $(x_I, 0, 0)$. Result (b) is due to the fact that the law for the propagation of light is given by $ds^2 = 0$, and is related to rate of clock ticking (a).

3-5. The Wu pseudo-group

It is well-established that the set of Poincaré transformations constitutes a unique one-to-one mapping between the whole of space-time in two inertial frames. The set of Poincaré transformations satisfies all group properties and form the Poincaré group. However, when the Poincaré transformations are generalized to the Wu transformations that map coordinates between inertial and non-inertial frames, the whole of space-time in the inertial frame is mapped to only a portion of the space-time in the accelerated frame, as was shown in the previous section, and the usual group properties are not completely satisfied. Nevertheless, as we will show, the Wu transformations satisfy a weaker set of criteria and form a "pseudo-group," as defined by Veblen and Whitehead.[9]

We first demonstrate explicitly that the subset of Wu transformations consisting of the transformations for all CLA frames with relative velocities along a particular direction (such as parallel x/x'-axes) satisfy the properties of a normal group. These transformations can be defined as non-linear transformations that preserve the quadratic form,

$$ds^2 = P_{\mu\nu}dx^\mu dx^\nu = P'_{\mu\nu}dx'^\mu dx'^\nu = P''_{\mu\nu}dx''^\mu dx''^\nu, \tag{3.20}$$

$$P_{\mu\nu} = (W^2, -1, -1, -1), \quad W = \gamma^2(\gamma_0^{-2} + \alpha_0 x),$$

$$P'_{\mu\nu} = (W'^2, -1, -1, -1), \quad W' = \gamma'^2(\gamma'^{-2}_0 + \alpha'_0 x'), \quad etc.$$

where $P_{\mu\nu}$, $P'_{\mu\nu}$ and $P''_{\mu\nu}$ are Poincaré metric tensors in the CLA frames $F(x)$, $F'(x')$ and $F''(x'')$, respectively. Each frame has its own constant acceleration, α_0, α'_0, and α''_0 with velocities $\beta(w) = \alpha_0 w + \beta_0$, $\beta'(w') = \alpha'_0 w' + \beta'_0$, and $\beta''(w'') = \alpha''_0 w'' + \beta''_0$.[10] With the help of (3.4), the transformations between an inertial frame F_I and the CLA frames $F(x)$, $F'(x')$ and $F''(x'')$ can be written as

$$w_I = \gamma\beta\left(x + \frac{1}{\alpha_0\gamma_0^2}\right) - \frac{\beta_0}{\alpha_0\gamma_0} = \gamma'\beta'\left(x' + \frac{1}{\alpha'_0\gamma'^2_0}\right) - \frac{\beta'_0}{\alpha'_0\gamma'_0} \tag{3.21}$$

$$= \gamma''\beta''\left(x'' + \frac{1}{\alpha''_0\gamma''^2_0}\right) - \frac{\beta''_0}{\alpha''_0\gamma''_0},$$

$$x_I = \gamma\left(x + \frac{1}{\alpha_0\gamma_0^2}\right) - \frac{1}{\alpha_0\gamma_0} = \gamma'\left(x' + \frac{1}{\alpha'_0\gamma'^2_0}\right) - \frac{1}{\alpha'_0\gamma'_0}$$

$$= \gamma''\left(x'' + \frac{1}{\alpha''_0\gamma''^2_0}\right) - \frac{1}{\alpha''_0\gamma''_0},$$

where $\gamma = 1/\sqrt{1 - \beta^2}$, $\gamma' = 1/\sqrt{1 - \beta'^2}$, and $\gamma'' = 1/\sqrt{1 - \beta''^2}$ and we have ignored the trivial transformations of y and z for simplicity. From (3.21), the coordinate transformations between the CLA frames F and F' are then

$$w = \frac{1}{\alpha_0}\left[\frac{\gamma'\beta'Q' - \beta'_0/(\alpha'_0\gamma'_0) + \beta_0/(\alpha_0\gamma_0)}{\gamma'Q' - 1/(\alpha'_0\gamma'_0) + 1/(\alpha_0\gamma_0)} - \beta_0\right], \tag{3.22}$$

$$x = \sqrt{\left(\gamma'Q' - \frac{1}{\alpha'_0\gamma'_0} + \frac{1}{\alpha_0\gamma_0}\right)^2 - \left(\gamma'\beta'Q' - \frac{\beta'_0}{\alpha'_0\gamma'_0} + \frac{\beta_0}{\alpha_0\gamma_0}\right)^2 - \frac{1}{\alpha_0\gamma_0^2}},$$

$$Q' \equiv \left(x' + \frac{1}{\alpha'_0\gamma'^2_0}\right), \quad \beta' = \alpha'_0 + \beta'_0,$$

with the inverse transformations

$$w' = \frac{1}{\alpha_o'} \left[\frac{\gamma\beta Q - \beta_0/(\alpha_o\gamma_o) + \beta_0'/(\alpha_o'\gamma_o')}{\gamma Q - 1/(\alpha_o\gamma_o) + 1/(\alpha_o'\gamma_o')} - \beta_0' \right], \qquad (3.23)$$

$$x' = \sqrt{\left(\gamma Q - \frac{1}{\alpha_o\gamma_o} + \frac{1}{\alpha_o'\gamma_o'} \right)^2 - \left(\gamma\beta Q - \frac{\beta_o}{\alpha_o\gamma_o} + \frac{\beta_o'}{\alpha_o'\gamma_o'} \right)^2 - \frac{1}{\alpha_o'\gamma_o'^2}},$$

$$Q \equiv \left(x + \frac{1}{\alpha_o\gamma_o^2} \right), \quad \beta = \alpha_o w + \beta_o.$$

Similarly from (3.21), the coordinate transformations between F and F'' are

$$w = \frac{1}{\alpha_o} \left[\frac{\gamma''\beta''Q'' - \beta_0''/(\alpha_o''\gamma_o'') + \beta_0/(\alpha_o\gamma_o)}{\gamma''Q'' - 1/(\alpha_o''\gamma_o'') + 1/(\alpha_o\gamma_o)} - \beta_0 \right], \qquad (3.24)$$

$$x = \sqrt{\left(\gamma''Q'' - \frac{1}{\alpha_o''\gamma_o''} + \frac{1}{\alpha_o\gamma_o} \right)^2 - \left(\gamma''\beta''Q'' - \frac{\beta_o''}{\alpha_o''\gamma_o''} + \frac{\beta_o}{\alpha_o\gamma_o} \right)^2 - \frac{1}{\alpha_o\gamma_o^2}},$$

$$Q'' \equiv \left(x'' + \frac{1}{\alpha_o''\gamma_o''^2} \right), \quad \beta'' = \alpha_o''w'' + \beta_o''.$$

From (3.23) and (3.24), the transformations between $F'(x')$ and $F''(x'')$ are

$$w' = \frac{1}{\alpha_o'} \left[\frac{\gamma''\beta''Q'' - \beta_0''/(\alpha_o''\gamma_o'') + \beta_0'/(\alpha_o'\gamma_o')}{\gamma''Q'' - 1/(\alpha_o''\gamma_o'') + 1/(\alpha_o'\gamma_o')} - \beta_0' \right], \qquad (3.25)$$

$$x' = \sqrt{\left(\gamma''Q'' - \frac{1}{\alpha_o''\gamma_o''} + \frac{1}{\alpha_o'\gamma_o'} \right)^2 - \left(\gamma''\beta''Q'' - \frac{\beta_o''}{\alpha_o''\gamma_o''} + \frac{\beta_o'}{\alpha_o'\gamma_o'} \right)^2 - \frac{1}{\alpha_o'\gamma_o'^2}}.$$

We see explicitly from (3.23), (3.24), and (3.25) that the transformations between any two CLA frames with relative motion along the $x/x'/x''$-axes have the same form, satisfying the group property of closure. Equation (3.23) demonstrates the existence of an inverse transformation that has the same form as the transformations themselves. Substituting (3.23) into (3.22) results in the identity transformation, $w = w$ and $x = x$. Finally, one can also show that this subset of Wu transformations satisfies the associative rule. Thus, this subset of Wu transformations satisfies all the properties of the usual continuous group.

However, the situation is different when transformations for non-parallel directions are considered. In the case of the Poincaré transformations, the combination of two Lorentz boosts in different directions is equivalent to a suitable combination of a boost and a rotation, and thus this compound boost still exists within the (usual) finite-dimensional set of Poincaré transformations. In the case of the Wu transformations however, the combination of two transformations for CLA frames with accelerations in different directions cannot be expressed as a combination of a single such transformation with a rotation. Thus, it does not belong to the usual finite-dimensional set of Wu transformations. In fact, the set obtained from generators of the Wu transformations using the Lie bracket, and the brackets of brackets, and so on, never closes. That is, the Wu transformations define a group with infinitely-many generators.[11] This difference is likely related to the distortion of space-time coordinates in accelerated frames relative to an inertial frame, as shown by the metric tensors $P_{\mu\nu}$ in (3.9). Also, the Wu transformations map only a portion of the space in a CLA frame F to a portion of the space in another CLA frame F'.

To deal with these types of transformations, mathematicians O. Veblen and J. H. C. Whitehead developed the concept of a pseudo-group[9]. A set of transformations forms a pseudo-group if

(i) the resultant of two transformations in the set is also in the set,

(ii) the set contains the inverse of every transformation in the set.

where "set" is defined as an infinite-dimensional set, rather than the usual Lie set. The complete set of Wu transformations for constant-linear-accelerations (in all directions) forms a pseudo-group we will call the Wu pseudo-group.[12 e] These pseudo-group properties are determined by the physical requirements of non-inertial frames and their accelerations, including the limiting continuation of physical laws and space-time transformations, within the taiji symmetry framework.

3-6. Relationship between the Wu and Møller transformations

The solutions of the integrability conditions (3.3) are not unique. Instead of assuming (or defining) the velocity β to be a linear function of time w

---

[e]In fact, the properties of the Wu pseudo-group are richer than those of traditional pseudo-groups because the complete set of Wu transformations also satisfies the inverse and associative properties of normal groups.

(i.e., $\beta = \alpha_o w + \beta_o$) to derive (3.4), suppose one assumes that $W(w, x)$ is independent of time, i.e., $W = W_x$ and that the velocity β is a function of time $\beta = \beta(w)$ only. We obtain a different solution

$$W_x = k_1 x + k_2, \quad \beta = \tanh(k_1 w + k_3), \tag{3.26}$$

in which the velocity β is a complicated hyperbolic tangent function of time. In this case, the time is in general different from that in the Wu transformations (3.4), so we will use w^* to denote time in the new solution (3.26). If the initial velocity is β_o, $\beta(0) = \beta_o$, then we have $k_3 = \tanh^{-1} \beta_o$.

Suppose we follow the approach of Wu and Lee[7] to impose a boundary condition that the velocity β is related to a constant acceleration α^* by the relation $\beta = \alpha^* w_I + \beta_o$ at $x_I = 0$. This condition enables us to determine $k_1 = \gamma_o \alpha^*$. In the limit of zero acceleration, where α^* or k_1 vanishes, W_x must be 1 for inertial frames. Thus, we have $k_2 = 1$ in (3.26). With the help of the principle of limiting continuation, we obtain

$$w_I = \left(x + \frac{1}{\gamma_o \alpha^*}\right) \sinh\left(\gamma_o \alpha^* w^* + \tanh^{-1} \beta_o\right) - \frac{\beta_o}{\alpha^*}$$

$$x_I = \left(x + \frac{1}{\gamma_o \alpha^*}\right) \cosh\left(\gamma_o \alpha^* w^* + \tanh^{-1} \beta_o\right) - \frac{1}{\alpha^*}, \tag{3.27}$$

$$y_I = y, \quad z_I = z,$$

which we call the generalized Møller–Wu–Lee (MWL) transformation for an inertial frame F_I and a CLA frame F with non-zero initial velocity. The Møller time w^* in the transformation (3.27) can take all values from $-\infty$ to $+\infty$. From the preceding discussions, we have $k_1 = \gamma_o \alpha^*$, $k_2 = 1$, $k_3 = \tanh^{-1} \beta_o$ and (3.27), so W_x and the velocity β in (3.26) (with w replaced by w^*), take the form

$$W_x = \gamma_o \alpha^* x + 1,$$

$$\beta = \tanh\left(\gamma_o \alpha^* w^* + \tanh^{-1} \beta_o\right) = \frac{w_I + \beta_o/\alpha^*}{x_I + 1/\alpha^*} \tag{3.28}$$

$$\rightarrow \alpha^* w_I + \beta_o, \quad \text{for } x_I \rightarrow 0.$$

In the special case of zero initial velocity, $\beta_o = 0$, the transformations (3.27) reduce to Møller's accelerated transformation (3.5) with $\alpha^* = \alpha_o$. This transformation was also derived by Wu and Lee using a purely kinematical approach in flat space-time. One can verify that in the limit of zero acceleration $\alpha^* \rightarrow 0$, (3.27) reduces to the Lorentz transformation

$$w_I = \gamma_0(w^* + \beta_0 x), \quad x_I = \gamma_0(x + \beta_0 w^*), \quad y_I = y, \quad z_I = z, \tag{3.29}$$

where

$$\sinh(\tanh^{-1}\beta_0) = \gamma_0\beta_0, \quad \cosh(\tanh^{-1}\beta_0) = \gamma_0.$$

Although the generalized MWL transformations (3.27) appear to be quite different from the Wu transformations (3.4), they are in fact related by a simple change of time variables. From (3.4), we have

$$\alpha_0 w + \beta_0 = \frac{w_I + \beta_0/(\gamma_0\alpha_0)}{x_I + 1/(\gamma_0\alpha_0)}. \tag{3.30}$$

Comparing (3.30) and (3.28) with the identification $\alpha^* = \alpha_0\gamma_0$, we see the following unique relation between the times w and w^*:

$$w = \frac{1}{\alpha_0}\left[\tanh\left(\gamma_0\alpha^*w^* + \tanh^{-1}\beta_0\right) - \beta_0\right], \quad \alpha^* = \alpha_0\gamma_0. \tag{3.31}$$

The generalized MWL transformation (3.27) can be obtained by substituting (3.31) into the Wu transformations (3.4). Both (3.27) and (3.4) reduce to the Lorentz transformation in the limit of zero acceleration and both have the same group properties. Although it may seem that only one of the two transformations can be correct because the readings on the clocks in the CLA frame will differ depending on whether one chooses to use (3.27) or (3.4), in actuality, both transformations are correct, in the sense of predicting the same physical outcomes. For example, let us re-derive the result for the change in energy per unit length of a particle undergoing CLA by following the steps from (3.11) to (3.16), but this time using the generalized MWL transformations (3.27). The metric tensor of the CLA frame is still $P_{\mu\nu} = (W^2, -1, -1, -1)$, but this time, the Wu factor $W = W_x = (1 + \alpha^*\gamma_0 x)$ is different. If a particle is at rest in the CLA frame ($\mathbf{p} = 0$), located at $\mathbf{r} = (x, y, z) = $ constant, then its change in energy per unit length, as measured in an inertial frame, would be given by

$$\left(\frac{dp_{I0}}{dx_I}\right)_x = m\gamma^3\frac{d\beta}{dx_I} = \frac{m\gamma_0\alpha^*}{(1 + \alpha^*\gamma_0 x)} = \frac{m\alpha_0}{\left(\gamma_0^{-2} + \alpha_0 x\right)}; \tag{3.32}$$

$$\beta = \tanh\left(\gamma_0\alpha^*w^* + \tanh^{-1}\beta_0\right), \quad \alpha^* = \alpha_0\gamma_0, \tag{3.33}$$

which is the same as the result (3.16) obtained using the Wu transformations (3.4) because we have the relation $\alpha^* = \alpha_0\gamma_0$. Thus, the Wu and the generalized MWL transformations have the same physical implications. The only quantities that would be different between the two are those that depend directly on clock readings, such as velocities and time intervals. However, physical predictions such as scattering cross sections, energies, and even the lifetime dilations of accelerated particles remain the same. This is a reflection of the fact that the choice of clock system for measuring physical quantities is purely a human convention and has no bearing on the physics itself. As long as the theory retains the essential limiting Lorentz and Poincaré invariance, its predictions will be consistent with experimental results. Within an experimentally consistent theoretical framework, a choice of clock synchronizations is analogous to a choice of coordinate systems. Different choices are convenient for different situations, but all lead to the same physics and physical predictions.

If one uses the generalized MWL transformations instead of the Wu transformations, one can set up a corresponding computer clock system in a CLA frame to show (w^*, x, y, z). Is this time w^* physical time for CLA frames?

It is reasonable to define 'physical time' as 'any clock system in CLA frames that is consistent with the lifetime or decay-length dilation of an accelerated unstable particle in flight'. In this sense, the 'times' or 'evolution variables' in both inertial and non-inertial frames have equal physical significance, as long as they can describe the physical phenomena of lifetime dilations of particles in flight. As we shall demonstrate in Section 3-7 below, the Wu transformation and the generalized MWL transformation predict the same lifetime dilation, as measured in an inertial frame. Therefore, both times w and w^* in the Wu and the generalized MWL transformations respectively are physical time for CLA frames. The situation is similar to relativistic and common times in relativity theory.[f] Our results suggest that any space-time transformations that preserve the fundamental metric tensor of the form $(W^2, -1, -1, -1)$ are physical space and time coordinates for CLA frames.

3-7.   Experimental tests of the Wu transformations

We now discuss two possible experimental tests of the Wu transformations. The first is the lifetime dilation of accelerated particles. The second is the Doppler effect on radiation from an accelerated source.

---

[f] See Ref. 5, Chapter 8.

Suppose a particle is at rest in an inertial frame F_I' and has a lifetime $\tau_I'(rest) = w_{I2}' - w_{I1}'$, as measured by an observer in F_I'. Its lifetime is $\tau_I = w_{I2} - w_{I1}$, as measured by observers in F_I. The Lorentz transformation (3.6) with $x^\mu = 0$ gives the usual lifetime dilation

$$\tau_I = \gamma_o \tau_I'(rest). \tag{3.34}$$

Experimentally, it is difficult to measure $\tau_I'(rest)$ because it is almost impossible to arrange for an observer to be in a frame that is co-moving with unstable particles in a high-energy laboratory. Fortunately, according to the principle of relativity, all inertial frames are equivalent. Thus, the lifetime $\tau_I(rest)$ of a particle at rest in F_I and measured by observers in F_I is the same as the lifetime $\tau_I'(rest)$ of the same kind of particle at rest in F_I' and measured by observers in F_I',

$$\tau_I(rest) = \tau_I'(rest). \tag{3.35}$$

Thus, the lifetime dilation (3.34) in standard special relativity can be expressed in terms of observable quantities in an inertial frame F_I,

$$\tau_I = \gamma_o \tau_I(rest),$$

which has been confirmed by high-energy experiments in the laboratory frame F_I.

An inertial frame F_I and a CLA frame F are not equivalent. However, in order to calculate the lifetime of an accelerated particle to compare it with experiment, one needs to postulate a relation similar to (3.35). Let us consider observable vectors such as space-time differential dx^μ, the wave vector k_μ, or a momentum vector. All of these physically observable vector quantities (e.g., energy–momentum of a particle, wave vector of the radiation emitted from a source, etc.) can be denoted by contravariant or covariant vectors, $V^\mu(rest)$ and $V_\mu(rest)$, if the particle or the source of light waves is 'at rest' in a CLA frame F and these quantities are measured by observers in the same CLA frame F. Following the principle of limiting continuation for physical laws, it is natural to postulate the following relation for two different CLA frames, F and F',

$$V^0(rest)\sqrt{P_{00}} = V'^0(rest)\sqrt{P_{00}'}, \tag{3.36}$$

$$or \quad V^0(rest)W = V'^0(rest)W',$$

which can also be expressed in terms of a covariant component

$$V_0(rest)\sqrt{P^{00}} = V'_0(rest)\sqrt{P'^{00}}, \tag{3.37}$$

$$or \quad \frac{V_0(rest)}{W} = \frac{V'_0(rest)}{W'},$$

where we have used $V_\mu = P_{\mu\nu}V^\nu$ or $V_0 = W^2 V^0$. This postulate (3.36) or (3.37) may be termed the 'weak equivalence of non-inertial frames.' Relationships (3.36) and (3.37) cannot be derived from the Wu transformations (3.21). However, they are consistent with the principle of limiting continuation of physical laws.

Now, let us consider a specific example of the lifetime of a particle at rest in a CLA frame $\tau(rest)$. Equation (3.36) implies that this lifetime is related to the lifetime of the same kind of particles at rest in inertial frame F_I (for which the Wu factor W is 1) by

$$\tau(rest)W = \tau_I(rest), \quad W = \gamma^2(\gamma_o^{-2} + \alpha_o x), \tag{3.38}$$

where the lifetime of a particle is usually sufficient small and can be considered as the zeroth component of a contravariant differential vector dx^μ. This relationship is consistent with the principle of limiting continuation and can be tested indirectly by experiments of measuring the lifetime dilation of accelerated particles.

Since the lifetime of unstable particles is usually very short, those lifetimes τ_I and τ, as measured by observed in F_I and F, can be identified with the infinitesimal time intervals dw_I and dw in (3.2), respectively. Suppose a particle is at rest with $\mathbf{r} = 0$ and $d\mathbf{r} = 0$ in a CLA frame, and has a position $x_I = L$ as measured in F_I. From (3.2) with $d\mathbf{r} = 0$ and (3.38), we have

$$\tau_I = \frac{1}{\sqrt{1-\beta^2}}\tau(rest)W = \gamma_o(1 + \alpha_o L)\tau_I(rest), \tag{3.39}$$

where all quantities are measured in the inertial laboratory frame F_I.

According to the Wu transformations, a constant-linear-acceleration α_o corresponds physically to a uniform increase in the energy of the particle per unit length traveled. This prediction (3.39) for the dilation of the lifetime of an unstable particle undergoing such a constant-linear-acceleration can be tested experimentally by measurements of the decay-length (or equivalently, the "lifetime") of an unstable particle in a linear accelerator.[13]

Given the discussion regarding the relationship between the Wu transformations and the generalized MWL transformations in Section 3-6, one might wonder whether the generalized MWL transformations predict the same result (3.39). The answer is yes. Differentiating the generalized MWL transformation (3.27), one obtains (3.2) with $W = W_x = 1 + \alpha^* \gamma_o x$ and β given by (3.28). The Wu factor W in (3.38) would be replaced by W_x. Substituting in these results and simplifying, one again obtains the same result (3.39), supporting the equivalence of the Wu and the generalized MWL transformations.

The Wu transformations (3.4) can also be tested experimentally by measuring the shift of the wavelength of light emitted from a source undergoing a constant-linear-acceleration. From Equation (3.14) one can obtain the Wu transformations of the covariant wave 4-vector $k_\mu = p_\mu$ between an inertial frame F_I and a CLA frame F:

$$k_{I0} = \gamma \left(\frac{k_0}{W} - \beta k_1 \right), \quad k_{I1} = \gamma \left(k_1 - \frac{\beta k_0}{W} \right), \tag{3.40}$$

and

$$k_{I0}^2 - k_{I1}^2 = \left(\frac{k_0}{W} \right)^2 - (k_1)^2 = 0, \tag{3.41}$$

where $k_{I2} = k_2 = 0, k_{I3} = k_3 = 0$. The quantities k_{I0} and k_0 are the frequencies (or equivalently, energies) of the same wave measured by observers in F_I and F respectively.

According to the weak equivalence of non-inertial frames (3.37), $k_0(rest)$ and $k_{I0}(rest)$ satisfy

$$\frac{k_0(rest)}{W} = k_{I0}(rest), \quad W = \gamma^2 \left(\gamma_o^{-2} + \alpha_o x \right), \tag{3.42}$$

where k_0 is the zeroth component of the covariant wave vector. From (3.40), (3.42) and $\beta = \alpha_o w + \beta_o$, we obtain the shift in frequency k_{I0} and wavelength λ_I

$$k_{I0} = k_{I0}(rest)\gamma(1 - \beta) \approx k_{I0}(rest)\gamma_o(1 - \beta_o)\left(1 - \alpha_o w \gamma_o^2\right), \tag{3.43}$$

$$\frac{1}{\lambda_I} = \frac{1}{\lambda_I(rest)}\gamma(1 - \beta) \approx \frac{1}{\lambda_I(rest)}\gamma_o(1 - \beta_o)\left(1 - \alpha_o w \gamma_o^2\right), \tag{3.44}$$

for radiation emitted from a source undergoing constant-linear-acceleration. In (3.43) and (3.44), we have again used the approximation $\gamma \approx \gamma_o(1 + \beta_o \alpha_o w \gamma_o^2)$, obtained by expanding γ to the first order in α_o (which is assumed to be small), while keeping $\beta_o < 1$ to all orders. We call the results (3.43) and (3.44) the Wu-Doppler shift or the Wu-Doppler effect.

For comparison with experiments, it is more convenient to express the quantities in (3.43) and (3.44) in terms of distances, rather than the time w of the CLA frame. For this purpose, the following relation may be used to express $\alpha_o w$ in (3.43) and (3.44) in terms of distances,

$$\alpha_o w + \beta_o = \frac{w_I + \beta_o/(\alpha_o \gamma_o)}{x_I + 1/(\alpha_o \gamma_o)} = \sqrt{1 - \left(\frac{\gamma_o^{-1} + \alpha_o \gamma_o x}{1 + \alpha_o \gamma_o x_I}\right)^2}. \qquad (3.45)$$

Suppose that the radiation source is located at $\mathbf{r} = 0$ in the CLA frame, and enters an accelerated potential at $x_I = 0$ in the laboratory frame F_I with an initial velocity β_o (measured from F_I). The radiation is then measured when the source has been accelerated to the point $x_I = L$. In this case, Equation (3.45) with small α_o can be approximated as

$$\alpha_o w \approx \sqrt{2\alpha_o L + \beta_o^2} - \beta_o. \qquad (3.46)$$

From Equations (3.44) and (3.46), we obtain

$$\delta\lambda_I = \lambda_I - \lambda_I(rest) \approx \lambda_I(rest)\left[-\frac{3}{2}\beta_o^2 - (1 - 2\beta_o)\sqrt{2\alpha_o L + \beta_o^2}\right], \qquad (3.47)$$

to first order in α_o and to second order in β_o. As expected, in the limit of zero acceleration, $\alpha_o \to 0$, the Wu-Doppler shift (3.47) reduces to the usual relativistic formula (to order β_o^2) for the Doppler shift.

As with our result for the lifetime dilation of an accelerated particle, we can also check our result (3.47) using the generalized MWL transformations rather than the Wu transformations. Following the derivation from (3.40) to (3.44) using the generalized MWL transformations (3.27), one again has the same Equations for (3.40) through (3.42), provided the Wu factor W is replaced by W_x and the velocity β is given by (3.28), where $\alpha^* = \alpha_o \gamma_o$. Using (3.28), we also obtain (3.43) and (3.44), but with $\alpha_o w \gamma_o$ replaced by $\alpha^* \gamma_o w^*$. Following steps (3.45) through (3.47), one can express the time w^* in the expression $\alpha^* \gamma_o w^*$ in terms of the distance L, and again arrive at the same expression (3.47) for the Wu-Doppler effect. Thus, the generalized MWL transformations are physically equivalent to the Wu transformations.

The predictions (3.44) and (3.47) can be tested in the laboratory frame F_I using a method similar to that in the Ives–Stilwell experiment.[14]

References

1. V. A. Brumberg and S. M. Kopeikin, in *Reference Frames*, (Astrophysics and Space Science Library, Vol. 154, edited by J. Kovalevsky, I. I. Mueller, and B. Kolaczek, Springer, 1989), pp. 115–141.

2. C. Møller, *Danske Vid. Sel. Mat-Fys.* **20**, No. 19 (1943); and *The Theory of Relativity* (Oxford Univ. Press, London, 1952), pp. 253–258; H. Lass, *Am. J. Phys.* **31**, 274 (1963); W. Rindler, *Am. J. Phys.* **34**, 1174 (1966); R. A. Nelson, *J. Math. Phys.* **28**, 2379 (1987), S. G. Turyshev, O. L. Minazzoli, and V. T. Toth, *J. Math. Phys.* **53**, 032501 (2012).

3. The name 'Poincaré group' was probably first used in a paper by E. Wigner and in a set of lecture notes by A. S. Wightman (informed by R. F. Streater). E. Wigner, *Proc. Nat'l Acad. Sci.* **51**, No. 5 (May, 1964), and *Symmetries and Reflections, Scientific Essays* (MIT Press, 1967) pp. 18–19. Wigner said that the geometrical principles of invariance "were recognized by Poincaré first, and I like to call the group formed by these invariables the Poincaré group." See also, H. Poincaré, *Compt. Rend. Acad. Sci. Paris,* **140**, 1504 (1905); *Rend. Circ. Mat. Palermo,* **21**, 129 (1906).

4. This is analogous to the case in which gauge symmetry does not uniquely determine the electromagnetic action and one must also postulate a minimal electromagnetic coupling. See J. J. Sakurai, *Invariance Principles and Elementary Particles* (Princeton Univ. Press, 1964) p. v, pp. 3–5, and p. 10.

5. Jong-Ping Hsu and Leonardo Hsu, Nuovo Cimento **112**, 575 (1997) and *Chin. J. Phys.* **35**, 407 (1997); *A Broad View of Relativity, General Implications of Lorentz and Poincaré Invariance* (World Scientific, Singapore, New Jersey, 2006) Chapters 18–19. (Available online at 'google books, Leonardo Hsu'.) The integrability conditions lead to a general solution for $W(w,x)$, i.e., $W(w,x) = \gamma^2 \alpha_0 x + f(w)$, where $f(w)$ is an arbitrary function of w and must approach 1 in the limit of zero acceleration. For example, if one chooses $f(w) = 1$, one has the following transformation

$$w_I = \gamma \beta x + \frac{1}{\alpha_0}(\sin^{-1}\beta - \sin^{-1}\beta_0), \quad x_I = \gamma x - \frac{1}{\alpha_0}\left(\frac{1}{\gamma} - \frac{1}{\gamma_0}\right), \quad y_I = y, \ldots$$

where $\beta = \alpha_0 w + \beta_0$. Thus far, it appears that no physical principle can determine $f(w)$ or the Wu factor $W(w,x)$ and $\beta(w)$ in (3.3) uniquely.

6. It is possible to make a different assumption for the function $\beta(w)$. In this case, one has a different time in CLA frames. See ref. 5, sec(20a) in chapter 20.

7. Ta-You Wu and Y. C. Lee, Int'l. *J. Theore. Phys.* **5**, 307–323 (1972); Ta-You Wu, *Theoretical Physics, vol. 4, Theory of Relativity* (Lian Jing Publishing Co., Taipei, 1978) pp. 172–175,

and references therein. The authors also made an exact calculation of the clock paradox problem, including the effects of linear accelerations and decelerations. Jong-Ping Hsu and Leonardo Hsu, in *JingShin Theoretical Physics Symposium in Honor of Professor Ta-You Wu* (Ed. J. P. Hsu and L. Hsu, World Scientific, Singapore, New Jersey, 1998) pp. 393–412.

8. J. P. Hsu and S. M. Kleff, *Chin. J. Phys.* **36**, 768 (1998); Silvia Kleff and J. P. Hsu, *JingShin Theoretical Physics Symposium in Honor of Professor Ta-You Wu* (Editors J. P. Hsu and L. Hsu, World Scientific Singapore, 1998), pp. 348–352. (Available online at 'google books, Leonardo Hsu'.)

9. See, for example, O. Veblen and J. H. C. Whitehead, *The Foundations of Differential Geometry* (Cambridge Univ. Press, 1953), pp. 37–38.

10. The transformations for the differentials in accelerated frames F and F' take the form
$\gamma(Wdw + \beta dx) = \gamma'(W'dw' + \beta'dx')$, $\gamma(dx + \beta Wdw) = \gamma'(dx' + \beta'W'dw')$, $dy = dy'$,
and $dz = dz'$. (cf. eq. (3.2)).

11. D. Fine, Private correspondence. (2012).

12. The Wu transformations appear to be also related to the pseudo-group discussed by Cartan, who required the identity transformation to exist in the pseudo-group he defined. (cf. online: 'wiki, pseudogroup'.)

13. J. P. Hsu and L. Hsu, ref. 5, pp. 310–317.

14. H. E. Ives and G. R. Stilwell, *J. Opt. Soc. Am.* **28**, 215 (1938).

4.

Coordinate Transformations for Frames with Arbitrary Linear Accelerations and the Taiji Pseudo-Group

4-1. Arbitrary linear accelerations: The Taiji transformations

In the previous chapter, we used the principle of limiting continuation of physical laws to develop the Wu transformations, a minimal generalization of the Lorentz transformations that give the coordinate transformations between an inertial frame and a non-inertial frame with a constant acceleration, where the acceleration and the velocity are both along the same line (they may point in the same or opposite directions).[1,2] In this chapter, we make a further generalization, again using the principle of limiting continuation, to develop a set of coordinate transformations between an inertial frame and a non-inertial frame with an 'arbitrary linear acceleration,' i.e., where the acceleration and velocity are both along the same line, but the acceleration is an arbitrary function of time. We will call these non-inertial frames "arbitrary linear acceleration frames" or ALA frames, and the transformations the "taiji transformations". As we shall see, the set of taiji transformations, like the Wu transformations, also forms a pseudo-group[3] and reduces to the Wu transformations in the limit of constant acceleration.

Consider a non-inertial frame $F(w, x, y, z)$ that has an acceleration $\alpha(w)$ along the x-axis that is an arbitrary function of time, rather than a constant. We consider the case where the velocity $\beta(w)$ is also taken to be along the x-axis,

$$\beta(w) = \beta_1(w) + \beta_o, \quad \alpha(w) = \frac{d\beta(w)}{dw} = \frac{d\beta_1}{dw},$$

$$\beta(0) = \beta_o, \quad \alpha(0) = \alpha_o.$$

$$(4.1)$$

The last two conditions in (4.1) imply that when $w = x = 0$, w_I and x_I need not also be zero, in general.

One of the simplest generalizations of the Wu transformations for the constant-acceleration case is to write the infinitesimal coordinate transformations between $F(x)$ and $F_I(x_I)$ in the following form,

$$dw_I = \gamma(W_a dw + \beta dx), \quad dx_I = \gamma(dx + \beta W_b dw),$$

$$dy_I = dy, \quad dz_I = dz; \tag{4.2}$$

$$\gamma = \sqrt{1 - \beta^2}, \quad \beta = \beta_1(w) + \beta_0, \quad \beta^2 < 1, \tag{4.3}$$

where $W_a = W_a(w, x)$ and $W_b = W_b(w, x)$ are two different unknown functions, in contrast to (3.2). For the same reasons as in the development of (3.2), the terms involving dx in (4.2) do not gain an extra coefficient.

The principle of limiting continuation requires that the two unknown functions W_a and W_b in (4.2), (a) reduce to the same function W in (3.8) in the limit of constant acceleration, and (b) satisfy the following two integrability conditions for the differential equations in (4.2)

$$\frac{\partial(\gamma W_a)}{\partial x} = \frac{\partial(\gamma\beta)}{\partial w}, \quad \frac{\partial(\gamma\beta W_b)}{\partial x} = \frac{\partial\gamma}{\partial w}. \tag{4.4}$$

Analogous to the derivation of the Wu transformations, γ and β are assumed to be functions of w only, leading to the results

$$W_a = \gamma^2\alpha(w)x + A(w), \quad W_b = \gamma^2\alpha(w)x + B(w). \tag{4.5}$$

Substituting (4.5) into (4.2), we obtain

$$w_I = \gamma\beta x + \int \gamma A(w)dw, \quad x_I = \gamma x + \int \gamma\beta B(w)dw, \tag{4.6}$$

while the transformations of the y and z coordinates remain

$$y_I = y + y_0, \quad z_I = z + z_0. \tag{4.7}$$

The principle of limiting continuation of physical laws now dictates that (4.6) must reduces to (3.4) in the limit,

$$\alpha(w) \to \alpha_0, \quad or \quad \beta_1(w) \to \alpha_0 w. \tag{4.8}$$

Thus, we have the following relations for the integrals in (4.6) involving an arbitrary linear acceleration $\alpha(w)$,[4]

$$\int \gamma A(w)dw = \gamma\beta\frac{1}{\alpha(w)\gamma_0^2} + a_0, \tag{4.9}$$

$$\int \gamma\beta B(w)dw = \gamma\frac{1}{\alpha(w)\gamma_0^2} + b_0. \tag{4.10}$$

By differentiation of (4.9) and (4.10), we can determine the two unknown functions $A(w)$ and $B(w)$,

$$A(w) = \frac{\gamma^2}{\gamma_0^2} - \frac{\beta J_e}{\alpha^2(w)\gamma_0^2}, \quad B(w) = \frac{\gamma^2}{\gamma_0^2} - \frac{J_e}{\beta\alpha^2(w)\gamma_0^2},$$

$$J_e(w) = \frac{d\alpha}{dw}, \tag{4.11}$$

where $J_e = J_e(w)$ is the 'jerk,' which is the third-order time derivative of the coordinates.

It thus follows from (4.6), (4.7) and (4.11), that the most general space-time coordinate transformations for the case where the acceleration and velocity of the non-inertial frame are both along parallel x and x_I axes are

$$w_I = \gamma\beta\left(x + \frac{1}{\alpha(w)\gamma_0^2}\right) - \frac{\beta_o}{\alpha_o\gamma_o} + w_o,$$

$$x_I = \gamma\left(x + \frac{1}{\alpha(w)\gamma_0^2}\right) - \frac{1}{\alpha_o\gamma_o} + x_o,$$

$$y_I = y + y_o, \quad z_I = z + z_o; \tag{4.12}$$

$$\gamma = \frac{1}{\sqrt{1 - \beta^2}}, \quad \gamma_o = \frac{1}{\sqrt{1 - \beta_o^2}}, \quad \beta = \beta_1(w) + \beta_o.$$

A constant space-time shift x_o^μ has been included so that both the Lorentz and Poincaré transformations are special limiting cases of (4.12). We call (4.12) the "taiji transformations" for a frame F with an acceleration $\alpha(w)$ that is an arbitrary function of w.

The inverse of the taiji transformation (4.12) is found to be

$$\beta(w) = \frac{w_I + \beta_o/(\alpha_o\gamma_o) - w_o}{x_I + 1/(\alpha_o\gamma_o) - x_o} \equiv \beta_1(w_I, x_I),$$

$$x + \frac{1}{\alpha(w)\gamma_0^2} = \sqrt{\left(x_I + \frac{1}{\alpha_o\gamma_o} - x_o\right)^2 - \left(w_I + \frac{\beta_o}{\alpha_o\gamma_o} - w_o\right)^2}, \tag{4.13}$$

$$y = y_I - y_o, \quad z = z_I - z_o.$$

Although the arbitrary velocity function $\beta(w) = \beta_I(w_I, x_I)$ can always be expressed in terms of w_I and x_I, the arbitrary acceleration function $\alpha(w)$ on the left-hand-side of (4.13) cannot, in the most general case. Thus, the inverse taiji transformations (4.13) cannot necessarily be written explicitly in the usual form $x^\mu = x^\mu(w_I, x_I)$.

However, if a specific function for $\beta(w)$ is given and one can solve for the time w in terms of $\beta(w)$, then it will be possible for the first equation in (4.13) to be written in the usual form $w = w(w_I, x_I)$. For example, if the jerk J_{eo} is constant, so that $\beta(w) = J_{eo}w^2/2 + \alpha_o w + \beta_o$, then when w is positive,

$$w = \frac{1}{J_{eo}}\left[-\alpha_o + \sqrt{\alpha_o^2 + 2J_{eo}[\beta_I(w_I, x_I) - \beta_o]}\right] = w(w_I, x_I), \qquad (4.14)$$

where $\beta(w) = \beta_I(w_I, x_I)$, is a function of x_I and w_I and is given by (4.13). Moreover, the result (4.14) enables $\alpha(w)$ to be expressed in terms of w_I and x_I

$$\alpha(w) = \alpha(w_I, x_I),$$

so that in this case, we can write the inverse taiji transformations (4.13) in the usual form, $w = w(w_I, x_I)$ and $x = x(w_I, x_I)$.

The taiji transformations can also be generalized to the case in which the velocity $\boldsymbol{\beta}(w)$ and the linear acceleration $\boldsymbol{\alpha}(w)$ are both in the same arbitrary direction, and not necessarily along the x and x_I axes. In this case, $\boldsymbol{\beta}(w)$, $\boldsymbol{\beta}_o$ and $\boldsymbol{\alpha}(w)$ are in the same fixed direction and the resultant taiji transformations take the form

$$w_I = \gamma\left(\frac{\beta}{\alpha(w)\gamma_o^2} + \boldsymbol{\beta}\cdot\mathbf{r}\right) - \frac{\beta_o}{\alpha_o\gamma_o} + w_o,$$

$$\mathbf{r}_I = \mathbf{r} + (\gamma - 1)(\boldsymbol{\beta}\cdot\mathbf{r})\frac{\boldsymbol{\beta}}{\beta^2} + \left(\frac{\gamma}{\alpha(w)} - \frac{\gamma_o}{\alpha_o}\right)\frac{\boldsymbol{\beta}}{\beta\gamma_o^2} + \mathbf{r}_o. \qquad (4.15)$$

The corresponding 'inverse' taiji transformation is

$$\beta(w) = \left[\frac{w_I - w_o + \beta_o/(\alpha_o\gamma_o)}{\mathbf{n}\cdot(\mathbf{r}_I - \mathbf{r}_o) + 1/(\alpha_o\gamma_o)}\right] \equiv \beta_I(w_I, \mathbf{r}_I),$$

$$\mathbf{r} + \frac{\mathbf{n}}{\alpha(w)\gamma_o^2} = \mathbf{r}_I - \mathbf{r}_o + \left(\frac{\gamma_o}{\alpha_o}\right)\frac{\mathbf{n}}{\gamma_o^2}$$

$$-\frac{(\gamma-1)\mathbf{n}}{\gamma\beta}\left(w_I - w_o + \frac{\beta_o}{\alpha_o\gamma_o}\right),\tag{4.16}$$

$$\mathbf{n} = \frac{\boldsymbol{\beta}}{\beta} = \frac{\boldsymbol{\beta_o}}{\beta_o}.$$

Since $\gamma = 1/\sqrt{1-\beta^2(w)} = 1/\sqrt{1-\beta_I^2(w_I,\mathbf{r}_I)}$ and $\beta(w) = \beta_I(w_I,\mathbf{r}_I)$ can be expressed in terms of w_I and $\mathbf{r}_I$, the right-hand-side of the 'inverse' taiji transformations (4.16) consists solely of quantities measured from inertial frame F_I. However, the left-hand-side of (4.16) can only be expressed in terms of $\beta(w)$ and $\mathbf{r} + \mathbf{n}/\alpha(w)\gamma_o^2$, because $\beta(w)$ and $\alpha(w) = d\beta(w)/dw$ are arbitrary (and unspecified) functions. Thus, in general, one cannot write the inverse transformations in analytical form. However, mathematically, (4.15) defines a function from one neighborhood of the origin in space-time to another. The implicit function theorem guarantees the existence of an inverse in the neighborhood of any point where the Jacobian is non-zero. For example, a simple calculation says this holds at the origin, and indeed in the open set defined by $x > -1/[\alpha(w)\gamma_o^2]$ for the special case (4.12). In practical terms, methods going back to Newton give a numerical approximation scheme which converges (quadratically) to the inverse of any given point where the Jacobian is non-zero.[5]

The taiji transformations appear to be the most generalized form of the Lorentz transformations from a constant velocity to the case of a frame with an arbitrary linear acceleration, as defined previously. Throughout the rest of this chapter, we shall set $x_o^\mu = (w_o, x_o, y_o, z_o) = 0$ and $\boldsymbol{\beta} = (\beta, 0, 0)$ in the taiji transformations (4.12) and (4.16) for simplicity.

4-2. Poincaré metric tensors for arbitrary-linear-acceleration frames

The invariant interval ds and Poincaré metric tensors in ALA frames can be obtained from (4.2), (4.5), and (4.11). We find

$$ds^2 = \eta_{\mu\nu}dx_I^\mu dx_I^\nu = P_{\mu\nu}dx^\mu dx^\nu,\tag{4.17}$$

$$P_{00} = W_\alpha^2,\quad P_{01} = P_{10} = U,\quad P_{11} = P_{22} = P_{33} = -1;\tag{4.18}$$

$$W_\alpha^2 = \left[\gamma^2\left(\gamma_o^{-2} + \alpha(w)x\right)\right]^2 - \left[\frac{J_e(w)}{\alpha^2(w)\gamma_o^2}\right]^2 > 0,\quad U = \frac{J_e(w)}{\alpha^2(w)\gamma_o^2},$$

where $J_e(w)$ is defined in (4.11).[6]

Mathematically, transformation (4.2) can be obtained by making the replacement $(dw, dx) \rightarrow (dw^*, dx^*) = (W_o dw, dx - U dw)$, where $W_o = \gamma^2$ $(\alpha x + 1/\gamma_o^2)$, and then performing a 4-dimensional rotation in the w–x plane with y and z axes fixed:

$$dw_I = \gamma(dw^* + \beta dx^*), \quad dx_I = \gamma(dx^* + \beta dw^*),$$

$$dy_I = dy, \quad dz_I = dz;$$

$$dw^* \equiv W_o dw, \quad dx^* \equiv dx - U dw. \tag{4.19}$$

This is analogous to the four-dimensional rotation of coordinates in special relativity.

The contravariant metric tensors for a general non-inertial frame can be obtained from (4.18):

$$P^{00} = \frac{1}{W_o^2}, \quad P^{01} = P^{10} = \frac{U}{W_o^2}, \tag{4.20}$$

$$P^{11} = \frac{-W_\alpha^2}{W_o^2}, \quad P^{22} = P^{33} = -1,$$

$$W_o^2 = W_\alpha^2 + U^2 = [\gamma^2(\gamma_o^{-2} + \alpha(w)x)]^2.$$

Using (4.18) and (4.20), one can verify that $P^{\alpha\gamma} P_{\gamma\beta} = \delta_\beta^\alpha$.

4-3. New properties of the Taiji transformations

As might be expected, the taiji transformations have many of the same mathematical properties as the Wu transformations. Similar to the Wu transformations, the taiji transformations (4.12) also imply a singular wall in the non-inertial frame at $x = -1/\alpha(w)\gamma_o \equiv x_s$, the location of which depends on the acceleration function $\alpha(w)$. This singular wall separates physical space ($x > x_s$ for $\alpha(w) > 0$) from unphysical space ($x < x_s$), which does not correspond to any physical part of inertial frames. Thus, the taiji transformations map the entire space-time of an inertial frame to just a portion of space-time in an ALA frame. The taiji transformations between two ALA frames, say F and F' would map a portion of the space-time of the F frame to a portion of the space-time in the F' frame.

Also like the Wu transformations, the taiji transformations form a pseudo-group rather than a usual group.[3] As in the previous chapter, we first demonstrate that the subset of taiji transformations for all ALA frames moving along a particular direction (such as parallel x/x'-axes) satisfies all the properties of a normal group.

In the limit $\alpha(w) = \alpha_o \to 0$ and $\beta_o \to 0$, one can verify that the taiji transformation (4.12) reduces to the identity transformation (identity property) and that the taiji transformation (4.12) has an inverse (as given by (4.16)) (inverse property). To demonstrate the other group properties of the taiji transformations for an arbitrary acceleration along the x-axis, let us consider two other ALA frames F' and F'', characterized by arbitrary velocities $\beta'(w')$, $\beta(w'')$, initial accelerations α'_o, α''_o, and initial velocities β'_o, β''_o, respectively. Using (4.12), the taiji transformations among F_I, F, F' and F'' are

$$
w_I = \gamma\beta\left(x + \frac{1}{\alpha(w)\gamma_o^2}\right) - \frac{\beta_o}{\alpha_o\gamma_o} = \gamma'\beta'\left(x' + \frac{1}{\alpha'(w')\gamma_o'^2}\right)
$$

$$
-\frac{\beta'_o}{\alpha'_o\gamma'_o} = \gamma''\beta''\left(x'' + \frac{1}{\alpha''(w'')\gamma_o''^2}\right) - \frac{\beta''_o}{\alpha''_o\gamma''_o}, \qquad (4.21)
$$

$$
x_I = \gamma\left(x + \frac{1}{\alpha(w)\gamma_o^2}\right) - \frac{1}{\alpha_o\gamma_o} = \gamma'\left(x' + \frac{1}{\alpha'(w')\gamma_o'^2}\right)
$$

$$
-\frac{1}{\alpha'_o\gamma'_o} = \gamma''\left(x'' + \frac{1}{\alpha''(w'')\gamma_o''^2}\right) - \frac{1}{\alpha''_o\gamma''_o},
$$

where we have ignored the y and z transformations for simplicity.

The taiji transformations between F and F' can be obtained from (4.21):

$$
\beta(w) = \frac{\gamma'\beta'Q' - \beta'_o/(\alpha'_o\gamma'_o) + \beta_o/(\alpha_o\gamma_o)}{\gamma'Q' - 1/(\alpha'_o\gamma'_o) + 1/(\alpha_o\gamma_o)}, \qquad (4.22)
$$

$$
x + \frac{1}{\alpha(w)\gamma_o^2}
$$

$$
= \sqrt{\left(\gamma'Q' - \frac{1}{\alpha'_o\gamma'_o} + \frac{1}{\alpha_o\gamma_o}\right)^2 - \left(\gamma'\beta'Q' - \frac{\beta'_o}{\alpha'_o\gamma'_o} + \frac{\beta_o}{\alpha_o\gamma_o}\right)^2},
$$

$$
y = y', \quad z = z'; \quad Q' = \left(x' + \frac{1}{\alpha'(w')\gamma_o'^2}\right).
$$

From (4.21), one can also obtain the inverse of (4.22)

$$\beta'(w') = \frac{\gamma\beta Q - \beta_0/(\alpha_0\gamma_0) + \beta_0'/(\alpha_0'\gamma_0')}{\gamma Q - 1/(\alpha_0\gamma_0) + 1/(\alpha_0'\gamma_0')}, \tag{4.23}$$

$$x' + \frac{1}{\alpha'(w')\gamma_0^2}$$

$$= \sqrt{\left(\gamma Q - \frac{1}{\alpha_0\gamma_0} + \frac{1}{\alpha_0'\gamma_0'}\right)^2 - \left(\gamma\beta Q - \frac{\beta_0}{\alpha_0\gamma_0} + \frac{\beta_0'}{\alpha_0'\gamma_0'}\right)^2},$$

$$y' = y, \quad z' = z; \quad Q = \left(x + \frac{1}{\alpha(w)\gamma_0^2}\right).$$

Similarly, the taiji transformations between F and F'' can be obtained from (4.21):

$$\beta(w) = \frac{\gamma''\beta''Q'' - \beta_0''/(\alpha_0''\gamma_0'') + \beta_0/(\alpha_0\gamma_0)}{\gamma''Q'' - 1/(\alpha_0''\gamma_0'') + 1/(\alpha_0\gamma_0)}, \tag{4.24}$$

$$x + \frac{1}{\alpha(w)\gamma_0^2}$$

$$= \sqrt{\left(\gamma''Q'' - \frac{1}{\alpha_0''\gamma_0''} + \frac{1}{\alpha_0\gamma_0}\right)^2 - \left(\gamma''\beta''Q'' - \frac{\beta_0''}{\alpha_0''\gamma_0''} + \frac{\beta_0}{\alpha_0\gamma_0}\right)^2},$$

$$y = y'', \quad z = z''; \quad Q'' = \left(x'' + \frac{1}{\alpha''(w'')\gamma_0''^2}\right).$$

From (4.23) and (4.24), one can show that the taiji transformation between $F'(x')$ and $F''(x'')$ is given by

$$\beta'(w') = \frac{\gamma''\beta''Q'' - \beta_0''/(\alpha_0''\gamma_0'') + \beta_0'/(\alpha_0'\gamma_0')}{\gamma''Q'' - 1/(\alpha_0''\gamma_0'') + 1/(\alpha_0'\gamma_0')}, \tag{4.25}$$

$$x' + \frac{1}{\alpha'\gamma_0'^2}$$

$$= \sqrt{\left(\gamma''Q'' - \frac{1}{\alpha_0''\gamma_0''} + \frac{1}{\alpha_0'\gamma_0'}\right)^2 - \left(\gamma''\beta''Q'' - \frac{\beta_0''}{\alpha_0''\gamma_0''} + \frac{\beta_0'}{\alpha_0'\gamma_0'}\right)^2},$$

$$y' = y'', \quad z' = z'', \quad \alpha' = \alpha'(w').$$

Thus the transformation (4.25) between two ALA frames F' and F'', has the same form as (4.22) for F and F' frames. Substituting (4.23) into (4.22), we obtain the 'identity transformation'

$$\beta(w) = \beta(w), \quad x + \frac{1}{\alpha(w)\gamma_0^2} = x + \frac{1}{\alpha(w)\gamma_0^2}, \quad y = y, \quad z = z.$$

$$(4.26)$$

Thus, just as with the Wu transformations, all the usual group properties are satisfied by this subset of taiji transformations. Also as was the case with the Wu transformations, when the full set of taiji transformations is considered, including those for non-inertial frames with accelerations in different directions, the combination of two taiji transformations with accelerations in different directions cannot be expressed as a combination of a single such transformation combined with a rotation. The set obtained from generators of the taiji transformations using the Lie bracket, the brackets of brackets, and so on, never closes. Therefore, the taiji transformations define a group with infinitely-many generators.[7]

Because the group properties of the taiji transformations appear to be somewhat more general than those of traditional pseudo-groups, it may be appropriate to denote such a pseudo-group as a 'taiji pseudo-group.' The taiji pseudo-group includes the Wu pseudo-groups (with constant accelerations) and the Lorentz and Poincaré groups (with constant velocities) as special cases. We observe that the fundamental metric tensors (4.18) for ALA frames appear to depend on both the coordinates x^μ and the differentials dx^μ (see Fig. 4.1).[8]

As a final note regarding the mathematical properties of the taiji transformations, we see that it is, in general, not possible to write the ALA transformations in the usual form

$$x^\mu = x^\mu(w', x', y', z') \quad \text{and} \quad x'^\mu = x'^\mu(w, x, y, z). \qquad (4.27)$$

Instead, the best that can be done is to express them in the symmetric form,

$$f^\mu(w, x, y, z, \alpha, \alpha_o, \beta, \beta_o) = f^\mu(w', x', y', z', \alpha', \alpha'_o, \beta', \beta'_o), \qquad (4.28)$$

as shown in (4.21)–(4.26). However, using this symmetric form, one can still discuss the space-time transformations for the differentials dx^μ and dx'^ν (4.2), the fundamental space-time metric tensors $P_{\mu\nu}$ ((4.17) and (4.18)), and the physics of particle dynamics and field theory in ALA frames.

If F is an ALA frame and F' is a different type of non-inertial frame, say, a rotating frame. (to be discussed in the next chapter), then even the symmetric form (4.28) does not hold.

Spacetime transformation between two inertial frames
$(\beta = \beta_0 = \text{constant}; \ \gamma_0 = (1 - \beta_0{}^2)^{-1/2})$

$$w_I = \gamma_0 \, (w - \beta_0 x), \quad x_I = \gamma_0 \, (x - \beta_0 w), \quad y' = y, \quad z' = z;$$

$$\beta = \beta_0 \ \rightarrow \ \beta = \beta_0 + \alpha_0 w$$

Spacetime transformation between an inertial frame and a frame with a Constant-Linear-Acceleration
$[\ \beta = \beta_0 + \alpha_0 w, \ \ \gamma = (1 - \beta^2)^{-1/2}\ ]$

$$w_I = \gamma\beta\left(x + \frac{1}{\alpha_0\gamma_0^2}\right) - \frac{\beta_0}{\alpha_0\gamma_0}, \quad x_I = \gamma\left(x + \frac{1}{\alpha_0\gamma_0^2}\right) - \frac{1}{\alpha_0\gamma_0}, \quad y_I = y, \quad z_I = z;$$

$$\beta = \beta_0 + \alpha_0 w \ \rightarrow \ \beta = \beta(w) = \text{arbitrary}$$

Spacetime transformation between an inertial frame and a frame with an Arbitrary-Linear-Acceleration
$[\ \beta = \beta(w) = \text{arbitrary}, \ \ \gamma = (1 - \beta^2)^{-1/2}\ ]$

$$w_I = \gamma\beta\left(x + \frac{1}{\alpha(w)\gamma_0^2}\right) - \frac{\beta_0}{\alpha_0\gamma_0}, \quad x_I = \gamma\left(x + \frac{1}{\alpha(w)\gamma_0^2}\right) - \frac{1}{\alpha_0\gamma_0}, \quad y_I = y, \quad z_I = z;$$

Fig. 4.1   The hierarchy of spacetime transformations between inertial and accelerated frames from least to most general.

4-4.   Physical implications

Some physical implications of the taiji transformations for ALA frames worthy of note are:

(A) Existence of a 'preferred' coordinate system

Similar to the way in which Cartesian coordinates are the preferred coordinates for writing down the Lorentz transformation, the coordinates x^μ with the Poincaré metric tensors given in (4.17) and (4.18) are the preferred coordinates for expressing the taiji transformations. No other choice of coordinates satisfies the principle of limiting continuation of physical laws. Thus, the taiji symmetry framework based on a flat space-time for both inertial and non-inertial frames does not have the property of general coordinate invariance, in contrast to general relativity.[9]

(B) Space-time-dependent speed of light in ALA frames

In an inertial frame, the law for the propagation of light is $ds = 0$. Analogously, the propagation of light in a non-inertial frame is described by the same invariant law (4.17) with $ds = 0$. Let us consider some specific and simple cases. Suppose a light signal travels along the x-axis, i.e., $dy = dz = 0$. The speed of light β_{Lx} is then

$$\beta_{Lx} = \frac{dx}{dw} = \gamma^2 \left(\alpha x + \frac{1}{\gamma_o^2} \right) + \frac{J_e(w)}{\alpha^2(w)\gamma_o^2}, \qquad \frac{dy}{dw} = \frac{dz}{dw} = 0, \qquad (4.29)$$

which is certainly different from the speed of light $\beta_L = 1$ (derived from $ds^2 = dw_I^2 - dx_I^2 = 0$) in an inertial frame. If the light signal moves in the y-direction, i.e., $dx = dz = 0$, eq. (4.14) with $ds = 0$ leads to the speed of light β_{Ly},

$$\beta_{Ly} = \frac{dy}{dw} = \sqrt{\left[\gamma^2 \left(\alpha x + \frac{1}{\gamma_o^2} \right) \right]^2 - \left[\frac{J_e(w)}{\alpha^2(w)\gamma_o^2} \right]^2},$$

$$\frac{dx}{dw} = \frac{dz}{dw} = 0. \qquad (4.30)$$

(C) Operational definition of the space-time coordinates in ALA frames

Since the speed of light in ALA frames is not constant, as shown in (4.29), it is very complicated to use light signals to synchronize clocks in an ALA frame $F(w, x, y, z)$. However, as discussed in Chapter 3 in reference to the Wu transformations, one can synchronize a set of clocks in the ALA frame F by using a grid of 'computerized space-time clocks' that are programmed to accept information concerning their positions x_I relative to the F_I frame, obtain w_I from the nearest F_I clock, and then compute and display w and x using the inverse transformation (4.16) with given parameters α_o, β_o and x_o^μ and functions $\beta(w)$ and $\alpha(w)$.[10] If this cannot be done for a given $\beta(w)$ in the taiji transformations, then the space-time coordinates have no operational meaning and we would consider such an ALA frame to be unphysical.

In an ALA frame, the values the physical time w can take on are restricted by the condition $\beta^2(w) < 1$ and the coordinates of physical space are limited by $x > -1/(\alpha(w)\gamma_o^2)$, as shown in (4.12).

(D) The velocity-addition laws in ALA frames

In general, the law for velocity addition can be obtained from (4.2),

$$\frac{dx_I}{dw_I} = \frac{dx/dw + \beta W_b}{W_a + \beta dx/dw}, \tag{4.31}$$

$$\frac{dy_I}{dw_I} = \frac{dy/dw}{\gamma(W_a + \beta dx/dw)}, \qquad \frac{dz_I}{dw_I} = \frac{dz/dw}{\gamma(W_a + \beta dx/dw)}.$$

The condition $dx = 0$ implies that an object is at rest in F and its velocity as measured by observers in F_I is then dx_I/dw_I. Similarly, the condition $dx_I = 0$ in (4.31) corresponds to the case where an object is at rest in F_I. Its velocity as measured by observers in the ALA frame F is dx/dw.

If $\beta_L = dx_I/dw_I = 1$, the velocity-addition law (4.31) leads to the same result for the space-time-dependent speed of light in an ALA frame as Equation (4.29). Similarly, (4.31) with $dx = dz = 0$ is consistent with (4.30).

(E) Classical electrodynamics in inertial and non-inertial frames

For a continuous charge distribution in space, the invariant action for electromagnetic fields and their interactions is postulated to be

$$S_{em} = -\int \left[a_\mu j^\mu + \frac{1}{4} f_{\mu\nu} f^{\mu\nu} \right] \sqrt{-P} d^4 x, \tag{4.32}$$

$$\sqrt{-P} = \sqrt{-\det P_{\mu\nu}} = \sqrt{W_\alpha^2 + U^2} = \gamma^2 \left(\alpha(w)x + \frac{1}{\gamma_o^2} \right) > 0,$$

$$f_{\mu\nu} = D_\mu a_\nu - D_\nu a_\mu = \partial_\mu a_\nu - \partial_\nu a_\mu,$$

where D_μ denotes the partial covariant derivative with respect to the metric tensor $P_{\mu\nu}$. The most general Maxwell equations for both inertial and non-inertial frames can be derived from the action S_{em}. The Lagrange equation of a_μ in a general frame characterized by the Poincaré tensor $P_{\mu\nu}$ leads to

$$\frac{1}{\sqrt{-P}}\partial_\mu\left(\sqrt{-P}f^{\mu\nu}\right) + j^\nu = 0. \tag{4.33}$$

This implies the continuity equation of charge,

$$\partial_\mu\left(\sqrt{-P}j^\mu\right) = 0, \tag{4.34}$$

which implies charge conservation. This general continuity Equation (4.34) also implies that the electric charge is constant in any frame of reference based on taiji symmetry framework.

(F) Generalized Klein–Gordon and Dirac equations for non-inertial frames

The general Klein–Gordon and Dirac equations for both inertial and non-inertial frames. are

$$\left[P^{\mu\nu}(D_\mu - iea_\mu)(D_\nu - iea_\nu) + m^2\right]\phi = 0, \quad e > 0, \tag{4.35}$$

$$i\Gamma^\mu(\partial_\mu - iea_\mu)\psi + \frac{1}{2}i(\partial_\mu\Gamma^\mu)\psi + \frac{1}{2}\left(\partial_\mu \ln\sqrt{-P}\right)i\Gamma^\mu\psi - m\psi = 0, \tag{4.36}$$

where $\{\Gamma^\mu, \Gamma^\nu\} = 2P^{\mu\nu}(x)$. The Lagrangians for these equations are

$$L_\phi = \sqrt{-P}\left(P^{\mu\nu}\left[(D_\mu - iea_\mu)\phi\right]^*(D_\nu - iea_\nu)\phi - m^2\phi^*\phi\right), \tag{4.37}$$

$$L_\psi = \sqrt{-P}\left[\frac{i}{2}\overline{\psi}\Gamma^\mu(D_\mu - iea_\mu)\right.$$
$$\left. -\frac{i}{2}\left[(D_\mu - iea_\mu)^*\overline{\psi}\right]\Gamma^\mu - m\overline{\psi}\right]\psi, \tag{4.38}$$

$$D_\mu\phi = \partial_\mu\phi, \quad D_\mu\psi = \partial_\mu\psi, \quad D_\mu U_\nu = \partial_\mu U_\nu - \Gamma^\lambda_{\mu\nu}U_\lambda,$$

$$\Gamma^\lambda_{\mu\nu} = \frac{1}{2}P^{\lambda\sigma}(\partial_\mu P_{\sigma\nu} + \partial_\nu P_{\sigma\mu} - \partial_\sigma P_{\mu\nu}),$$

where the asterisk * denotes complex conjugate. The invariance of the action involving the Lagrangians (4.37) and (4.38) implies that the quantity corresponding to the electric charge in the Klein–Gordon and Dirac equations has the same value in an arbitrary frame of reference and hence, is a true universal constant, as opposed to the speed of a light signal.

4-5.   Experimental tests of the Taiji transformations

Because it is extremely difficult to construct an experimental set up in which objects moving at relativistic speeds also have a non-constant or arbitrary acceleration, obtaining experimental support for the taiji transformations will be difficult. However, we can still use the taiji transformations to predict the shift in the frequency of radiation from a source undergoing an arbitrary linear acceleration.

With the help of the Poincaré metric tensors (4.18) and (4.20), the taiji transformations for covariant vectors can be obtained from the transformations for the contravariant coordinate vectors dx^{μ}. Thus, we can derive the transformations for the covariant wave vectors k_{μ} between an inertial frame F_I and an ALA frame F

$$k_{I0} = \gamma \left(\frac{k_0}{W} - \left[\beta - \frac{U}{W_o} \right] k_1 \right),$$

$$k_{I1} = \gamma \left(k_1 \left[1 - \frac{\beta U}{W_o} \right] - \frac{\beta k_0}{W_o} \right), \qquad (4.39)$$

and

$$k_{I0}^2 - k_{I1}^2 = \left(\frac{k_0}{W_o} \right)^2 - (k_1)^2 \left(\frac{W_\alpha}{W_o} \right)^2 + 2 k_1 k_0 \frac{U}{W_o^2} = 0, \qquad (4.40)$$

where $k_{I2} = k_2 = 0$ and $k_{I3} = k_3 = 0$. The quantities k_{I0} and k_0 are the frequencies (or equivalently, energies) of the same wave measured by observers in F_I and F respectively.

Consider a radiation source at rest in an ALA frame F. Suppose $k_0(rest)$ denotes the frequency of the waves emitted from this source as measured by observer in the F frame and $k_{I0}(rest)$ denotes the frequency of the radiation from the same source at rest in an inertial frame F_I as measured by observers in F_I. In practice, it is difficult to determine the relationship between $k_0(rest)$ and $k_{I0}(rest)$ experimentally because of the difficulty of putting observers in the ALA frame. However, the postulate of the weak equivalence of non-inertial frames (3.37) gives the relationship between $k_0(rest)$ and $k_{I0}(rest)$ as

$$k_0(rest)\sqrt{P^{00}} = \frac{k_0(rest)}{W_o} = k_{I0}(rest), \quad W_o = \gamma^2(\gamma_o^{-2} + \alpha(w)x).$$

$$(4.41)$$

Based on (4.39), (4.40), and (4.41), we obtain the shift in the frequency of the radiation emitted from a source undergoing an arbitrary linear acceleration as

$$k_{I0} = k_{I0}(rest)\frac{\gamma(1-\beta)}{[1 - U/W_o]}, \tag{4.42}$$

$$\frac{1}{\lambda_I} = \frac{1}{\lambda_I(rest)}\frac{\gamma(1-\beta)}{[1 - U/W_o]}, \tag{4.43}$$

where we have used (4.40) to find the relation $k_0 = k_1(W_o - U)$ in the ALA frame F, which holds for the case when the radiation source is at rest in F. Results (4.42) and (4.43) may be called the taiji-Doppler effect. The taiji Doppler shift includes the Wu-Doppler shift (3.43) and (3.44) in the limit where the jerk vanishes $J_e = 0$ or $U = 0$, or equivalently, when the acceleration is constant, $\alpha(w) = \alpha_o$.

References

1. For early discussions of space-time transformations for accelerated frames, see A. Einstein, *Jahrb. Rad. Elektr.* **4**, 411 (1907); L. Page, *Phys. Rev.* **49**, 254 (1936); C. Møller, *Danske Vid. Sel. Mat. Fys.* **20**, No. 19 (1943); T. Fulton, F. Rohrlich and L. Witten, *Nuovo Ciment*, **XXVI**, 652 (1962) and *Rev. Mod. Phys.* **34**, 442 (1962); E. A. Desloge and R. J. Philpott, *Am. J. Phys.* **55**, 252 (1987).

2. The Wu transformations also lead to space-time transformations between any two frames with a constant linear accelerations, where the limit of zero acceleration exists. For more recent discussions based on limiting Lorentz and Poincaré invariance (which is a special case of the principle of limiting continuation of physical laws), see Jong-Ping Hsu and Leonardo Hsu, *Nuovo Cimento B* **112**, 575 (1997) and *Chin. J. Phys.* **35**, 407 (1997); **40**, 265 (2002). Jong-Ping Hsu, *Einstein's Relativity and Beyond — New Symmetry Approaches*, (World Scientific, Singapore, 2000), chapters 21–23. (Available online at 'google books, Jong-Ping Hsu'.)

3. E. Cartan, *Annales scientifiques de l'I.N.S.* **21**, 153206 (1904); O. Veblen and J. Whitehead, *Proc. Natl. Acad. Sci. USA* **17**, 551 (1931); O. Veblen and J. Whitehead, *The Foundations of Differential Geometry*, (Cambridge Univ. Press. 1932), pp. 37–38.

4. The generalization with the replacement $\alpha_o \rightarrow \alpha(w)$ (for the accelerations in the denominators of (4.9) and (4.10)) is crucial. Mathematically, this replacement implies that the two variables w and x in $W(w, x)$ cannot be separated, in general. In contrast, a general transformation for ALA frames was discussed in a previous paper based on the separation of w and x in $W(w, x)$. See J. P. Hsu, in *Frontiers of Physics at the Millennium*

Symposium (Ed. Y. L. Wu and J. P. Hsu, World Scientific, 2001). (Available online at 'google books, Jong-Ping Hsu'.) However, the generality in this conference paper turns out to be restricted, and hence not completely satisfactory because an additional technical assumption of the separation of variables, $W(w, x) = W_1(w) W_2(x)$, prevents the arbitrary linear acceleration transformations from being fully realized.

5. We would like to thank Dana Fine for helpful discussions.

6. For $\beta(w) = \mathbf{n}\beta$ in an arbitrary direction, the invariant interval $ds^2 = P_{\mu\nu}dx^\mu dx^\nu = W^2 dw^2 + 2U\, dw\mathbf{n} \cdot \mathbf{r} - d\mathbf{r}^2$ with $W = W(w, \mathbf{r})$ given by

$$W_\alpha^2(w, \mathbf{r}) = \left[\gamma^2 \left(\alpha(w)(\mathbf{n} \cdot \mathbf{r}) + \frac{1}{\gamma_0^2} \right) \right]^2 - \left[\frac{J_e(w)}{\alpha^2(w)\gamma_0^2} \right]^2 .$$

7. Dana Fine, private correspondence (2012).

8. Although the space-time of all ALA frames has a zero Riemann–Christoffel curvature tensor, the metric tensors $P_{\mu\nu}$ in (4.17) and (4.18), etc. have unusual structures. The quantity $P_{00} = W^2$, for example, depends on x^μ, $\alpha(w) = d\beta(w)/dw$, and $J_e(w) = d\alpha(w)/dw$. This property of P_{00} is interesting because in Riemannian geometry the metric tensors $g_{\mu\nu}(x)$ depend only on the position coordinates x^μ. However, the metric tensors of space-time in ALA frames appear to be more general than that of Riemannian geometry. In the Finsler geometry, the metric tensor, $g_{\mu\nu}(x^\rho, dx^\sigma)$ depends on the coordinates and their differentials (or the velocities). In this sense, the geometry of space-time in ALA frames appears to be a generalized Riemannian geometry. Possible applications of the Finsler geometry in physics have been suggested by the mathematician S. S. Chern and merit further investigation.

9. The generalization of the Lorentz transformations to the taiji transformations (4.12) does not change the curvature of space-time. Thus, the curvature of space-time in ALA frames is the same as that of inertial frames, i.e., all these frames have vanishing Riemann–Christoffel curvature tensor. Cf. C. Møller, *The Theory of Relativity* (Oxford Univ. Press, London, 1952), p. 285.

10. See the operational meaning of space-time coordinates, discussed in section 3-3.

5.

Coordinate Transformations for Rotating Frames
and Experimental Tests

5-1.   Rotational taiji transformations

In the previous two chapters, we applied the principle of limiting continuation to derive coordinate transformations between an inertial frame and a frame with a velocity and acceleration pointing along the same line (either parallel or anti-parallel). We now use the same approach to derive coordinate transformations between an inertial frame and a frame that rotates with a constant angular velocity.[1] As we shall see, our results are not only consistent with the results of high energy experiments involving unstable particles in a circular storage ring, but also support Pellegrini and Swift's analysis of the Wilson experiment,[2] in which they point out that rotational transformations cannot be locally replaced by Lorentz transformations.

Suppose $F_I(w_I, x_I, y_I, z_I)$ is an inertial frame and $F(w, x, y, z)$ (which we subsequently refer to as $F(\Omega)$) is a frame that rotates with a constant angular velocity Ω (to be defined more precisely below). The origins of both frames coincide at all times and we use a Cartesian coordinate system in both frames. The usual classical transformation equations between F_I and F are

$$w_I = w, \quad x_I = x \cos(\Omega w) - y \sin(\Omega w),$$

$$y_I = x \sin(\Omega w) + y \cos(\Omega w), \quad z_I = z; \tag{5.1}$$

In order to derive a set of coordinate transformations between F_I and F that satisfy the requirements of limiting Lorentz and Poincaré invariance (i.e., limiting four-dimensional symmetry),[a] we first consider a slightly more general case. In this more general case, there is an inertial reference frame F_I and a non-inertial frame $F_R(\Omega)$ whose origin orbits the origin of the inertial frame at a constant distance R, with a constant angular velocity Ω. A Cartesian coordinate system is used in

---

[a]This is a special case of the principle of limiting continuation of physical laws when the accelerated frame becomes an inertial frame in the limit of zero acceleration. There is no additional constant velocity term here.

both frames, set up in such a way that the positive portion of the y-axis of the $F_R(\Omega)$ frame always extends through the origin of F_I. This is useful because in the limit $R \to \infty$ and $\Omega \to 0$ such that the product $R\Omega = \beta_o$ is a finite non-zero constant velocity, the two frames become inertial frames with coordinates related by the familiar Lorentz transformations. With that in mind, the classical coordinate transformations between F_I and $F_R(\Omega)$ (the orbiting frame) are

$$w_I = w, \quad x_I = x \cos(\Omega w) - (y - R) \sin(\Omega w),$$
$$y_I = x \sin(\Omega w) + (y - R) \cos(\Omega w), \quad z_I = z. \tag{5.2}$$

According to the principle of limiting continuation, we postulate that the transformations between F_I and $F_R(\Omega)$ that satisfy the requirements of limiting Lorentz and Poincaré invariance (i.e., the transformations satisfy Lorentz and Poincaré invariance in the limit of zero acceleration) should have the form

$$w_I = Aw + B\boldsymbol{\rho} \cdot \boldsymbol{\beta}, \quad x_I = Gx \cos(\Omega w) + E(y - R) \sin(\Omega w),$$
$$y_I = Ix \sin(\Omega w) + H(y - R) \cos(\Omega w), \quad z_I = z; \tag{5.3}$$

where $\boldsymbol{\rho} = (x, y)$, $\mathbf{S} = (x, y - R)$, $\boldsymbol{\beta} = \boldsymbol{\Omega} \times \mathbf{S}$, $\boldsymbol{\Omega} = (0, 0, \Omega)$, and the functions A, B, E, G, H, and I may, in general, depend on the coordinates x^μ.

One unusual feature of the transformation equations (5.3) is that we will treat both the constant angular velocity Ω (defined as $\Omega = d\phi/dw$) and the orbital radius R as quantities that are measured by observers in the non-inertial frame $F_R(\Omega)$. This is counter to the usual procedure of measuring such parameters from the 'lab' or 'inertial' reference frame, but will simplify the following discussion since all quantities on the right side of the transformation equations (5.3) are measured with respect to $F_R(\Omega)$ observers. Thus, when $w = w_I = 0$, the y and y_I axes overlap and the origin of F_I, i.e., $x_I = y_I = 0$, is found at the coordinates $(x, y) = (0, +R)$. Equivalently, the origin of the rotating frame $F_R(\Omega)$, i.e., $x = y = 0$, is found at the inertial coordinates $(x_I, y_I) = (0, -HR)$, where H is an unknown function to be determined.

To determine the unknown functions A, B, E, G, H, and I, we consider the following limiting cases. First, when $R = 0$, transformations (5.3) must have x/y symmetry (i.e., be symmetric under an exchange of x and y). This implies that

$$-E = G = H = I; \quad \text{when } R = 0. \tag{5.4}$$

Second, in the limit of small Ω (or $|\boldsymbol{\Omega} \times \mathbf{S}| \ll 1$) with $R \to 0$, transformations (5.3) should reduce to the classical rotational transformations (5.1). Thus,

$$-E \approx G \approx H \approx I \approx 1, \quad \text{for small } \Omega \text{ with } R = 0. \tag{5.5}$$

Finally, and perhaps most importantly, in the limit where $R \to \infty$ and $\Omega \to 0$ such that their product $R\Omega = \beta_o$ is a non-zero constant velocity parallel to the x_I axis, the finite and differential forms of (5.3) must reduce to the Lorentz transformations, in both its finite form

$$w_I = \gamma_o(w + \beta_o x), \quad x_I = \gamma_o(x + \beta_o w), \quad y_I = y, \quad z_I = z; \tag{5.6}$$

$$\gamma_o = \frac{1}{\sqrt{1 - \beta_o^2}},$$

and its differential form,

$$dw_I = \gamma_o(dw + \beta_o dx), \quad dx_I = \gamma_o(dx + \beta_o dw),$$

$$dy_I = dy, \quad dz_I = dz. \tag{5.7}$$

(Strictly speaking, $y_I = y$ should be replaced by $y_I = -\infty$ in (5.6), because $R \to \infty$ in this limit.) This is necessary for (5.3) to satisfy the limiting Lorentz and Poincaré invariance, as specified by the principle of limiting continuation of physical laws. The taiji time w in (5.2) reduces to the usual relativistic time in this limit. Thus, we have

$$A = B = G = -E = \gamma_o, \quad H = 1, \tag{5.8}$$

$$\text{when } R \to \infty \quad \text{and} \quad \Omega \to 0 \text{ such that } \beta_o = R\Omega.$$

The requirements put on A, B, E, G, H, and I by considering these three limiting cases do not lead to a unique solution for the unknown functions. This is analogous to the case in which gauge symmetry does not uniquely determine the electromagnetic action[3] and one must also postulate a minimal electromagnetic coupling. Here, as we did in the previous two chapters in deriving the Wu and taiji transformations, we postulate a minimal generalization of the classical rotational transformations (5.2).

Based on the limiting cases considered above, it is not unreasonable for transformations (5.3) to have the following two properties: (i) For non-zero Ω and finite R, the functions A, B, G, I and $-E$ are simply generalized from $\gamma_o = (1 - \beta_o^2)^{-1/2}$ to $\gamma = (1 - \beta^2)^{-1/2}$, where $\beta = S\Omega$, and (ii) Only H depends on R

and it is required to be the simplest function involving only the first power of γ. Combining (5.4), (5.5), and (5.8) with the two properties just named leads to the following solutions

$$A = B = G = I = \gamma = (1 - \beta^2)^{-1/2}, \quad E = -\gamma, \tag{5.9}$$

$$H = (\gamma + R/R_o)(1 + R/R_o), \quad \gamma = (1 - \beta^2)^{-1/2}.$$

The quantity R_o in (5.9) is an undetermined length parameter that seems to be necessary for the transformations to satisfy the limiting criteria specified in (5.4), (5.5), and (5.8), as well as the limiting Lorentz and Poincaré invariance. From a comparison of rotational transformations and their classical approximation, R_o should be very large value. However, since it is unsatisfactory to have such undetermined parameters in a theory and there are no known experimental results that can help to determine a value for R_o, we shall take the limit $R_o \to \infty$ in the following discussions for simplicity.

Thus, a simple rotational taiji transformation corresponding to the classical transformations (5.2) is

$$w_I = \gamma(w + \boldsymbol{\rho} \cdot \boldsymbol{\beta}), \quad x_I = \gamma[x \cos(\Omega w) - (y - R) \sin(\Omega w)],$$

$$y_I = \gamma[x \sin(\Omega w) + (y - R) \cos(\Omega w)], \quad z_I = z; \tag{5.10}$$

$$\beta = |\boldsymbol{\Omega} \times \mathbf{S}| = \Omega\sqrt{x^2 + (y - R)^2} = \Omega S < 1, \quad \boldsymbol{\rho} \cdot \boldsymbol{\beta} = xR\Omega.$$

Because all known experiments can be analyzed in the $R = 0$ case, for the rest of this chapter, we set $R = 0$ and concentrate solely on the implications of the transformations under that condition. With $R = 0$ the coordinate transformations between an inertial frame F_I and a non-inertial frame $F(\Omega)$ whose origin coincides with that of F_I and that rotates with a constant angular velocity Ω about the origin are

$$w_I = \gamma(w + \boldsymbol{\rho} \cdot \boldsymbol{\beta}) = \gamma w, \quad x_I = \gamma[x \cos(\Omega w) - y \sin(\Omega w)],$$

$$y_I = \gamma[x \sin(\Omega w) + y \cos(\Omega w)], \quad z_I = z, \tag{5.11}$$

where $\gamma = 1/\sqrt{1 - \rho^2\Omega^2}$ and $\boldsymbol{\rho} \cdot \boldsymbol{\beta} = \Omega Rx = 0$ for $R = 0$. Again, we take the constant angular velocity Ω (defined as $\Omega = d\phi/dw$) as a quantity that is measured by observers in the non-inertial frame $F(\Omega)$.

To derive the inverse transformations of (5.11), we must first find a way to express Ωw and $\gamma = 1/\sqrt{1 - \rho^2\Omega^2}$ in terms of quantities measured in the inertial frame F_I. While the coordinate transformations (5.10) for the non-zero R case can only be written in Cartesian coordinates (so that they satisfy the limiting Lorentz and Poincaré invariance), the transformations in the $R = 0$ case (5.11) can be written in terms of cylindrical coordinates. We introduce the relations $x_I = \rho_I \cos \phi_I, y_I = \rho_I \sin \phi_I, x = \rho \cos \phi$, and $y = \rho \sin \phi$ so that (5.11) can be written as

$$w_I = \gamma w, \quad \rho_I = \sqrt{x_I^2 + y_I^2} = \gamma \rho, \quad \phi_I = \phi + \Omega w, \quad z_I = z;$$

with

$$\Omega_I \equiv \frac{d\phi_I}{dw_I} = \frac{d\phi_I}{dw}\frac{dw}{dw_I} = \Omega \frac{1}{\gamma}, \quad (\rho \text{ and } \phi \text{ fixed}).$$

The conditions that ρ and ϕ are fixed mean that an object is at rest in the rotating frame, so that an observer in F_I can measure its angular velocity Ω_I and identify it with the angular velocity of the rotating frame. These equations give the operational definitions of Ω_I and Ω. Thus, we have

$$w_I \Omega_I = w\Omega, \quad \rho_I \Omega_I = \rho\Omega, \quad \gamma = \frac{1}{\sqrt{1 - \rho^2 \Omega^2}} = \frac{1}{\sqrt{1 - \rho_I^2 \Omega_I^2}}. \quad (5.12)$$

From (5.11) and (5.12), we can then derive the inverse rotational transformations

$$w = \frac{w_I}{\gamma}, \quad x = \frac{1}{\gamma}[x_I \cos(\Omega_I w_I) + y_I \sin(\Omega_I w_I)],$$

$$y = \frac{1}{\gamma}[-x_I \sin(\Omega_I w_I) + y_I \cos(\Omega_I w_I)], \quad z = z_I; \quad (5.13)$$

where

$$\gamma = \frac{1}{\sqrt{1 - \Omega_I^2(x_I^2 + y_I^2)}}.$$

If one imagines the frame $F(\Omega)$ as a carousel, then for any given angular velocity, objects at rest relative to $F(\Omega)$ that are sufficiently far from the origin would have a classical linear velocity that exceeds the speed of light in the inertial frame F_I. As expected, the taiji rotational transformations display some unusual behavior near that region. From the relationship between Ω and Ω_I, i.e., $\Omega_I = \Omega/\gamma = \Omega\sqrt{1 - \rho^2\Omega^2}$, one can see that for a given constant angular velocity Ω, the corresponding angular velocity Ω_I depends on ρ as well as Ω. In particular, when $\rho = 1/\Omega$, the value of Ω_I becomes zero. Furthermore, the reading of $F(\Omega)$

clocks at that radius, as viewed by F_I observers, stops changing, much like the case in special relativity where the rate of ticking of clocks slows down and stops as the speed of those clocks approaches the speed of light.

These effects are not unexpected because the rotational transformations (5.11) map only a portion of the space in the rotating frame $F(\Omega)$ ($\rho < 1/\Omega$) to the entire inertial frame F_I. Similar to the singular walls found in frames with a linear acceleration, there is a cylindrical singular wall at $\rho = 1/\Omega$ in the rotating frame. $F(\Omega)$.

The rotational transformations (5.11) have the inverse transformations (5.13), but when $R \neq 0$ in (5.10), it appears that there is no algebraic expression for the inverse of the transformations (5.10). However, mathematically, (5.10) defines a function from one neighborhood of the origin in space-time to another. The implicit function theorem then guarantees the existence of an inverse in a neighborhood of any point where the Jacobian is non-zero.[4] The situation is similar to the inverse of the space-time transformations (4.15) involving arbitrary linear accelerations in chapter 4.

5-2.   Metric tensors for the space-time of rotating frames

We note that the transformations of the contravariant 4-vectors $dx_I^\mu = (dw_I, dx_I, dy_I, dz_I)$ and $dx^\mu = (dw, dx, dy, dz)$ can be derived from (5.11). We have

$$dw_I = \gamma \left[dw + \left(\gamma^2 \Omega^2 wx \right) dx + \left(\gamma^2 \Omega^2 wy \right) dy \right],$$

$$dx_I = \gamma \left\{ \left[\cos\left(\Omega w \right) + \gamma^2 \Omega^2 x^2 \cos\left(\Omega w \right) - \gamma^2 \Omega^2 xy \, \sin\left(\Omega w \right) \right] dx \right.$$

$$- \left[\sin\left(\Omega w \right) + \gamma^2 \Omega^2 y^2 \sin\left(\Omega w \right) - \gamma^2 \Omega^2 xy \cos\left(\Omega w \right) \right] dy$$

$$\left. - \left[\Omega x \sin\left(\Omega w \right) + \Omega y \cos\left(\Omega w \right) \right] dw \right\},$$

$$dy_I = \gamma \left\{ \left[\sin\left(\Omega w \right) + \gamma^2 \Omega^2 x^2 \sin\left(\Omega w \right) + \gamma^2 \Omega^2 xy \cos\left(\Omega w \right) \right] dx \right.$$

$$+ \left[\cos\left(\Omega w \right) + \gamma^2 \Omega^2 y^2 \cos\left(\Omega w \right) + \gamma^2 \Omega^2 xy \sin\left(\Omega w \right) \right] dy$$

$$\left. + \left[\Omega x \cos\left(\Omega w \right) - \Omega y \sin\left(\Omega w \right) \right] dw \right\}, \tag{5.14}$$

$$dz_I = dz.$$

To find the Poincaré metric tensor $P_{\mu\nu}$ for the rotating frame. $F(\Omega)$, it is convenient to use (5.11) to first write $ds^2 = dw_I^2 - dx_I^2 - dy_I^2 - dz_I^2$ as

$$ds^2 = d(\gamma w)^2 - \left(x^2 + y^2\right)\gamma^2\Omega^2 dw^2 - d(\gamma x)^2 - d(\gamma y)^2 - dz^2$$
$$+ 2\gamma\Omega y\, d(\gamma x)dw - 2\Omega\gamma x\, d(\gamma y)dw. \tag{5.15}$$

Then, with the help of the relation $d\gamma = \gamma^3\Omega^2(xdx + ydy)$, (5.15) can be written as

$$ds^2 = P_{\mu\nu}dx^\mu dx^\nu, \quad \mu,\nu = 0,1,2,3, \tag{5.16}$$

where the non-vanishing components of $P_{\mu\nu}$ are given by

$$P_{00} = 1, \quad P_{11} = -\gamma^2\left[1 + 2\gamma^2\Omega^2 x^2 - \gamma^4\Omega^4 x^2\left(w^2 - x^2 - y^2\right)\right],$$
$$P_{22} = -\gamma^2\left[1 + 2\gamma^2\Omega^2 y^2 - \gamma^4\Omega^4 y^2\left(w^2 - x^2 - y^2\right)\right], \quad P_{33} = -1,$$
$$P_{01} = \gamma^2\left[\Omega y + \gamma^2\Omega^2 wx\right], \quad P_{02} = \gamma^2\left[-\Omega x + \gamma^2\Omega^2 wy\right], \tag{5.17}$$
$$P_{12} = -\gamma^4\Omega^2 xy\left[2 - \gamma^2\Omega^2\left(w^2 - x^2 - y^2\right)\right].$$

The contravariant metric tensor $P^{\mu\nu}$ is found to be[b]

$$P^{00} = \gamma^{-2}\left[1 - \Omega^4 w^2\left(x^2 + y^2\right)\right], \quad P^{33} = -1,$$
$$P^{11} = -\gamma^{-2}\left[\gamma^{-2}\left(1 - \Omega^2 x^2\right) - 2\gamma^{-2}\Omega^3 wxy + \Omega^6 w^2 y^2\left(x^2 + y^2\right)\right],$$
$$P^{22} = -\gamma^{-2}\left[\gamma^{-2}\left(1 - \Omega^2 y^2\right) + 2\gamma^{-2}\Omega^3 wxy + \Omega^6 w^2 x^2(x^2 + y^2)\right], \tag{5.18}$$
$$P^{01} = -\gamma^{-2}\left[-\Omega y - \gamma^{-2}\Omega^2 wx + \Omega^5 w^2 y\left(x^2 + y^2\right)\right],$$
$$P^{02} = -\gamma^{-2}\left[\Omega x - \gamma^{-2}\Omega^2 wy - \Omega^5 w^2 x\left(x^2 + y^2\right)\right],$$
$$P^{12} = \gamma^{-2}\left[\gamma^{-2}\Omega^2 xy - \gamma^{-2}\Omega^3 w\left(x^2 - y^2\right) + \Omega^6 w^2 xy\left(x^2 + y^2\right)\right].$$

All other components in (5.17) and (5.18) are zero. Indeed, one can verify $P_{\mu\lambda}P^{\lambda\nu} = \delta_\mu^\nu$.

5-3. The rotational pseudo-group

As mentioned previously, the rotational transformations (5.11) imply the existence of a cylindrical singular wall at $\rho = \sqrt{x^2 + y^2} = 1/\Omega \equiv \rho_s$, which

---

[b]The components of this contravariant tensor could also be obtained using the momentum transformations (5.30) below and the invariant relation, $p_{I\nu}p_I^\nu = P^{\mu\nu}p_\mu p_\nu$.

depends on the angular velocity Ω. Like the Wu and taiji transformations, the rotational transformations map only the portion of space-time. of the rotating frame. within the singular wall to the entire space-time of an inertial frame and the 'group' of rotating transformations is a pseudo-group.

Two of the group properties of the rotational transformations are straightforward to verify. One can easily see that in the limit $\Omega \to 0$, the rotational transformations (5.11) reduce to the identity transformation. Also, the inverse transformations of (5.11) are given by (5.13).

In order to see other group properties of the rotational transformations, let us consider two other rotating frames, $F'(\Omega')$ and $F''(\Omega'')$, which are characterized by two different constant angular velocities Ω' and Ω'', respectively. With the help of (5.11), we can derive the rotational transformations between $F(0) = F_I, F(\Omega), F'(\Omega')$ and $F''(\Omega'')$,

$$w_I = \gamma w = \gamma' w' = \gamma'' w'',$$

$$x_I = \gamma \left[x \cos(\Omega w) - y \sin(\Omega w) \right] = \gamma' \left[x' \cos(\Omega' w') - y' \sin(\Omega' w') \right]$$

$$= \gamma'' \left[x'' \cos(\Omega'' w'') - y'' \sin(\Omega'' w'') \right], \qquad (5.19)$$

$$y_I = \gamma \left[x \sin(\Omega w) + y \cos(\Omega w) \right] = \gamma' \left[x' \sin(\Omega' w') + y' \cos(\Omega' w') \right]$$

$$= \gamma'' \left[x'' \sin(\Omega'' w'') + y'' \cos(\Omega'' w'') \right],$$

$$\gamma' = \frac{1}{\sqrt{1 - \beta'^2}}, \quad \beta' = \Omega' \rho', \quad \gamma'' = \frac{1}{\sqrt{1 - \beta''^2}}, \quad \beta'' = \Omega'' \rho''.$$

where we have neglected to write the trivial transformation equations for the z direction. The rotational transformations between $F(\Omega)$ and $F'(\Omega')$ can be obtained from (5.19):

$$w = \frac{\gamma'}{\gamma} w',$$

$$x = \gamma' \left\{ \left[x' \cos(\Omega' w') - y' \sin(\Omega' w') \right] \frac{\cos(\Omega w)}{\gamma} \right.$$

$$\left. + \left[x' \sin(\Omega' w') + y' \cos(\Omega' w') \right) \frac{\sin(\Omega w)}{\gamma} \right] \right\}, \qquad (5.20)$$

$$y = \gamma' \left\{ [x' \sin(\Omega' w') + y' \cos(\Omega' w')] \frac{\cos(\Omega w)}{\gamma} \right.$$
$$\left. - [x' \cos(\Omega' w') - y' \sin(\Omega' w')] \frac{\sin(\Omega w)}{\gamma} \right\}.$$

If one compares these transformations with those in equations (5.11) to (5.13), one can see that the transformations between two rotating frames are more algebraically complicated than those between an inertial frame and a rotating frame. In general, it does not seem possible to write the inverse of the transformations (5.20) in analytic form. However, as was the case for the taiji transformations (4.15) and (4.16), the implicit function theorem guarantees the existence of an inverse in the neighborhood of any point where the Jacobian is non-zero.[4]

5-4. Physical implications

Some physical implications of the rotational taiji transformations for rotating frames worthy of note are:

(A) The invariant action for electrodynamics in rotating frames

We are now able to write the invariant action S_{em} in natural units in a rotating frame for a charged particle with mass m and charge $e > 0$ moving in the 4-potential A_μ:

$$S_{em} = \int \left[-m\,ds + eA_\mu\,dx^\mu - \frac{1}{4} F_{\mu\nu} F^{\mu\nu} \sqrt{-P}\,d^4x \right], \qquad (5.21)$$

$$F_{\mu\nu} = D_\mu A_\nu - D_\nu A_\mu = \partial_\mu A_\nu - \partial_\nu A_\mu, \quad P = \det P_{\mu\nu}, \qquad (5.22)$$

where ds is given by (5.16).

The Lagrange equation of motion of a charged particle can be derived from (5.21). We obtain

$$m\frac{Du_\mu}{ds} = -eF_{\mu\nu}u^\nu, \qquad (5.23)$$

$$Du_\mu = D_\nu u_\mu\,dx^\nu, \quad u^\nu = \frac{dx^\nu}{ds}, \quad u_\mu = P_{\mu\nu}u^\nu, \qquad (5.24)$$

where D_ν denotes the partial covariant derivative associated with the rotational metric tensors $P_{\mu\nu}$ in (5.17).

Starting with the invariant action (5.21) and replacing the second term with

$$\int A_\mu j^\mu \sqrt{-P} d^4 x \qquad (5.25)$$

for a continuous charge distribution in space, we obtain the invariant Maxwell's equations in a rotating frame

$$D_\nu F^{\mu\nu} = -j^\mu, \qquad \partial_\lambda F_{\mu\nu} + \partial_\mu F_{\nu\lambda} + \partial_\nu F_{\lambda\mu} = 0. \qquad (5.26)$$

Based on gauge invariance and the rotational invariance of the action (5.21), the electromagnetic potential must be a covariant vector A_μ in non-inertial frames. Since the force **F** and the fields **E** and **B** are naturally related to a change of the potential A_μ with respect to a change of coordinates (i.e., x^μ, by definition), the fields **E** and **B** are naturally identified with components of the covariant tensor $F_{\mu\nu}$ as given by (5.22) in non-inertial frames. The metric tensor $P_{\mu\nu}$ behaves like a constant under covariant differentiation, $D_\alpha P_{\mu\nu} = 0$.

(B) Absolute contraction of a rotating radius and absolute slow-down of a rotating clock

Comparing the taiji rotational transformations (5.11) and the classical rotational transformations (5.1), one can see that the rotational taiji transformations (5.11) predict that the length of a rotating radius $\sqrt{x^2 + y^2}$ is contracted by a factor of γ

$$\sqrt{x_I^2 + y_I^2} = \gamma \sqrt{x^2 + y^2}. \qquad (5.27)$$

This contraction is absolute, meaning that both observers in the inertial frame F_I and in the rotating frame $F(\Omega)$ agree that the radius, as measured in the rotating frame $F(\Omega)$ is shorter, because there is no relativity between an inertial frame and a rotating (non-inertial) frame.

In some discussions of phenomena involving circular motion at high speeds, such as the lifetime dilation of unstable particles traveling in a circular storage ring, the argument is made that during a very short time interval, one can approximate the true rotational transformations using the Lorentz transformations.[2] However, making this approximation leads to a completely different conclusion regarding the radius, namely, that it does not contract because it is always perpendicular to the direction of motion.

Furthermore, for a given ρ, (5.11) implies $\Delta w_I = \gamma \Delta w$, which is independent of the spatial distance between two events.[c] In other words, clocks at rest relative to a rotating frame. and located at a distance $\rho = \sqrt{x^2 + y^2}$ from the center of rotation slow down by a factor of $\gamma = \sqrt{(1 - \beta^2)}$ in comparison with clocks in the inertial frame F_I. Analogous to the absolute contraction of radial lengths as shown in (5.27), this time dilation is also an absolute effect in that observers in both F_I and $F(\Omega)$ agree that it is the accelerated clocks that are slowed.

Both the contraction of radial distances and the slowing down of clocks are consequences of requiring the rotational transformations to satisfy the limiting Lorentz and Poincaré invariance.

(C) The covariant four-momentum in a rotating frame

Although the speed of light is not a universal constant in a rotating frame F, one can still formulate a covariant four-momentum $p_\mu = (p_0, p_i)$, $i = 1, 2, 3$ using the time w as the evolution variable in the Lagrangian formalism. From the invariant 'free' action $S_f = -\int m\,ds = \int L\,dw$ for a 'non-interacting particle' with mass m in the rotating frame $F(\Omega)$, the spatial components of the covariant four-momentum are

$$p_i = -\frac{\partial L}{\partial v^i} = mP_{vi}\frac{dx^\nu}{ds} = P_{vi}p^\nu, \qquad L = -m\sqrt{P_{\mu\nu}v^\mu v^\nu}, \qquad (5.28)$$

where $i = 1, 2, 3$. Both L and p_i have the dimension of mass and $v^\mu \equiv dx^\mu/dw = (1, v^i)$. The zeroth component p_0 (or the Hamiltonian) with the dimension of mass is defined as usual

$$p_0 = v^i \frac{\partial L}{\partial v^i} - L = mP_{v0}\frac{dx^\nu}{ds} = P_{v0}p^\nu. \qquad (5.29)$$

The rotational taiji transformations of the differential operators $\partial/\partial x_I^\mu$ and $\partial/\partial x^\mu$ can be calculated from (5.13). Just as in quantum mechanics, the covariant momentum p_μ has the same transformation properties as the covariant differential

---

[c]For comparison with the usual time t, if one were to define $w = ct$ and $w_I = ct_I$, then one would have the usual relation $\Delta t_I = \gamma \Delta t$. However, the constant 'speed of light' c in a non-inertial frame is not well-defined because the speed of a light signal is no longer isotropic or constant.

operator $\partial/\partial x^\mu$. From (5.13), we obtain

$$p_{I0} = \gamma^{-1}(p_0 + \Omega y p_1 - \Omega x p_2),$$

$$p_{I1} = \left[-\gamma^{-2}\Omega^2 w x_I\right] p_0 + \gamma^{-2}\left[\gamma \cos(\Omega w) - \Omega^2 x_I x - \Omega^3 w x_I y\right] p_1$$
$$+ \gamma^{-2}\left[-\gamma \sin(\Omega w) - \Omega^2 x_I y + \Omega^3 w x_I x\right] p_2, \qquad (5.30)$$

$$p_{I2} = \left[-\gamma^{-2}\Omega^2 w y_I\right] p_0 + \gamma^{-2}\left[\gamma \sin(\Omega w) - \Omega^2 y_I x - \Omega^3 w y_I y\right] p_1$$
$$+ \gamma^{-2}\left[\gamma \cos(\Omega w) - \Omega^2 y_I y + \Omega^3 w y_I x\right] p_2,$$

$$p_{I3} = p_3, \quad \gamma = \frac{1}{\sqrt{1 - \rho^2\Omega^2}},$$

where x_I and y_I can be expressed in terms of x, y, and w of the rotating frame $F(\Omega)$ using (5.11).

Consider the case of a particle traveling at high speed in a circular ring. The particle can be considered to be at rest in a rotating frame, so that $dx^i = 0$ and $ds = dw$. Using $p^\nu = mdx^\nu/ds$ in (5.28), we have $p^i = 0$, $i = 1, 2, 3$, and $p^0 = m$. In this case ρ and γ are constant, so (5.14), (5.17), (5.18) and (5.30) will be simplified. We have

$$p_0 = P_{0\mu}p^\mu = m, \quad p_1 = m\gamma^2\Omega y, \quad p_2 = -m\gamma^2\Omega x, \quad p_3 = 0; \qquad (5.31)$$
$$(\rho^2 = x^2 + y^2 = constant.)$$

This difference between p_μ and p^μ is due to the presence of the metric tensor components P_{01} and P_{02}. The covariant momenta of the particle, as measured in an inertial frame F_I, are given by simplified (5.30) and (5.31),

$$p_{I0} = \gamma m, \quad p_{I1} = m\gamma\left[\Omega x \sin(\Omega w) + \Omega y \cos(\Omega w)\right],$$

$$p_{I2} = -m\gamma\left[\Omega x \cos(\Omega w) - \Omega y \sin(\Omega w)\right], \quad p_{I3} = 0. \qquad (5.32)$$

Thus, the expression for the energy of a rotating particle $p_{I0} = \gamma m$ agrees with the well-established results of high energy experiments performed in an inertial laboratory frame F_I.

5-5. Experimental tests of the rotational taiji transformations

Because particles moving in a straight line at relativistic speeds travel large distances in very short times, designing experiments to test relativistic effects with linear motion, where the Lorentz transformations are most directly applicable, can be challenging. Experiments in which objects move in a circular path at relativistic speeds can be performed with much more compact apparatus. However, as was discussed briefly in the previous section, applying the Lorentz transformations to analyzing such experiments is problematic because of the non-relativity between inertial frames and rotating frames. In this section, we discuss three experiments that can test the usefulness of the rotational taiji transformations.

(A) Absolute dilation of decay length for particle decay in circular motion

Let us first consider the lifetime dilation of unstable particles traveling in a circular storage ring.[5] If the particle's rest lifetime in $F(\Omega)$ is denoted by $\Delta w(rest)$, then the rotational taiji transformation (5.11) gives

$$\Delta w_I = \gamma \Delta w(rest), \quad \gamma = \frac{1}{\sqrt{1 - \beta^2}}, \quad \beta = \Omega \rho. \tag{5.33}$$

As mentioned in the previous two chapters, the rest lifetime of the particle $\Delta w(rest)$ in the rotating frame F cannot be directly measured. However, according to the weak equivalence of non-inertial frames (3.36),

$$\Delta w(rest)\sqrt{P_{00}} = \Delta w_I(rest)\sqrt{\eta_{00}}, \tag{5.34}$$

where we have taken the limit (5.8) for the non-inertial frame F' in (3.36), so that F' becomes F_I in the limit and $P'_{00} = \eta_{00}$. Since the lifetime is typically very short, $\Delta w(rest)$ can be considered as the zeroth component of the contravariant differential coordinate vector V^μ:

$$\Delta w = dw = dx^0 = V^0.$$

It follows from (5.33) and (5.34) that

$$\Delta w_I = \gamma \Delta w_I(rest), \tag{5.35}$$

where we have used $P_{00} = 1$ in (5.17).

Since Δw_I has the dimension of length, (5.35) can be understood as a dilation of the decay length, i.e., the distance traveled by the particle before it decays. Result (5.35) is consistent with well-established experimental results of the decay lifetime dilation of muons in flight in a circular storage ring.[5] The lifetime dilation (5.35) is an absolute effect in that observers in both F_I and $F(\Omega)$ agree that it is the accelerated muons whose mean decay length is dilated.

(B) Davies–Jennison experiment

A second experiment that can test the usefulness of the taiji rotational transformations is the Davies–Jennison experiment,[6] in which the shift in frequency of transversely-emitted radiation from an orbiting source was measured. The arrangement of the apparatus for this experiment is roughly as follows: A laser beam is directed straight downward onto a mirror at the center of a horizontal rotating table. The mirror sends the beam radially outward, parallel to the surface of the table, where it strikes a second mirror mounted at the edge of the table. This second mirror reflects the beam in a direction perpendicular to its velocity, thus acting as a source emitting radiation in the transverse direction. This resultant beam is then mixed with light from the original laser in order to produce interference fringes and measure any shift in its frequency compared to the original laser light. In the experiments of Davies and Jennison, no such shift was observed. Let us analyze this experiment using the rotational taiji transformations.

Consider a radiation source at rest in a rotating frame $F(\Omega)$, located at $\vec{\rho}_s = (x_s, y_s)$, that emits a wave with a frequency $k_0 = k_0(rest)$, as measured in $F(\Omega)$. If this wave, travels along the $-\vec{\rho}_s$ direction to the origin, then $k_1/k_2 = k_x/k_y = x/y$, because the two vectors $-\vec{\rho}$ and $\vec{k}$ are on the x–y plane and parallel to each other. Because the rotational taiji transformations for the covariant wave vector k_μ are the same as those for the covariant momentum $p_\mu = \hbar k_\mu$, the frequency shift of transversely-emitted radiation as measured by observers in an inertial laboratory frame F_I is given by (5.30). To detect the transverse effect, the observer is assumed to be at rest at the origin $x_I = y_I = x = y = 0$. Equation (5.30) with $p_\mu = \hbar k_\mu$ gives

$$k_{I0} = \gamma^{-1}(k_0 + \Omega[yk_1 - xk_2]) = \gamma^{-1}k_0, \qquad (5.36)$$

$$xk_2 = yk_1, \quad \gamma = \frac{1}{\sqrt{1 - \rho_s^2\Omega^2}}, \quad \rho_s^2 = x_s^2 + y_s^2 = constant,$$

where $k_0 = k_0(rest)$ is the frequency of the source at rest in the rotating frame $F(\Omega)$, as measured by observers in $F(\Omega)$.

Just as in our discussion of the lifetime dilation of unstable particles traveling in a circle, it is extremely difficult to measure the frequency $k_0(rest)$

directly because we cannot move our measuring apparatus to the rotating frame $F(\Omega)$. Furthermore, because of the non-equivalence of inertial and rotating frames, we cannot assume that $k_0(rest)$ is equal to $k_{I0}(rest)$, the frequency of the radiation from the same source at rest in an inertial frame F_I as measured by observers in F_I. However, we can again use the weak equivalence of non-inertial frames (3.36) to write

$$k_0(rest)\sqrt{P^{00}} = k_{I0}(rest)\sqrt{\eta^{00}}, \quad P^{00} = \gamma^{-2}, \quad \eta^{00} = 1. \qquad (5.37)$$

Recall that $k_0(rest)$ is the frequency of radiation emitted from the source at rest in the rotating frame $F(\Omega)$, located at the constant radius, $\rho = fixed$ on the x–y plane. When this condition is used in the derivation of (5.14), then $dw_I = \gamma dw$, etc. and the contravariant component P^{00} in (5.18) reduces to $P^{00} = \gamma^{-2} = constant$ in (5.37). It follows from (5.36) and (5.37) that the relationship between the frequency of the transversely-emitted radiation emitted from the orbiting source as measured in the inertial laboratory frame k_{I0}, and the frequency of the radiation emitted from the same source at rest in the inertial laboratory frame as measured in the inertial laboratory frame $k_{I0}(rest)$ is

$$k_{I0} = \gamma^{-1}k_0(rest) = k_{I0}(rest). \qquad (5.38)$$

Thus, the rotational taiji transformation predicts that no frequency shift for transverse radiation should be observed for an orbiting radiation source, as measured by observers in inertial frames. This result (5.38) appears to be consistent with the experiments of Davies and Jennison.[6] It is interesting to note that, based on experimental results, Jennison concluded that rotating radii are contracted and that angular velocities increase, as measured from the rotating frame. Those conclusions are consistent with those in (5.12), derived from the rotational taiji transformations (5.11). These results thus lend support to the taiji rotational transformations and the concept of limiting Lorentz–Poincaré invariance.

(C) Thim's experiment with orbiting receivers

In an experiment by H. W. Thim,[7] a radiation source is located at the center of a rotating disk and eight detectors (monopole antennas, in this case) are mounted at the edges of the disk so that they move in a circle with a radius ρ. This experiment is similar to that of Davies and Jennison, but with the source and receiver interchanged. Like Davies and Jennison, Thim found no shift in the radiation frequency. However, because of the absolute difference between an inertial and a rotating frames, the two null results cannot be explained by a simple relativity argument.

In Thim's experiment, the source at rest at the center of the rotating frame can also be considered to be at rest in the inertial laboratory frame F_I. Thus, the frequency of its radiation k_{I0} measured in F_I is the same as $k_0(rest)$,

$$k_{I0} = k_{I0}(rest) = k_0(rest).	(5.39)$$

Consider one specific detector located at the rim of the rotating disk. The wave with the wave vector $\vec{k} = (k_1, k_2)$ propagates from the center of the disk to this detector along the radius $\vec{\rho} = (x, y)$, so that $k_1/k_2 = x/y$. Thus, similar to (5.36), we have

$$k_{I0} = \gamma^{-1} k_0, \quad xk_2 = yk_1, \quad \gamma = \frac{1}{\sqrt{1 - \rho_a^2 \Omega^2}}.	(5.40)$$

From (5.39) and (5.40), we derive the transverse frequency shift,

$$k_0 = \gamma k_{I0} = \gamma k_0(rest),	(5.41)$$

as measured by an observer at rest in the rotating frame $F(\Omega)$. Result (5.41), derived from the rotational space-time transformations (5.11) holds for each of the eight detectors and implies that if the frequency k_0 is measured in the rotating frame $F(\Omega)$, there is a shift in frequency by a factor of γ as a result of the orbiting motion of the detectors.

Although Thim measured no shift in the radiation frequency, result (5.41) actually does not contradict Thim's null result. The reason is as follows: In Thim's experiment,[7] the frequency of the radiation k_0 received by the orbiting detector is not directly measured in the rotating frame $F(\Omega)$. Instead, the signal received by the detector on the rotating disk is transferred through a second rotating disk into a stationary detector and its frequency is measured in the inertial laboratory F_I.

According to the rotational transformations (5.11), clocks in the rotating frame located at a radius ρ are slowed by a factor of γ, resulting in an increase in the frequency k_0 by a factor of γ. This is shown in the first equation in (5.36) with $xk_2 = yk_1$, because the wave vector points in the radial direction in the rotating frame. Conversely, the frequency of the radiation $k_{I0}(detector)$ received by the detector, as measured in the laboratory, will be smaller by a factor of γ because the clocks in the inertial frame run faster by the factor of γ, $k_{I0}(detector) = k_0/\gamma$.

This relation and Equation (5.41) lead to the result

$$k_{I0}(detector) = k_o(rest) = k_{I0}, \qquad (5.42)$$

in the inertial frame F_I. Thus, the rotational transformations (5.11) imply that when the frequency of the radiation $k_{I0}(detector)$ received by the detector is measured by standard mixer and interferometer techniques in the inertial laboratory frame and compared with the frequency $k_o(rest)$ of the source in (5.39), no shift should be measured.

Furthermore, according to the rotational transformations (5.11) and (5.36), conclusion (5.42) should be independent of the angular velocity Ω' of the second disk. This property can be seen by regarding the second disk as a second rotating frame $F(\Omega')$. The rotational transformations for F_I, $F(\Omega)$ and $F(\Omega')$ relevant to the experiment are given by

$$w_I = \gamma w = \gamma' w' = w_I, \quad etc. \qquad (5.43)$$

$$k_{I0} = \gamma^{-1} k_0 = \gamma'^{-1} k_0' = k_{I0}, \quad etc. \qquad (5.44)$$

$$\gamma = \frac{1}{\sqrt{1 - \rho_a^2 \Omega^2}}, \quad \gamma' = \frac{1}{\sqrt{1 - \rho'^2 \Omega'^2}};$$

where ρ_a and ρ' are constant. Thus, if the frequency of the radiation were to be measured by an apparatus in either the $F(\Omega)$ (disk 1) or the $F(\Omega')$ (disk 2) frame, it would differ from its original value (measured in the inertial laboratory frame). However, because this radiation is eventually transferred back and measured by a detector that is at rest in the same inertial laboratory frame as the source from which it was emitted, there will be no overall frequency shift.

We note that the transverse frequency shift given by (5.41) should be detectable if one were to measure the frequency received by the orbiting detector directly, in the rotating frame, rather than transferring the signal back to a stationary detector to be measured in the inertial laboratory frame. It would be an interesting challenge to redesign the experiment to measure directly the frequency of the signal received by the orbiting antenna by mounting a standard mixer and interferometer on the rotating disk. Assuming that the centrifugal effects on the apparatus are small and negligible, one can test the transverse frequency shift predicted in (5.41). In fact, one could test for centrifugal effects by repeating the experiment with different angular velocities.

In conclusion, the three experiments discussed here are important because they suggest new principles in physics and novel space-time properties of non-inertial frames, which are beyond the realm of special relativity. They can help to develop a deeper and broader understanding of physics in non-inertial frames based on the general principle of limiting continuation of physical laws and, in particular, limiting Lorentz–Poincaré invariance.

References

1. Leonardo Hsu and Jong-Ping Hsu, Nuovo Cimento B **112**, 1147 (1997); Jong-Ping Hsu and Leonardo Hsu, in *JingShin Theoretical Physics Symposium in Honor of Professor Ta-You Wu* (Ed. J. P. Hsu and L. Hsu, World Scientific, Singapore; New Jersey.) pp. 393–412; Nuovo Cimento B **112**, 575 (1997).
2. G. N. Pellegrini and A. R. Swift, Am. J. Phys. **63**, 694–705 (1995) and references therein.
3. J. J. Sakurai, *Invariance Principles and Elementary Particles*, (Princeton Univ. Press, 1964) p. v, pp. 3–5, and p. 10.
4. We would like to thank Dana Fine for helpful discussions.
5. F. J. M Farley, J. Bailey and E. Picasso, Nature **217**, 17–18 (1968).
6. P. A. Davies and R. C. Jennison, Nature, vol. 248, Apil. 1974; J. Phys. A. Math. vol 8, p. 1390 (1975). The reflected radiation from the moving mirror is considered to be transversely-emitted radiation from an orbiting radiation source. The Davies–Jennison experiment 2 sends the reference beam of the interferometer directly to the detector while the other beam is reflected in turn from a mirror moving towards the beam and then from a mirror moving away from the beam, where both beams are parallel to the x_I. Suppose the mirror moving away from the beam is at $(x, y) = (0, -R_s)$ and the mirror moving towards the beam is at $(x, y) = (0, +R_s)$. The shifts of frequencies at these two locations of orbiting mirrors are respectively given by $k_{I0} = k_{I0}(rest)[1 - \Omega R_s(k_1/k_0)]$, and $k_{I0} = k_{I0}(rest)[1 + \Omega R_s(k_1/k_0)]$, where, the ratio k_1/k_0 is determined by the law $k_\mu k_\nu P^{\mu\nu} = 0$ and depends on the constant radius $\rho > 0$. These results show that the blue-shift is completely cancelled by the red-shift, $k_{I0} = k_{I0}(rest)$, as confirmed by the Davies–Jennison experiment 2. The physical reasons for this null frequency shift are the rotational transformations and the weak equivalence of non-inertial frames, as expressed by (5.37).
7. H. W. Thim, IEEE Trans. Instrumentation and Measurement, vol. 52, no. 5, pp. 1660–1664 (2003).

6.

Conservation Laws and Symmetric Energy–Momentum Tensors

6-1. Conservation laws in the Taiji symmetry framework

Traditionally, theories of physics have been formulated and discussed in inertial frames based on flat space-time and are consistent with the principle of Lorentz and Poincaré invariance. In particular, conservation laws are usually derived from global symmetries in inertial frames through Noether's well-known Theorem I.[1] Einstein's theory of gravity however, is based on the principle of general coordinate invariance in curved space-time. Despite its success, this theory of gravity fails to preserve certain fundamental elements of physics such as the law of conservation of energy because the symmetry group of general coordinate transformations involves a continuously infinite number of generators, as shown by Noether's Theorem II.[1] Although the covariant continuity equation for the energy–momentum tensor and the usual continuity equation for the energy–momentum pseudo-tensor can be found within general relativity, they are not conservation laws in the usual sense and are thus not ideal. Furthermore, since non-inertial frames in general relativity are assumed to involve curved space-time, it is not obvious that energy and momentum are conserved in such reference frames. Although there is a well-known general result that states that any equation in inertial frames can be expressed in a covariant form, so that it holds in arbitrary coordinate system,[2] the framework of general relativity does not allow for inertial frames.[3] In this chapter, we show how the taiji symmetry framework, which is based on a flat space-time for both inertial and non-inertial frames of reference, avoids all the complications caused by the curved space-time of general relativity and naturally accommodates all known conservation laws.

6-2.   Symmetric energy–momentum tensors and variations of metric tensors in taiji space-time

We assume that the space-time of all frames of reference, both inertial and non-inertial, is described by coordinates x^μ and characterized by symmetric Poincaré metric tensors $P_{\mu\nu} = P_{\nu\mu}$. As we have discussed in previous chapters, we apply the principle of limiting continuation and postulate that in the limit of zero acceleration, all non-inertial frames reduce to inertial frames and the Poincaré metric tensors $P_{\mu\nu}$ reduce to the Minkowski metric tensor $\eta_{\mu\nu} = (1, -1, -1, -1)$. In general, in order for x^0 and $x^i = (x, y, z)$ to have the character of temporal and spatial coordinates respectively, it is necessary and sufficient that

$$P_{00} > 0, \quad P_{ik} dx^i dx^k < 0, \tag{6.1}$$

where i and k are summed over 1, 2 and 3 and the components P_{10}, P_{20} and P_{30} may be positive or negative.

Noether's theorem allows one to obtain the conserved energy–momentum tensor $t_{\mu\nu}$ for a physical system in a flat space-time with global translational symmetry. Such a tensor $t_{\mu\nu}$ is, in general, not symmetric. Although, it is possible to symmetrize the energy–momentum tensor, the procedure can be quite complicated. A simple and direct way to obtain a symmetric energy–momentum tensor was first discussed by Hilbert in 1915 when he combined Einstein's idea of general covariance and the variation of an invariant action involving scalar curvature to derive the gravitational field equation. The relationship between the symmetric energy–momentum tensor and the variation of the metric tensor have been discussed by Hilbert, Fock, and others.[4,5]

Hilbert's symmetric energy–momentum tensor $T_{HF}^{\mu\nu}$ in arbitrary coordinates is given by[4,5]

$$T_{HF}^{\mu\nu} = \frac{-2}{\sqrt{-P}} \left(\frac{\partial L\sqrt{-P}}{\partial P_{\mu\nu}} - \partial_\lambda \frac{\partial L\sqrt{-P}}{\partial(\partial_\lambda P_{\mu\nu})} \right). \tag{6.2}$$

This definition can be applied in a general frame with the Poincaré metric tensor $P_{\mu\nu}$. Because the ten elements of $\delta P_{\mu\nu}$ are constrained by the condition that the Riemann–Christoffel curvature tensor must vanish, they are not independent.

To obtain a symmetric energy–momentum tensor $U_{\alpha\beta}$ for an electromagnetic field within the taiji space-time framework, let us consider the U(1) gauge invariant action for electromagnetic fields,

$$S_f = \int L\sqrt{-P}\, d^4x, \quad L = -\frac{1}{4}F_{\mu\nu}F^{\mu\nu}. \tag{6.3}$$

Under the variation δ_p of the metric tensor $P_{\mu\nu}$, we have

$$\delta_p S_f = -\frac{1}{4}\int \left[F_{\mu\nu}\delta_p\left(F^{\mu\nu}\sqrt{-P}\right)\right] d^4x$$

$$= -\frac{1}{2}\int U^{\alpha\beta}\delta P_{\alpha\beta}\sqrt{-P}\, d^4x, \tag{6.4}$$

$$U^{\alpha\beta} = \frac{1}{4}F_{\mu\nu}F^{\mu\nu}P^{\alpha\beta} - P_{\mu\nu}F^{\mu\alpha}F^{\nu\beta}, \tag{6.5}$$

where we have used the relations,

$$F_{\mu\nu}\delta_p F^{\mu\nu} = -2F_{\mu\nu}F^{\mu\alpha}P^{\nu\beta}\delta P_{\alpha\beta}, \quad \delta_p P_{\alpha\beta} = \delta P_{\alpha\beta}, \tag{6.6}$$

$$\delta_p\sqrt{-P} = \frac{1}{2}\sqrt{-P}P^{\alpha\beta}\delta P_{\alpha\beta}. \tag{6.7}$$

Note that A_μ and $F_{\mu\nu}$ are not varied because A_μ is originally introduced through the U(1) gauge covariant derivative, $\partial_\mu - ieA_\mu$, $e > 0$. Only the contravariant potential $A^\mu = P^{\mu\nu}A_\nu$ depends on the metric tensors.

In the Fock–Hilbert approach to the energy–momentum tensor, the law of conservation of energy–momentum is derived on the basis of a local transformation in flat space-time.[2] Specifically, that the action (6.3) is invariant under local translations in flat space-time,

$$x'^\mu = x^\mu + \Lambda^\mu(x), \tag{6.8}$$

where $\Lambda^\mu(x)$ are arbitrary infinitesimal functions. Transformation (6.8) can also be viewed as a general coordinate transformation. The invariance of actions under this local translation is the mathematical expression of the space-time 'taiji symmetry.'[a] In this case, Noether's theorem assumes a global space-time translational-symmetry of the action to derive the conservation of momentum.

---

[a]This taiji symmetry is essential for Yang–Mills gravity. It enables one to introduce the gauge covariant derivative involving the generators of the space-time translation group. All the properties of the gravitational interaction appear to be manifestations of this taiji symmetry, as we shall see in Part II of this book.

According to the general transformation formula for tensors, the transformation of $P_{\mu\nu}$ is given by

$$P'^{\mu\nu}(x') = P^{\alpha\beta}(x)\frac{\partial x'^{\mu}}{\partial x^{\alpha}}\frac{\partial x'^{\nu}}{\partial x^{\beta}}, \tag{6.9}$$

which can be calculated to first order using (6.8) and

$$P'^{\mu\nu}(x') = P'^{\mu\nu}(x) + \Lambda^{\lambda}(\partial_{\lambda}P'^{\mu\nu}) = P'^{\mu\nu}(x) + \Lambda^{\lambda}(\partial_{\lambda}P^{\mu\nu}).$$

The difference between $\Lambda^{\lambda}(\partial_{\lambda}P'^{\mu\nu})$ and $\Lambda^{\lambda}(\partial_{\lambda}P^{\mu\nu})$ is of the second order and negligible. We obtain expressions for $\delta P^{\mu\nu}(x)$ and $\delta P_{\mu\nu}$

$$\delta P^{\mu\nu}(x) = P'^{\mu\nu}(x) - P^{\mu\nu}(x)$$

$$= -(\partial_{\lambda}P^{\mu\nu})\Lambda^{\lambda} + P^{\mu\alpha}\partial_{\alpha}\Lambda^{\nu} + P^{\nu\alpha}\partial_{\alpha}\Lambda^{\nu}$$

$$= D^{\mu}\Lambda^{\nu} + D^{\nu}\Lambda^{\mu}, \tag{6.10}$$

$$\delta P_{\mu\nu} = -D_{\mu}\Lambda_{\nu} - D_{\nu}\Lambda_{\mu}, \tag{6.11}$$

where

$$D_{\mu}\Lambda_{\nu} = \partial_{\mu}\Lambda_{\nu} - \Gamma^{\lambda}_{\mu\nu}\Lambda_{\lambda},$$

$$\Gamma^{\lambda}_{\mu\nu} = \frac{1}{2}P^{\lambda\alpha}(\partial_{\nu}P_{\mu\alpha} + \partial_{\mu}P_{\nu\alpha} - \partial_{\alpha}P_{\mu\nu}). \tag{6.12}$$

Since the action S_f is invariant under an infinitesimal local space-time translation (6.8),

$$\delta_p S_f = -\frac{1}{2}\int U^{\alpha\beta}\delta P_{\alpha\beta}\sqrt{-P}d^4x$$

$$= -\frac{1}{2}\int U^{\mu\nu}(-D_{\mu}\Lambda_{\nu} - D_{\nu}\Lambda_{\mu})$$

$$= -\int (D_{\mu}U^{\mu\nu})\Lambda_{\nu}\sqrt{-P}d^4x = 0, \tag{6.13}$$

where we have used (6.10) and (6.11). From (6.13) with the arbitrary infinitesimal functions Λ_{ν}, we obtain the following equation for the symmetric energy–momentum tensor $U^{\alpha\beta}$ in (6.5),

$$D_{\alpha}U^{\alpha\beta} = 0, \tag{6.14}$$

which is the covariant continuity equation for all frames and arbitrary coordinates. The form of Equation (6.14) holds for all physical systems with space-time translational gauge symmetry. Combining the continuity Equation (6.14) with the metric tensors derived in the preceding chapters, one can obtain explicit transformation equations for energy and momentum between inertial frames, frames with an arbitrary linear acceleration, and rotating frames.

In Einstein's theory of gravity, there is a covariant equation analogous to (6.14) for the energy–momentum tensor, but the field equation does not necessarily imply that the energy–momentum tensor is conserved. In the next section, we discuss the conditions under which the covariant Equation (6.14) implies the conservation of energy–momentum in the usual sense.[6,7]

6-3. Integral forms of conservation laws in non-inertial frames

Understanding the conservation laws in non-inertial frames is not a simple thing. For example, consider a simple two-particle scattering process. In an inertial frame, it is easy to see that energy, momentum, and angular momentum are conserved because the energy–momentum 4-vector of each particle is constant. However, in a non-inertial frame, the energy–momentum 4-vector of each particle is a vector function of time because of the acceleration of the reference frame.

To see how the conservation laws are instantiated in the taiji symmetry framework, we first note the following three mathematical results.

(i) In taiji space-time, the symmetric energy–momentum tensor $T^{\mu\nu} = T^{\nu\mu}$ satisfies the covariant continuity condition given by (6.14),

$$D_\nu T^{\mu\nu} = \partial_\nu T^{\mu\nu} + \Gamma^\nu_{\rho\nu} T^{\mu\rho} + \Gamma^\mu_{\rho\nu} T^{\rho\nu} = 0, \qquad (6.15)$$

for any physical system whose action is invariant under a space-time translation. The first two terms can then be combined, so that

$$\frac{1}{\sqrt{-P}} \partial_\nu(\sqrt{-P} T^{\mu\nu}) + \Gamma^\mu_{\rho\nu} T^{\rho\nu} = 0. \qquad (6.16)$$

(ii) It can be shown that vanishing curvature, $R^\rho_{\mu,\nu\alpha} = 0$, is a necessary and sufficient condition to ensure that the following equation has a solution,[4,5]

$$\partial_\mu \partial_\nu \phi - \Gamma^\rho_{\mu\nu} \partial_\rho \phi = \partial_\mu \phi_\nu - \Gamma^\rho_{\mu\nu} \phi_\rho = 0, \qquad \phi_\rho = \partial_\rho \phi. \qquad (6.17)$$

(iii) Let us introduce a vector ϕ_μ and integrate the equation

$$D_\nu(T^{\mu\nu}\phi_\mu) = \frac{1}{\sqrt{-P}}\partial_\nu(\sqrt{-P}T^{\mu\nu}\phi_\mu) = 0 \qquad (6.18)$$

over a 3-dimensional volume. As usual, we assume the energy–momentum tensor $T^{\mu\nu}$ vanishes on the surface of this volume, obtaining

$$\int D_\nu(T^{\mu\nu}\phi_\mu)\sqrt{-P}d^3r = \frac{d}{dx^0}\int T^{\mu 0}\phi_\mu\sqrt{-P}d^3r$$

$$= \frac{1}{2}\int T^{\mu\nu}(D_\nu\phi_\mu + D_\mu\phi_\nu)\sqrt{-P}d^3r, \qquad (6.19)$$

where we have used $T^{\mu\nu} = T^{\nu\mu}$ and Equation (6.15). If the vector ϕ_μ satisfies

$$D_\nu\phi_\mu + D_\mu\phi_\nu = 0, \qquad (6.20)$$

then Equation (6.19) leads to

$$K \equiv \int T^{\mu 0}\phi_\mu\sqrt{-P}d^3r = constant, \qquad (6.21)$$

which is independent of time x^0.

To demonstrate the conservation laws in the taiji symmetry framework in an inertial frame F_I with coordinates $x_I^\mu = (x_I^0, x_I^1, x_I^2, x_I^3)$, we introduce the vector

$$\phi_\mu^{(0)} = \frac{\partial x_I^0}{\partial x^\mu}, \quad \phi_\mu^{(1)} = \frac{\partial x_I^1}{\partial x^\mu}, \quad \phi_\mu^{(2)} = \frac{\partial x_I^2}{\partial x^\mu}, \quad \phi_\mu^{(3)} = \frac{\partial x_I^3}{\partial x^\mu}, \qquad (6.22)$$

These quantities constitute a 4-vector with respect to the unprimed coordinates (x^0, x^1, x^2, x^3).[4] Using (6.17), i.e., $\partial_\mu\phi_\nu - \Gamma^\rho_{\mu\nu}\phi_\rho = D_\mu\phi_\nu = 0$, and (6.22) we have

$$D_\nu\phi_\mu^{(\alpha)} = 0, \quad (\alpha = 0, 1, 2, 3). \qquad (6.23)$$

Thus, each of the elements of the 4-vector in (6.22) also satisfies (6.23). One can also introduce the 6-vector

$$\phi_\mu^{\alpha\beta} = x_I^\alpha\partial_\mu x_I^\beta - x_I^\beta\partial_\mu x_I^\alpha, \qquad (6.24)$$

which satisfies

$$D_\nu\phi_\mu^{\alpha\beta} + D_\mu\phi_\nu^{\alpha\beta} = 0. \qquad (6.25)$$

This is the same as the relation $D_\nu \phi_\mu + D_\mu \phi_\nu = 0$ in (6.20). Substituting (6.22) and (6.24) into (6.21), we obtain two sets of constant integrals p_I^α and $M_I^{\alpha\beta}$,

$$p_I^\alpha = \int T^{\mu 0} \frac{\partial x_I^\alpha}{\partial x^\mu} \sqrt{-P} d^3 r \qquad (6.26)$$

and

$$M_I^{\alpha\beta} = \int T^{\mu 0} (x_I^\alpha \partial_\mu x_I^\beta - x_I^\beta \partial_\mu x_I^\alpha) \sqrt{-P} d^3 r, \qquad (6.27)$$

where p_I^α is the constant energy–momentum 4-vector and $M_I^{\alpha\beta}$ is the constant angular-momentum tensor in an inertial frame $F_I(x_I)$. Physically, the components $M_I^{0k}(k = 1, 2, 3)$ are the constants of the motion of the center of mass, and M_I^{ik} is the constant angular momentum in the inertial frame F_I. Comparison of (6.26) and (6.27) with (6.21) shows both of these quantities to be constant, as expected, demonstrating the conservation of the energy–momentum 4-vector and the angular momentum.[6]

To see the analogous result in non-inertial frames explicitly, consider a frame with an arbitrary linear acceleration (ALA) $F(x) \equiv F(x^\mu)$ with coordinates $x^\mu = (w, x, y, z)$. According to (4.12), the coordinate transformation between the inertial frame $F_I(x_I)$ and an ALA frame $F(x)$ is

$$x_I^0 = \gamma\beta \left(x^1 + \frac{1}{\alpha(x^0)\gamma_0^2} \right) - \frac{\beta_o}{\alpha_o \gamma_o} + x_o^0,$$

$$x_I^1 = \gamma \left(x^1 + \frac{1}{\alpha(x^0)\gamma_0^2} \right) - \frac{1}{\alpha_o \gamma_o} + x_o^1,$$

$$x_I^2 = x^2 + x_o^2, \quad x_I^3 = x^3 + x_o^3;$$

$$\gamma = \frac{1}{\sqrt{1 - \beta^2}}, \quad \gamma_0 = \frac{1}{\sqrt{1 - \beta_o^2}},$$

$$\beta = \beta_1(x^0) + \beta_o, \quad \alpha(x^0) = \frac{d\beta}{dx^0}. \qquad (6.28)$$

It follows from (6.28) that

$$\frac{\partial x_I^0}{\partial x^\nu} = (A, \gamma\beta, 0, 0), \quad A \equiv \gamma[W_o - \beta U],$$

$$\frac{\partial x_I^1}{\partial x^\nu} = (B, \gamma, 0, 0), \quad B \equiv \gamma\beta \left[W_o - \frac{U}{\beta} \right],$$

$$\frac{\partial x_I^2}{\partial x^\nu} = (0,0,1,0), \quad \frac{\partial x_I^3}{\partial x^\nu} = (0,0,0,1); \quad \alpha = \alpha(x^0);$$

$$W_o = \gamma^2 \left(\alpha x^1 + \frac{1}{\gamma_0^2} \right), \quad U = \frac{d\alpha/dx^0}{\alpha^2 \gamma_0^2}. \tag{6.29}$$

Thus, in an ALA frame, we have the energy–momentum vector

$$p_\nu = p_{I\alpha} \frac{\partial x_I^\alpha}{\partial x^\nu} = \eta_{\alpha\lambda} p_I^\lambda \frac{\partial x_I^\alpha}{\partial x^\nu}, \tag{6.30}$$

where $p_{I\alpha}$ is the constant covariant momentum 4-vector as measured in the inertial frame $F_I(x_I)$. Similarly, the angular momentum tensor in an ALA frame is

$$M_{\mu\nu} = \eta_{\alpha\lambda} \eta_{\beta\rho} M_I^{\lambda\rho} \frac{\partial x_I^\alpha}{\partial x^\mu} \frac{\partial x_I^\beta}{\partial x^\nu}. \tag{6.31}$$

The ten elements of p_ν and $M_{\mu\nu}$ are not constant. However, they satisfy the covariant continuity equations,

$$D_\lambda p_\nu = 0, \quad D_\lambda M_{\mu\nu} = 0. \tag{6.32}$$

The momentum p_ν and the anti-symmetric angular momentum $M_{\mu\nu}$ are functions of space-time and depend on ten constants, which are just the ten constants of motion $p_{I\alpha}$ and $M_I^{\lambda\rho}$ in the inertial frame F_I. Therefore, the covariant continuity Equations (6.32) in an ALA frame with a flat space-time imply the conservation of momentum and angular momentum. This is a specific demonstration of the general case in Equations (6.15), (6.21), (6.26) and (6.27) using the space-time transformations between an ALA frame and an inertial frame.

To see explicitly the relationship between the 4-momentum vectors p_ν and $p_{I\alpha}$ in (6.30), let us consider an ALA frame $F(x)$ and an inertial frame F_I. Equations (6.29) and (6.30) lead to the transformation between p_ν and $p_{I\alpha}$,

$$p_0 = \gamma \left[(W_o - \beta U) p_{I0} + \beta \left(W_o - \frac{U}{\beta} \right) p_{I1} \right],$$

$$p_1 = \gamma(p_{I1} + \beta p_{I0}), \quad p_2 = p_{I2}, \quad p_3 = p_{I3}, \tag{6.33}$$

where the functions W_o and U are given in (6.29) and $p_\nu(\nu = 0,1,2,3,)$ is the time-dependent energy and momentum vector as measured in the ALA frame $F(x)$. It can be verified that the transformation of the covariant 4-vector in (6.33) can also be derived from the covariant coordinate differentials dx_μ. These, in turn, can be obtained using the transformations (6.28) and the relation $dx^\mu = P^{\mu\nu} dx_\nu$, where $P^{\mu\nu}$ is given in Equation (4.20).

The inverse transformations of (6.33) are found to be

$$p_{I0} = \frac{\gamma}{W_o} \left[p_0 - \beta \left(W_o - \frac{U}{\beta} \right) p_1 \right],$$

$$p_{I1} = \frac{\gamma}{W_o} \left[(W_o - \beta U) p_1 - \beta p_0 \right], \quad p_{I2} = p_2, \quad p_{I3} = p_3. \tag{6.34}$$

It is clear that the momentum p_1 and energy p_0, as measured in the accelerated frame are not separately constant because β and γ are time-dependent, as shown in (6.28). However, their combinations to form the energy and momentum (p_{I0} and p_{I1}, respectively) in an inertial frame are time-independent constants, as shown in (6.26).

The results (6.34) are the same as those in (3.17) obtained on the basis of Wu transformations since p'^{μ} is the 4-momentum vector in an inertial frame, $p'_{\mu} = p_{I\mu}$.

6-4. Symmetry implications of global and local space-time translations

The previous discussions may be puzzling. How, one might ask, can the invariance of an action under one a single transformation (6.8) lead to two different conservation laws for energy–momentum and angular momentum? Conventionally, it is translational invariance that leads to the conservation of energy–momentum and rotational invariance that leads to the conservation of angular momentum.

In physics, one usually discusses the relations between symmetry and conservation laws based on global transformations and Noether's theorem in inertial frames. For example, the most general space-time transformations for two inertial frames $F_I(x_I)$ and $F'_I(x'_I)$ are the Poincaré transformations,

$$x'^{\mu}_I = x^{\mu}_I + \epsilon^{\mu}{}_{\nu} x^{\nu}_I + \epsilon^{\mu}, \tag{6.35}$$

where we have written the infinitesimal transformation for simplicity. The quantities ϵ^{μ}_{ν} and ϵ^{μ} denote the constant (or global) rotations and translations in a flat 4-dimensional space-time, respectively. Noether's theorem implies that if the Lagrangian of a physical system is invariant under an infinitesimal translation,

$$x'^{\mu}_I = x^{\mu}_I + \epsilon^{\mu}, \tag{6.36}$$

then one has the conserved energy–momentum 4-vector p_ν,

$$p_\nu = \int d^x T_{0\nu}, \quad \partial^\mu T_{\mu\nu} = 0, \tag{6.37}$$

where $T_{\mu\nu}$ may not be symmetric in general. If the Lagrangian is invariant under an infinitesimal 4-dimensional rotation (i.e., the Lorentz transformation),

$$x_I'^\mu = x_I^\mu + \epsilon^\mu{}_\nu x_I^\nu, \tag{6.38}$$

then one has the conserved angular momentum tensor,

$$M^{\nu\lambda} = \int d^3x M^{0\nu\lambda}, \quad \partial_\mu M^{\mu\nu\lambda} = 0. \tag{6.39}$$

Let us compare these two transformations, (6.36) and (6.38), with the single local translation (6.8). Since $\Lambda^\mu(x)$ in (6.8) are arbitrary infinitesimal functions, they include both the infinitesimal translation ϵ^μ in (6.36) and the infinitesimal Lorentz transformation $\epsilon^\mu{}_\nu x_I^\nu$ in (6.38) as two special cases. Thus, if the action of a physical system is invariant under the local space-time translations (6.8), the system actually has both the symmetries of global translations and infinitesimal Lorentz transformations.

Therefore, in using the Fock–Hilbert approach to considering the energy–momentum tensor (6.2) based on the local translation (6.8) and the variation of the metric tensor $P_{\mu\nu}$ (6.10), we can obtain both the conservation laws of the energy–momentum vector (6.26) and the angular momentum tensor (6.27). The Fock–Hilbert results have two interesting differences from Noether's theorem, as usually presented in physics:

(a) The Fock–Hilbert approach based on local symmetries leads directly to a symmetric energy–momentum tensor (6.2) in arbitrary coordinates in the taiji symmetry framework.
(b) The resultant conservation laws hold not only in inertial frames with arbitrary coordinates, but also in non-inertial frames based on the taiji symmetry framework.

Noether's well-known theorem shows the deep connection between symmetries and conservation laws. In her 1918 paper,[1] Noether discussed her Theorem II, which showed that general relativity does not include the usual law for the conservation of energy. The reason is that the group of general coordinate transformations in Einstein's gravity is a Lie group with a continuously infinite number of generators (an infinite continuous group).[8,9] The law of conservation of energy in classical mechanics and in special relativity holds because the groups have a finite number of generators. Thus, from the viewpoint of physics, general coordinate invariance in curved space-time is not a 'proper symmetry' because it is not associated with the law of conservation of energy.

References

1. E. Noether, Goett. Nachr., 235 (1918). English translation of Noether's paper by M. A. Tavel is online. Google search: M. A. Tavel, Noether's paper.

2. V. Fock, *The Theory of Space Time and Gravitation* (trans by N. Kemmer, Pergamon Press, 1959), p. 149.

3. V. Fock, ref. 2, p. xvii; E. P. Wigner, *Symmetries and Reflections, Scientific Essays* (The MIT Press, 1967), pp. 52–53; L. Landau and E. Lifshitz, *The Classical Theory of Fields* (Trans. by M. Hamermesh, Addison-Wesley, Cambridge, MA. 1951), pp. 248–289; C. Møller, *The Theory of Relativity* (Oxford, London, 1969), p. 226.

4. V. Fock, ref. 2, pp. 158–161; L. Landau and E. Lifshitz, pp. 293–296, pp. 316–318.

5. D. Hilbert, translated by D. Fine in *100 Years of Gravity and Accelerated Frames, The deepest Insights of Einstein and Yang–Mills* (Ed. J. P. Hsu and D. Fine, World Scientific, 2005) p. 120. (Available online: google books, 100 years of gravity).

6. V. Fock, ref. 2, pp. 163–165.

7. Mathematically, any space-time with a constant curvature K possesses maximal symmetry. Fock showed that the system of differential equations (6.14) is completely integrable if the Riemann–Christoffel curvature takes the form $R_{\mu\nu,\alpha\beta} = K(g_{\nu\alpha}g_{\mu\beta} - g_{\mu\alpha}g_{\nu\beta})$, where K is the constant of curvature. Thus, only in this case do the covariant continuity equations (6.14) imply the conservation laws (see ref. 4). However, if $K \neq 0$, we have neither inertial frames nor the principle of relativity for physical laws in such a space-time framework. This is the physical reason that the taiji symmetry framework is based on flat space-time, i.e., zero Riemann–Christoffel curvature, $R_{\mu\nu,\alpha\beta} = 0$.

8. Nina Byers, 'E. Noether's Discovery of the Deep Connection Between Symmetries and Conservation Laws'. Google search: N. Byers, Noether's discovery. Noether's results were consistent with Hilbert's assertion (1918) that the failure of the law of conservation of energy was a characteristic of Einstein's theory of gravity.

9. If one compares Noether's Theorem and Fock's results in Section 6-2, they seem to suggest some sort of equivalence between invariance under a finite (or countably infinite) number of infinitesimal generators of external space-time groups, and symmetry in a space-time with a constant curvature (Cf. ref. 7).

Part II

Quantum Yang–Mills Gravity

Jong-Ping Hsu and Leonardo Hsu

7.

The Yang–Mills–Utiyama–Weyl Framework for Internal and External Gauge Symmetries

7-1. The Yang–Mills–Utiyama–Weyl framework

Since the creation of the Yang–Mills theory in 1954, quantum field theories with gauge symmetries based on a flat space-time have been successfully applied to all fundamental interactions in nature except for the gravitational interaction. This problem is not necessarily surprising since it has been noted that one of "the most glaring incompatibility of concepts in contemporary physics" is that between the principle of general coordinate invariance and a quantum-mechanical description of all of nature.[1]

In contrast to the field-theoretic approach, Einstein's approach involving the geometrization of physics using Riemannian geometry appears to be successful only in the formulation of classical gravity. Einstein's principle of general coordinate invariance implies that space-time points can be given arbitrary labels. Because the presence of gravity implies a change in the space-time metric, there is no way to define a space-like separation between two regions.[1] Thus, it is difficult to construct quantum gravity because local commutativity is meaningless. Another difficulty with Einstein's principle of general coordinate invariance is the failure of the conservation law for energy-momentum, as was discussed in the Overview.

It is fair to say that general coordinate invariance, which is one of the most powerful ideas in Einstein's theory of the gravitational interaction, is also the origin of its incompatibility with quantum field theory and its non-conservation of the energy-momentum tensor. If a theory of gravity could be formulated in flat space-time, both the difficulties of the quantization of gravity and of conserving the energy-momentum tensor would automatically be resolved. In addition, the aforementioned conceptual difficulty related to the operational meaning of coordinates and momentum would also disappear.

From the viewpoint of established field theories such as quantum elec-trodynamics and unified electroweak theory, a geometrization of physics should be based on the geometry of fiber bundles with a flat space-time as the base manifold, rather than something like the Riemannian or Finsler geometries.[a] In such a Yang–Mills approach, the central roles are played by the connections (i.e., gauge potential fields) on a principal fiber bundle and the gauge curvatures on the connections. These gauge curvatures are closely related to the symmetry of groups. However, despite extensive investigation over several decades, it seems unlikely that one can develop a satisfactory formulation of gravity within the usual Yang–Mills framework. Previous such studies based on the Yang–Mills idea and curved space-time[4,5] have not shed light on the problems of quantization and energy-momentum conservation.

One reason is that the usual gauge fields in the Yang–Mills framework are 'phase fields' associated with internal symmetry groups, whose generators have constant matrix representations. Consequently, all gauge potential fields are vector fields in space-time with dimensionless coupling constants. However, tensor fields and a coupling constant with the dimension of length appear to be necessary for a description of gravity, and so a new framework to accommodate these different properties of the gravitational gauge field must be found. One effective and simple way that we describe here is to generalize the traditional Yang–Mills framework based on flat space-time so as to accommodate an external symmetry such as the space-time translational symmetry. This symmetry is what leads to the conserved energy-momentum tensor.

The generalized Yang–Mills framework we will describe in this chapter can accommodate both internal symmetry groups and external symmetry groups, enabling us to discuss all the interactions of nature, including gravity, within the same gauge symmetry framework. Because one can trace the idea of the unification of interactions with gauge symmetry to Weyl,[6] and the insight of gravity as a Yang–Mills gauge field to Utiyama,[4] we call this generalized Yang–Mills framework the Yang–Mills–Utiyama–Weyl (YMUW) framework.

Wigner divided symmetry in physics into two classes: (i) geometric symmetries and (ii) dynamical symmetries. It now appears clear that dynamical symmetries are the gauge symmetries. The generators of the dynamical groups, such as $SU_2 \times U_1$ and SU_3 for the electroweak theory and quantum chromodynamics, respectively, have constant matrix representations.[7] The gauge transformations of these dynamical groups and parallel transport correspond to phase changes.

---

[a]If a velocity-dependent interaction, such as the electromagnetic interaction, were to be treated using Einstein's geometric approach, one might reasonably employ the Finsler geometry, rather than the Riemannian geometry.[2,3]

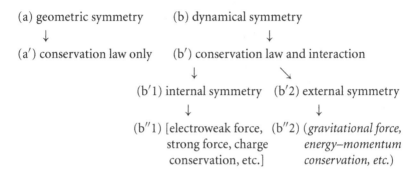

Fig. 7.1   Yang–Mills–Utiyama–Weyl framework for gauge symmetry

These are key properties of the usual gauge theories with internal symmetry groups. Analogously, Yang–Mills gravity suggests that an 'external' space-time translational group is also a dynamical group (see Fig. 7.1).[8] Parallel transport related to a space-time translational group corresponds to a scale change.

In Yang–Mills gravity, the flat space-time translational gauge symmetry insures that the wave equations of quarks, leptons, photons, and all other particles reduce to Hamilton-Jacobi type equations that explicitly display 'effective Riemannian metric tensors' in the geometric-optics (or classical) limit. This gives a new and interesting picture of the physical world, namely that in the presence of T_4 gauge fields, classical objects and light rays exhibit motions as if they were in a 'curved space-time,' even though the real physical space-time for fields, quantum particles, and their interactions is flat. The obvious challenge is to construct a gauge-invariant action in flat space-time that correctly predicts experimental results and in the following chapters, we do just that.

7-2.   The Levi-Civita connection and interpretations of Einstein gravity

In 1917, Levi-Civita introduced a connection on a Riemannian manifold by parallel transport.[9] In Euclidean space, a vector field whose direction at any point is that of the unit vector **i** is parallel if the derivative of **i** vanishes at every point and for all directions, but this cannot be true of a vector field in a Riemannian manifold in general. Instead, Levi-Civita defined the 'parallel transport' of vectors with respect to a given curve C, where one can express the coordinates of points on the curve C in terms of the arc-length t. A vector field $V^\nu(t)$ of constant magnitude is parallel with respect to the Riemannian manifold along the curve C, if its derived vector in the direction of the curve is zero at all points of C. The differential equation for the parallel transport of the vector field $V^\nu(t)$ is[9]

$$\frac{dV^\nu(t)}{dt} + \Gamma^\nu_{\mu\lambda}\frac{dx^\mu(t)}{dt} V^\lambda(t) = 0, \quad V^\nu(t) = V^\nu(x(t)). \tag{7.1}$$

It can also be written in the form

$$\frac{dx^\mu}{dt} D_\mu V^\nu = 0, \tag{7.2}$$

where

$$D_\mu V^\nu = \partial_\mu V^\nu + \Gamma^\nu_{\mu\lambda} V^\lambda \tag{7.3}$$

is the covariant derivative of V^ν in Riemannian geometry. As expected, equation (7.2) is independent of the choice of coordinates, i.e., it is invariant under general (or arbitrary infinitesimal) coordinate transformations

$$x'^\mu = g^\mu(x^\lambda) \quad \text{or} \quad x'^\mu = x^\mu + \Lambda^\mu(x), \tag{7.4}$$

where g^μ are single-valued functions whose functional determinant does not vanish, and $\Lambda^\mu(x)$ are arbitrary infinitesimal functions. The Levi-Civita connection is the Christoffel symbol $\Gamma^\nu_{\mu\lambda}$, which is expressed in terms of the Riemannian metric tensors. Although the connection is not a tensor, the difference between two connections or any infinitesimal variation of a connection is a tensor. As an aside, the Levi-Civita connection can also be determined by noting that it is the only connection that satisfies $\Gamma^\nu_{\mu\lambda} = \Gamma^\nu_{\lambda\mu}$ (i.e., it is torsionless) and preserves the length of the vector $V^\nu(x)$ in the parallel transport (7.1), i.e. $d(V_\mu V^\mu) = 0$. The Riemann–Christoffel curvature of the Riemannian manifold can be considered as the curvature on the Levi-Civita connection.

The Riemann–Christoffel curvature played a central role in the derivation of the gravitational field equation, as discussed by Hilbert.[10] Utiyama first interpreted Einstein's field equation as a gauge field equation in 1956. C. N. Yang proposed interpreting the Levi-Civita connection as the gravitational gauge potential and the Riemann–Christoffel curvature as the gauge curvature for gravity, constructing an action involving the quadratic gauge curvature and obtaining a new gravitational equation that contained the third-order derivative of the metric tensor.[7] However, these results did not shed light on the basic difficulties of Einstein gravity discussed earlier.

7-3. Weyl's parallel transport of scale and electromagnetic fields

In 1917, G. Hessenberg published a discussion of the vectorial foundations of differential geometry.[11] His idea led to a very broad generalization of Riemannian geometry by dispensing with the Riemannian metric tensor so that in Hessenberg's generalized geometry there is no fundamental metric tensor to define the length of a vector. In place of it, he introduced axiomatically the coefficients of affine connection — a certain function of the coordinates. Thus, Hessenberg and Weyl[12] laid down the mathematical foundations of gauge field theories by liberating the concept of connection from the bondage of metric tensors, so that one can axiomatically introduce connections associated with various internal and external symmetry groups.

Weyl's insight led to the application of parallel transport to any vector field attached to a point of the manifold and even the possible scales at various points as a 'non-geometric' vector bundle on space-time, using a prescribed connection. He then used these mathematical properties to investigate a unified field theory involving the electromagnetic and gravitational interactions.[12] The transformation law for the connection is determined uniquely by requiring that the parallel transport be independent of the choice of coordinates. The concepts of connection, parallel transport and covariant differentiation are essentially equivalent. These general properties, which are independent of the Riemannian metric tensors, turned out to be the cornerstones of modern gauge fields and the unification of the interactions.[13,14]

Weyl defined the parallel transport of 'scale' $s(t)$ by introducing the following corresponding relation to (7.1),

$$A_\mu \, dx^\mu \leftrightarrow \Gamma^\nu_{\mu\lambda} \, dx^\mu. \tag{7.5}$$

In analogy to (7.1), the differential equation that describes the parallel transport for scale $s(t)$ is given by

$$\frac{ds(t)}{dt} + \frac{dx^\mu(t)}{dt} A_\mu(t) s(t) = 0, \quad s(t) = s(x(t)), \tag{7.6}$$

which leads to the solution

$$s(t) = T \exp\left[-\int_0^t A_\mu \frac{dx^\mu}{dt} dt\right] s(0). \tag{7.7}$$

Here, T denotes that $s(t)$ is to be evaluated by path-ordering each term in the series expansion for the exponential,[15] and the line integral is along the path C from 0 to t. According to Weyl's view, the scales at various points can be considered as a vector bundle on space-time, connections can be introduced on this bundle, and their curvatures as 2-forms can be defined on space-time.[16] This curvature turned out to be the electromagnetic field strength, which plays a key role in the physics of electromagnetism.

After the birth of quantum mechanics, it was realized that the vector potential A_μ should be related to the phase change of a wave function rather than to the scale change. In this case, the parallel transport of phase and the solution are formally the same as (7.6) and (7.7), except that the A_μ in (7.6) and (7.7) is replaced by $-ieA_\mu$, where e is the electric charge associated with the coupling of the electromagnetic field A_μ. One can also generalize (7.6) and (7.7) to Yang–Mills fields by the replacement, $A_\mu \rightarrow ifb_\mu^a \tau^a/2$, where the SU_2 generators satisfy the relations, $[\tau^a/2, \tau^b/2] = ie^{abc}\tau^c/2$. Thus, the differential equation that describes the parallel transport for $s(t)$ is given by

$$\frac{ds(t)}{dt} = \frac{dx^\mu(t)}{dt} Z_\mu(x(t))s(t), \tag{7.8}$$

$$Z_\mu = ieA_\mu \quad \text{for } U_1, \quad \text{and} \quad Z_\mu = -ifb_\mu^a \frac{\tau^a}{2} \quad \text{for } SU_2.$$

This leads to the solution

$$s(t) = T \exp\left[\int_0^t Z_\mu(t') \frac{dx^\mu(t')}{dt'} dt'\right] s(0). \tag{7.9}$$

The equation of the parallel transport (7.8) corresponds to the gauge covariant derivative of the form,

$$\Delta_\mu = \partial_\mu - ieA_\mu \quad \text{for } U_1; \quad \Delta_\mu = \partial_\mu + ifb_\mu^a \frac{\tau^a}{2}, \quad \text{for } SU_2. \tag{7.10}$$

This can be seen by expressing ds/dt as $(\partial s/\partial x^\mu)(dx^\mu/dt)$ in (7.8), so that we have

$$\frac{dx^\mu}{dt} \Delta_\mu s(t) = 0, \quad \Delta_\mu = \partial_\mu - Z_\mu. \tag{7.11}$$

The parallel transport (7.9) in this case is the gauge-dependent or non-integral phase factor.

Thus, the definitions of the gauge covariant derivatives in (7.10) and (7.11) resemble the covariant derivative (7.3) in Riemannian geometry.

The U_1 and SU_2 gauge curvatures $F_{\mu\nu}$ and $B^a_{\mu\nu}$ are defined by

$$[\Delta_\mu, \Delta_\nu] = -ieF_{\mu\nu}, \quad F_{\mu\nu} = \partial_\mu A_\nu - \partial_\nu A_\mu, \quad \text{for } U_1, \tag{7.12}$$

$$[\Delta_\mu, \Delta_\nu] = +ifB^a_{\mu\nu}\frac{\tau^a}{2}, \quad \text{for } SU_2, \tag{7.13}$$

$$B^a_{\mu\nu} = \partial_\mu b^a_\nu - \partial_\nu b^a_\mu + f\epsilon^{abc} b^b_\mu b^c_\nu. \tag{7.14}$$

In physics, the generalization from the Abelian group U_1 for electromagnetic fields to the non-Abelian group SU_2 for Yang–Mills fields is highly non-trivial, especially with respect to their quantum properties. Yang and Mills attempted to make this generalization, motivated by the question of whether one could understand the conservation of isospin quantum number in a way similar to the conservation of electric charge. Although their original formulation of a gauge theory for the strong interaction was incorrect, their idea turned out to be very significant, gradually becoming the mainstream of particle physics and paving the way for the unification of the electromagnetic and weak interactions. In hindsight, the generalization of U_1 to SU_2 is mathematically simple or even trivial if one knows their relations to fiber bundles. Nevertheless, physicists initially struggled to understand non-Abelian gauge fields, without the help of fiber bundles.

7-4.   Curvatures on the connections

Gauge curvatures associated with various internal (dynamical) gauge groups are essential for our understanding of fundamental interactions in physics. To see the wonderful analogy of gauge curvatures and the Riemann–Christoffel curvature, let us calculate the relation between the gauge curvatures (e.g., given by (7.12), (7.13), and (7.14)), and the parallel transport along a small square with side c in the x^1–x^2 plane. For convenience of treating U_1 and SU_2 simultaneously, we use the following notations:

$$\Delta_\mu = \partial_\mu - Z_\mu, \quad Z_\mu = -ifb^a_\mu(x)\frac{\tau^a}{2}, \quad \text{for } SU_2, \tag{7.15}$$

$$\Delta_\mu = \partial_\mu - Z_\mu, \quad Z_\mu = ieA_\mu(x), \quad \text{for } U_1. \tag{7.16}$$

To calculate the parallel transport (7.9) along the x-direction on one side of the square from $(0, x, 0, 0)$ to $(0, x', 0, 0)$, let us use the following parametrization for $x^\mu(t)$:

$$f^\mu(t) = x^\mu + ct\delta_1^\mu, \quad df^\mu = c\delta_1^\mu \, dt. \tag{7.17}$$

We have

$$\exp\left[\int_x^{x'} Z_\lambda(f^\mu(t))df^\lambda\right] = \exp\left[\int_0^1 cZ_\lambda\left(x^\mu + ct\delta_1^\mu\right) dt\delta_1^\lambda\right]$$

$$= \exp\left[-c\int_0^1 \left[Z_1(x^\mu) + \partial_\mu Z_1 ct\delta_1^\mu\right] dt\right]. \tag{7.18}$$

For parallel transport along the $+y$ direction on the side of the square from $(0, 0, y, 0)$ to $(0, 0, y', 0)$, we use the parameterization

$$f^\mu(t) = x^\mu + c\delta_1^\mu + ct\delta_2^\mu, \quad df^\mu = c\delta_2^\mu \, dt. \tag{7.19}$$

We have

$$\exp\left[\int_y^{y'} Z_\lambda(f^\mu(t))df^\lambda\right] = \exp\left[\int_0^1 cZ_\lambda(x^\mu + c\delta_1^\mu + ct\delta_2^\mu)dt\delta_2^\lambda\right],$$

$$= \exp\left[-c\int_0^1 \left(Z_2(x^\mu) + (\partial_\mu Z_2)\left[c\delta_1^\mu + ct\delta_2^\mu\right]\right) dt\right],$$

$$= \exp\left[-c\left(Z_2(x^\mu) + (\partial_\mu Z_2)\left[c\delta_1^\mu + \frac{c}{2}\delta_2^\mu\right]\right)\right],$$

$$= \exp\left[-cZ_2\left(x^\mu + c\delta_1^\mu + \frac{c}{2}\delta_2^\mu\right)\right]. \tag{7.20}$$

Using (7.18) and (7.20), the parallel transport along the perimeter of a small square with side c is then

$$\Phi_{square} = \exp\left[cZ_2\left(x^\mu + \frac{c}{2}\delta_2^\mu\right)\right]\exp\left[cZ_1\left(x^\mu + \frac{c}{2}\delta_1^\mu + c\delta_2^\mu\right)\right]$$

$$\times \exp\left[-cZ_2\left(x^\mu + c\delta_1^\mu + \frac{c}{2}\delta_2^\mu\right)\right]$$

$$\times \exp\left[-cZ_1\left(x^\mu + \frac{c}{2}\delta_1^\mu\right)\right]$$

$$= \exp\left[cZ_2(x^\mu) + c^2\partial_\mu Z_2 \frac{1}{2}\delta_2^\mu\right]$$

$$\times \exp\left[cZ_1(x^\mu) + c^2\partial_\mu Z_1\left[\frac{1}{2}\delta_1^\mu + \delta_2^\mu\right]\right]$$

$$\times \exp\left[cZ_2(x^\mu) - c^2\partial_\mu Z_2\left[\delta_1^\mu + \frac{1}{2}\delta_2^\mu\right]\right]$$

$$\times \exp\left[cZ_1(x^\mu) - c^2\partial_\mu Z_1\left[\frac{1}{2}\delta_1^\mu\right]\right]. \tag{7.21}$$

Let us use the Baker–Campbell–Hausdorff formula for operators A and B,

$$e^A e^B = e^{A+B} e^{\frac{1}{2}[A,B]}, \tag{7.22}$$

which suffices for our purpose of calculating terms of order $O(c^2)$. The phase factor Φ_{square} for a small square with side c in the x^1–x^2 plane can be expressed as

$$\Phi_{square} = e^A e^B e^C e^D = e^{A+B+C+D+[A,B]/2+[C,D]/2} e^{[A+B,C+D]/2}$$

$$A = cZ_2(x^\mu) + c^2\partial_\mu Z_2 \frac{1}{2}\delta_2^\mu, \quad B = cZ_1(x^\mu) + c^2\partial_\mu Z_1\left[\frac{1}{2}\delta_1^\mu + \delta_2^\mu\right], \tag{7.23}$$

$$C = cZ_2(x^\mu) - c^2\partial_\mu Z_2\left[\delta_1^\mu + \frac{1}{2}\delta_2^\mu\right], \quad D = cZ_1(x^\mu) - c^2\partial_\mu Z_1 \frac{1}{2}\delta_1^\mu,$$

to the second order in c. We thus obtain the curvature $C_{\mu\nu}$ with $\mu = 2, \nu = 1$,

$$\Phi_{square} = \exp[c^2 C_{21} + O(c^3)],$$
$$C_{21} \equiv (\partial_2 Z_1 - \partial_1 Z_2 + Z_2 Z_1 - Z_1 Z_2), \tag{7.24}$$

where $\Delta_\mu = \partial_\mu - Z_\mu$.

For gauge theories with U_1 and SU_2 groups, we have the usual results

$$C_{21} = ie(\partial_2 A_1 - \partial_1 A_2) \equiv ieF_{21}, \quad \text{for } U_1, \tag{7.25}$$

$$C_{21} = -if(\partial_2 b_1^a - \partial_1 b_2^a + f\epsilon^{abc} b_2^b b_1^c)\frac{\tau^a}{2} \equiv -ifB_{21}^a \frac{\tau^a}{2}, \quad \text{for } SU_2. \tag{7.26}$$

Other components of $C_{\mu\nu}$ can also be obtained. We note again that the derivations of gauge curvatures in (7.24), (7.25) and (7.26) also resemble the

Riemann–Christoffel curvatures $R^{\beta}_{\alpha,\mu\nu}$ given by

$$[D_{\mu}, D_{\nu}]V_{\alpha} = R^{\beta}_{\alpha,\mu\nu}V_{\beta},$$

$$R^{\beta}_{\alpha,\mu\nu} = \partial_{\nu}\Gamma^{\beta}_{\alpha\mu} - \partial_{\mu}\Gamma^{\beta}_{\alpha\nu} + \Gamma^{\rho}_{\alpha\mu}\Gamma^{\beta}_{\rho\nu} - \Gamma^{\rho}_{\alpha\nu}\Gamma^{\beta}_{\rho\mu}. \tag{7.27}$$

7-5.   Taiji symmetry and the space-time translational symmetry group T_4

[1] *Taiji space-time symmetry*

Yang–Mills gravity[17] is based on the external space-time translational gauge group T_4 in flat space-time with arbitrary coordinates, which is applicable to both inertial and non-inertial frames.[8] Translational gauge symmetry can be expressed as a local space-time translation with an arbitrary infinitesimal vector gauge function $\Lambda^{\mu}(x)$,

$$x^{\mu} \to x'^{\mu} = x^{\mu} + \Lambda^{\mu}(x), \quad x \equiv x^{\lambda} = (w, x, y, z), \tag{7.28}$$

where $\Lambda^{\mu}(x)$ must not be affected by the transformations at a fixed point because they are gauge functions (i.e., group parameters). The Jacobian is assumed to be non-zero. The invariance of the action for a physical system under this *local translation in flat space-time* is the physical essence and the mathematical expression of taiji symmetry, and furthermore is intimately related to the gravitational field and its universal interaction with all other fields in nature.

[2] *T_4 gauge transformations and groups*

The local space-time translations (7.28) are also the arbitrary infinitesimal transformations of coordinates in flat space-time. This dual property embedded in (7.28) makes the formulation of gravity highly non-trivial. Nevertheless, we can accommodate these two mathematical implications (i.e., local space-time translations and arbitrary coordinate transformations) of the local transformations (7.28) by defining suitable gauge transformations of space-time translations for the various functions that appear in a Lagrangian.[8] For example, the infinitesimal T_4 gauge transformations for a scalar Q and a vector Γ^{μ} are given by

$$\Phi \to \Phi^{\$} = \Phi - \Lambda^{\lambda}\partial_{\lambda}\Phi, \quad \psi \to \psi^{\$} = \psi - \Lambda^{\lambda}\partial_{\lambda}\psi, \tag{7.29}$$

$$D_{\mu}Q \to (D_{\mu}Q)^{\$} = D_{\mu}Q - \Lambda^{\lambda}\partial_{\lambda}(D_{\mu}Q) - (D_{\lambda}Q)\partial_{\mu}\Lambda^{\lambda}. \tag{7.30}$$

Here, D_μ denotes the partial covariant derivative associated with a metric tensor $P_{\mu\nu}(x)$ in a general reference frame (inertial or non-inertial). We shall use $P_{\mu\nu}$ to denote the metric tensor in flat space-time with arbitrary coordinates in general and call it the Poincaré metric tensor, which includes the Minkowski metric tensor for inertial frames as a special case.

Mathematically, only the second terms, involving $\Lambda^\lambda \partial_\lambda$, on the right-hand-side of the gauge transformations (7.29) and (7.30) correspond to a local space-time translation (i.e., the T_4 group with generators $p_\lambda = i\partial_\lambda$). All other terms reflect a much more complicated group of arbitrary infinitesimal transformations of coordinates in flat space-time. This complicated group actually has a continuously infinite number of generators, which have little to do with the action in Yang–Mills gravity and the physics of the gravitational interaction. (See Section 13-4.)

[3] T_4 and Weyl's parallel transports of 'scale' $s(t)$ and T_4 gauge curvature

The following displays of the mathematical properties of the translational gauge symmetry (in Yang–Mills gravity) and the corresponding fiber bundle are based purely on algebraic analogy, rather than a rigorous derivation. The geometric picture of the external translational gauge symmetry in Yang–Mills gravity seems to be much more complicated than that of gauge symmetries of internal groups, the reason being that the local space-time translations (7.28) are also the general coordinate transformations. Thus, the T_4 gauge transformations (7.29) and (7.30) and the associated group generators related to arbitrary coordinate transformations are complicated and again, will be discussed only briefly in Section 13-4.

The tensor potential field $\phi_{\mu\nu}$ in Yang–Mills gravity is associated with the space-time translational group T_4 with the generators $p_\mu = i\partial/\partial x^\mu$ in inertial frames with the Minkowski metric tensor. Thus, the results for T_4 can be obtained from (7.8)–(7.14) with the replacement

$$ifb^a_\mu \frac{\tau^a}{2} \rightarrow ig\phi^\nu_\mu p_\nu \equiv Y_\mu, \quad p_\nu = i\partial/\partial x^\nu. \tag{7.31}$$

In Yang–Mills gravity we have the following corresponding results,

$$\frac{ds(t)}{dt} - ig\frac{dx^\mu(t)}{dt}\phi^\nu_\mu p_\nu s(t) = 0, \quad c = \hbar = 1, \tag{7.32}$$

$$s(t) = T \exp\left[\int_0^t dt' \frac{dx^\mu(t')}{dt'} Y_\mu(t') \right] s(0)$$

$$= \exp\left[-g \int_{C(t)} dx^\mu \, \phi_\mu^\nu \partial_\nu \right] s(0), \quad s(t) = s(x^\mu(t)), \quad (7.33)$$

where $C(t)$ is the path of integration. It is important to note that the generators of the T_4 group are $p_\mu = i\partial/\partial x^\mu$, so that $s(t)$ is a scale factor rather than a phase factor. These properties have important physical implications, namely that the gauge fields associated with scale factor lead to only one kind of force (e.g., attractive force), while those associated with phase factors lead to both attractive and repulsive forces.

Equations (7.32) and (7.33) for T_4 resemble Weyl's parallel transport in (7.6) and (7.7) for a scale. The equation for parallel transport (7.32) can be written as

$$\frac{dx^\mu(t)}{dt} \Delta_\mu s(t) = \frac{dx^\mu}{dt} [\partial_\mu - ig\phi_\mu^\nu p_\nu] s(x(t)) = 0. \quad (7.34)$$

We note that the derivative $\partial_\mu - ig\phi_\mu^\nu p_\nu = \partial_\mu + g\phi_\mu^\nu \partial_\nu$ is not independent of the choice of the coordinates. In order for the gravitational action in the Yang–Mills gravity to hold for both inertial and non-inertial frames (i.e., to be independent of the choice of coordinates) we must replace the ordinary derivative ∂_μ in $\partial_\mu + g\phi_\mu^\nu \partial_\nu$ by the covariant derivative D_μ associated with the Poincaré metric tensors $P_{\mu\nu}$ in general frames based on flat space-time. Thus the T_4 gauge covariant derivatives are

$$\Delta_\mu = D_\mu + g\phi_\mu^\nu D_\nu, \quad (7.35)$$

which can be written in the form

$$\Delta_\mu = J_\mu^\nu D_\nu, \quad J_\mu^\nu = \delta_\mu^\nu + g\phi_\mu^\nu. \quad (7.36)$$

Thus, the T_4 gauge curvature $C_{\mu\nu\alpha}$ is given by

$$[\Delta_\mu, \Delta_\nu] = g C_{\mu\nu\alpha} D_\alpha, \quad (7.37)$$

$$C_{\mu\nu\alpha} = J_\mu^\sigma D_\sigma J_{\nu\alpha} - J_\nu^\sigma D_\sigma J_{\mu\alpha} = C_{\nu\mu\alpha}, \quad (7.38)$$

in general frames of reference.

There is a difference between the invariance under the gauge transformations of internal symmetry groups and that of external symmetry groups. Equation (7.11) with $\Delta_\mu = \partial_\mu - ieA_\mu$ for parallel transport with the U_1 group or the Lagrangian $L = \left[\left(\partial_\mu + ieA_\mu\right)\Phi^*\right]\left[(\partial^\mu - ieA^\mu)\Phi\right]$ can be verified to be invariant under the U_1 gauge transformations

$$A'_\mu = A_\mu + \partial_\mu \Lambda(x), \quad s'(x) = e^{[ie\Lambda(x)]}s(x). \tag{7.39}$$

However, the corresponding equation (7.34) with $\Delta_\mu = \partial_\mu + g\phi^\nu_\mu\partial_\nu$ for parallel transport with the external T_4 group or the Lagrangian $L = [(\partial_\mu + g\phi^\nu_\mu)\Phi^*][(\partial^\mu + g\phi^\mu_\nu)\Phi]$ will be changed by a term due to an infinitesimal space-time shift $\Lambda^\lambda(x)$

$$\left[\frac{dx^\mu(t)}{dt}\Delta_\mu s(t)\right]^\$ = \frac{dx^\mu(t)}{dt}\Delta_\mu s(t) - \Lambda(x)\partial^\lambda\left[\frac{dx^\mu(t)}{dt}\Delta_\mu s(t)\right], \tag{7.40}$$

$$L^\$ = L - \Lambda^\lambda(x)\partial_\lambda L, \quad L = [(\partial_\mu + g\phi^\nu_\mu)\Phi^*][(\partial^\mu + g\phi^\mu_\nu)\Phi], \tag{7.41}$$

and will be invariant only after integrating over all space-time

$$\int d^4x\left[\sqrt{-P}W\right]^\$ = \int d^4x\sqrt{-P}W, \quad W = \frac{dx^\mu(t)}{dt}\Delta_\mu s(t) \quad \text{or} \quad L, \tag{7.42}$$

where we have used the T_4 gauge transformation for $P = \det\left(P_{\mu\nu}\right)$

$$\sqrt{-P}^\$ = [(1 - \Lambda^\sigma\partial_\sigma)\sqrt{-P}](1 - \partial_\lambda\Lambda^\lambda). \tag{7.43}$$

The YMUW framework in flat space-time can accommodate both gauge fields associated with SU_N bundle and T_4 bundle. It suggests that all interactions are dictated by the dynamical gauge symmetries in this framework. The internal (compact) symmetry groups such as SU_N lead to vector gauge fields, while the external (non-compact) space-time translation group T_4 leads to tensor gauge fields $\phi_{\mu\nu}$.[8] The T_4 gauge curvature $C_{\mu\nu\alpha}$ in (7.38) will play a central role in Yang–Mills gravity, in which the gravitational interaction is dictated by the T_4 gauge symmetry. In this sense, gravity is a manifestation of the space-time translation gauge symmetry or, equivalently, the taiji space-time symmetry. As a result, the difficulties of Einstein gravity discussed earlier are resolved because Yang–Mills gravity is based on flat space-time.

Table 7.1. Gauge fields in Yang–Mills–Utiyama–Weyl framework and fiber bundle terminology.

'phase' and 'scale' gauge fields	fiber bundle terminology
vector 'phase' gauge field b_μ^k	connection on an SU_N bundle
tensor 'scale' gauge field $\phi_{\mu\nu}$	connection on a T_4 bundle
phase factor	parallel transport (with SU_N)
scale factor	parallel transport (with T_4)
SU_N gauge curvature $f_{\mu\nu}^k$	curvature on SU_N connections
T_4 gauge curvature $C_{\mu\nu\lambda}$	curvature on T_4 connection
electromagnetism	connection on a U_1 bundle
chromodynamics	connection on $(SU_3)_{color}$ bundle
gravity	connection on a T_4 bundle
$(\partial_\mu - ieA_\mu)\Phi$, $(\partial_\mu + ieA_\mu)\Phi^*$ "attrac. and repul. electromagnetic force" with dimensionless coupling constants $\pm e$	U_1 covariant differentiation
$(\partial_\mu + g\phi_\mu^\nu \partial_\nu)\Phi$, $(\partial_\mu + g\phi_\mu^\nu \partial_\nu)\Phi^*$ "only attractive gravitational force" where g has the dimension of length.	T_4 covariant differentiation

Based on algebraic analogy, the correspondences between the physical gauge fields in the YMUW framework and mathematical fiber bundles are summarized in Table 7.1.[18,19]

References

1. F. J. Dyson, in *100 Years of Gravity and Accelerated Frames, The Deepest Insights of Einstein and Yang–Mills* (Ed. J. P. Hsu and D. Fine, World Scientific, 2005) p. 348.

2. The result of such a geometrization of electrodynamics was not satisfactory. See also discussions in section 0-3.

3. J. P. Hsu, *Nuovo Cimento*, **108B**, 183 (1993); J. P. Hsu, *Nuovo Cimento*, **108B**, 949 (1993); **109B**, 645 (1994).

4. R. Utiyama, *Phys. Rev.* **101**, 1597 (1956).

5. T. W. B. Kibble, *J. Math. Phys.* **2**, 212 (1961); R. Utiyama and T. Fukuyama, *Prog. Theor. Phys.* **45**, 612 (1971); Y. M. Cho, *Phys. Rev. D*, **14**, 2521 (1976); Y. M. Cho, *Phys. Rev. D*, **14**, 3341 (1976); K. Hayashi and T. Shirafuji, *Phys. Rev. D.* **19**, 3524 (1979); T. W. Hehl, P. von der Heyde, G. D. Kerlick and J. M. Nester, *Rev. Mod. Phys.* **48**, 393 (1976).

6. H. Weyl, *Preuss. Akad. Berlin*, 465 (1918).

7. C. N. Yang, *Phys. Rev. Lett.*, **32**, 445 (1974); 'Symmetry and Physics', talk at the centenary symposium of the Peking University, May 5, 1998.

8. J. P. Hsu, *Int. J. Mod. Phys. A* **21**, 5119 (2006).

9. T. Levi-Civita, *Rend. Circolo. Mat. Palermo.* **42**, 173 (1917); C. E. Weatherburn, *An Introduction to Riemannian Geometry and the Tensor Calculus* (Cambridge Univ. Press, 1938), pp. 85–88; V. S. Varadarajan, in *A Tribute to C.S. Seshadri Perspectives in Geometry and Representation Theory* (Ed. V. Lakshmibai, *et al.* Hindustan Book Agency, 2003) pp. 502–541.

10. D. Hilbert, in *100 Years of Gravity and Accelerated Frames, The Deepest Insights of Einstein and Yang–Mills* (Ed. J. P. Hsu and D. Fine, World Scientific, 2005) p. 120.

11. G. Hessenberg, *Math. Ann.* **78**, 187 (1917).

12. H. Weyl, Raum-Zeit-Materie (Berlin) § 14, (1918).

13. J. P. Hsu, *Mod. Phys. Lett. A*, **26**, 1707 (2011).

14. J. P. Hsu, *Chinese Phys. C*, **36**, 403 (2012); arXiv, 1108.2423.

15. S. Weinberg, *The Quantum Theory of Fields* vol. 1 (Cambridge University Press, 1995) pp. 142–144.

16. The expression ($\Gamma^{\nu}_{\mu\lambda} dx^{\mu}$) in (7.1) can be considered as the matrix A of 1-form. The matrix A and its transformation law define the connection on the manifold. The components of F in the matrix 2-form $dA = F$ are $F_{\mu\nu} = \partial_\mu A_\nu - \partial_\nu A_\mu$, which is the electromagnetic field strength.

17. J. P. Hsu and L. Hsu, *A Broader View of Relativity, General Implications of Lorentz and Poincaré Invariance* (World Scientific, 2006), Appendix D.

18. For fiber bundle terminology related to internal gauge symmetry groups, see T. T. Wu and C. N. Yang, *Phys. Rev. D*, **12**, 3845 (1975). The Levi-Civita connection does not play a role in Yang–Mills gravity based on flat space-time.

19. It seems that the phase factor or the parallel transport associated with compact Lie groups is related to gauge transformations and gauge curvatures, while the scale factor or the parallel transport associated with non-compact Lie groups such as T_4 is related only to gauge curvatures.

8.

Yang–Mills Gravity Based on Flat Space-time and Effective Curved Space-Time for Motions of Classical Objects

8-1. Translational gauge transformations in taiji space-time

Let us consider a particle-physics approach to gravity based on the Yang–Mills–Utiyama–Weyl (YMUW) framework with external space-time translational gauge symmetry. The formulation should hold for both inertial and non-inertial frames in taiji space-time. We first briefly review the basic replacement in the formulation of a gauge theory with internal groups. Then we generalize it to a gauge theory with external space-time translation group T_4 and discuss the associated T_4 gauge transformations in flat space-time.

The formulations of electromagnetic and Yang–Mills theories associated with internal gauge groups are all based on the replacement, $\partial_\mu \to \partial_\mu + igB_\mu$ in flat space-time where the field $B_\mu = B_\mu^a t_a$ involves constant matrix representations of the generators t_a of the gauge groups associated with internal symmetry. However, because the generators of the external space-time translation group T_4 are the displacement operators, $p_\mu = i\partial_\mu$ (in natural units where $c = \hbar = 1$), the replacement takes the form,

$$\partial_\mu \to \partial_\mu - ig\phi_\mu^\nu p_\nu \equiv J_\mu^\nu \partial_\nu, \quad J_\mu^\nu = \delta_\mu^\nu + g\phi_\mu^\nu, \qquad (8.1)$$

in inertial frames, where g is the coupling constant associated with the tensor gauge field $\phi_{\mu\nu}$. As usual, we assume the tensor gauge field to be symmetric,

$$\phi_{\mu\nu} = \phi_{\nu\mu}, \quad \phi_{\mu\nu} = \phi_\mu^\lambda \eta_{\lambda\nu}.$$

We interpret this field as a spin-2 field, which is related to the generators of the space-time translational group T_4 and follow the Yang–Mills approach to formulate a theory of gravity based on T_4 in taiji space-time. It is precisely this unique property (8.1) due to the displacement operators of the T_4 group that leads naturally to an 'effective' Riemannian metric tensor,[1][a] and a universal attractive force for all matter and anti-matter. The gravitational force is characterized by a coupling constant g with the dimension of length in natural units. This property is in sharp contrast with the dimensionless coupling constants in the usual Yang–Mills theories. For external gauge groups related to space-time, e.g., the de Sitter group or the Poincaré group, the gauge invariant Lagrangian involving fermions turns out to be richer in content (e.g., it also has a new gravitational spin force generated by the fermion spin density as well as genuine scale gauge fields to insure the gauge invariance of the Lagrangian). If the space-time translation group T_4 is replaced by the Poincaré group, there is a new gravitational spin force generated by the fermion spin density.[2] However, we will concentrate on the external gauge group of space-time translations T_4, which is the Abelian subgroup of the Poincaré group and is non-compact. This group T_4 is particularly interesting because it is the minimal group related to the conserved energy-momentum tensor.[3]

Translation gauge symmetry is based on the local space-time translation with an arbitrary infinitesimal vector gauge-function $\Lambda^\mu(x)$,

$$x^\mu \to x'^\mu = x^\mu + \Lambda^\mu(x), \tag{8.2}$$

as given by (7.28), which has two possible interpretations: (i) local translations of the space-time coordinates, and (ii) arbitrary infinitesimal coordinate transformations.[3]

In taiji space-time with arbitrary coordinates, both of these conceptual interpretations can be accommodated by defining an infinitesimal gauge transformation of space-time translations for physical quantities $Q^{\mu_1...\mu_m}_{\alpha_1...\alpha_n}(x)$ in the Lagrangian of fields[3]

$$Q^{\mu_1...\mu_m}_{\alpha_1...\alpha_n}(x) \to (Q^{\mu_1...\mu_m}_{\alpha_1...\alpha_n}(x))^\$ = \left(Q^{\nu_1...\nu_m}_{\beta_1...\beta_n}(x) - \Lambda^\lambda(x)\partial_\lambda Q^{\nu_1...\nu_m}_{\beta_1...\beta_n}(x)\right)$$
$$\times \frac{\partial x'^{\mu_1}}{\partial x^{\nu_1}} \cdots \frac{\partial x'^{\mu_m}}{\partial x^{\nu_m}} \frac{\partial x^{\beta_1}}{\partial x'^{\alpha_1}} \cdots \frac{\partial x^{\beta_n}}{\partial x'^{\alpha_n}}, \tag{8.3}$$

where μ_1, ν_1, α_1, β_1, etc. are space-time indices and $\Lambda^\mu(x)$ are infinitesimal functions. As usual, both the (Lorentz) spinor field ψ and the (Lorentz) scalar field Φ are treated as 'coordinate scalars' and have the same translational gauge

---

[a]The speculation that Einstein's theory of gravity may be an effective field theory has been around for decades among theorists. The idea of an effective Riemannian space due to the presence of the gravitational field in Minkowski space-time was discussed by Logunov, Mestvirishvili, N. Wu and others. (See ref. 15).

transformation. The gauge transformations for scalar, vector and tensor fields are given by

$$Q(x) \rightarrow (Q(x))^\$ = Q(x) - \Lambda^\lambda \partial_\lambda Q(x), \quad Q(x) = \psi, \overline{\psi}, \Phi, \tag{8.4}$$

$$\Gamma_\mu \rightarrow (\Gamma_\mu)^\$ = \Gamma_\mu - \Lambda^\lambda \partial_\lambda \Gamma_\mu - \Gamma_\lambda \partial_\mu \Lambda^\lambda, \tag{8.5}$$

$$\Gamma^\mu \rightarrow (\Gamma^\mu)^\$ = \Gamma^\mu - \Lambda^\lambda \partial_\lambda \Gamma^\mu + \Gamma^\lambda \partial_\lambda \Lambda^\mu,$$

$$S_{\mu\nu} \rightarrow (S_{\mu\nu})^\$ = S_{\mu\nu} - \Lambda^\lambda \partial_\lambda S_{\mu\nu} - S_{\mu\alpha} \partial_\nu \Lambda^\alpha - S_{\alpha\nu} \partial_\mu \Lambda^\alpha, \tag{8.6}$$

$$Q^{\mu\nu} \rightarrow (Q^{\mu\nu})^\$ = Q^{\mu\nu} - \Lambda^\lambda \partial_\lambda Q^{\mu\nu} + Q^{\lambda\nu} \partial_\lambda \Lambda^\mu + Q^{\mu\lambda} \partial_\lambda \Lambda^\nu,$$

where $T_{\mu\nu} = J_{\mu\nu}, P_{\mu\nu}$.

Suppose D_μ denotes the partial covariant derivative associated with a Poincaré metric tensor $P_{\mu\nu}(x)$ in a general reference frame (inertial or non-inertial) within flat space-time. We have, for example,

$$D_\mu V^\nu = \partial_\mu V^\nu + \Gamma^\nu_{\rho\mu} V^\rho, \quad D_\mu V_\nu = \partial_\mu V_\nu - \Gamma^\rho_{\nu\mu} V_\rho,$$

where $\Gamma^\nu_{\mu\rho} = \frac{1}{2} P^{\nu\sigma}(\partial_\mu P_{\sigma\rho} + \partial_\rho P_{\sigma\mu} - \partial_\sigma P_{\mu\rho})$. Also, the Riemann–Christoffel curvature tensor in (7.27) vanishes, i.e., $R^\nu_{\mu,\alpha\beta} = 0$. Note that the functions $D_\mu Q$ and $D_\mu D_\nu Q$ transform, by definition, as a covariant vector $\Gamma_\mu(x)$ and a covariant tensor $T_{\mu\nu}(x)$ respectively under the translational gauge transformation. The change of variables in the translational gauge transformations (8.4)–(8.6) in flat space-time is formally similar to the Lie variations in the coordinate transformations in Riemannian geometry.

The coordinate transformation (8.2) indicates that the theory of Yang–Mills gravity is formulated in a general frame of reference (inertial or non-inertial) characterized by an arbitrary coordinate and a Poincaré metric tensor $P_{\mu\nu}$. In general, the ordinary derivative ∂_μ in the basic replacement (8.1) should be expressed in terms of the partial covariant derivative D_μ associated with the Poincaré metric tensor $P_{\mu\nu}$:

$$D_\mu \rightarrow D_\mu + g\phi^\nu_\mu D_\nu \equiv J^\nu_\mu D_\nu, \tag{8.7}$$

$$J_{\mu\nu} = P_{\mu\nu} + g\phi_{\mu\nu}, \quad J_{\mu\nu} = J_{\nu\mu}, \quad J_{\mu\nu} = J^\lambda_\mu P_{\lambda\nu}, \tag{8.8}$$

for general frames with arbitrary coordinates.

All these equations (8.3)–(8.8) are necessary for the general formulation of Yang–Mills gravity in an arbitrary frame of reference based on taiji space-time with arbitrary coordinates.

8-2.   Translational gauge symmetry and the field-theoretic origin of effective metric tensors

To see the field-theoretic origin of the effective Riemannian metric tensors, let us consider a fermion field ψ. The kinetic energy term in the Lagrangian for a fermion in a general frame is given by

$$i\overline{\psi}\Gamma_\alpha D^\alpha\psi - m\overline{\psi}\psi = i\overline{\psi}\Gamma_\alpha\partial^\alpha\psi - m\overline{\psi}\psi,$$

$$\{\Gamma_\mu, \Gamma_\nu\} = 2P_{\mu\nu}(x), \quad \Gamma_\mu = \gamma_a e^a_\mu, \tag{8.9}$$

$$\{\gamma_a, \gamma_b\} = 2\eta_{ab}, \quad \eta_{ab}e^a_\mu e^b_\nu = P_{\mu\nu}.$$

Here we have the usual relation $D_\mu\psi = \partial_\mu\psi$ because the 'Lorentz spinor' ψ transforms as a 'coordinate scalar' and D_μ is the partial covariant derivative defined in terms of the Poincaré metric tensor $P_{\mu\nu}$ in a general frame. In a general frame, the gauge field $\phi_{\mu\nu}$ and the translational gauge symmetry dictates that the replacement be

$$D^\alpha \rightarrow D^\alpha + g\phi^{\alpha\beta}D_\beta \equiv J^{\alpha\beta}D_\beta, \quad J^{\alpha\beta} = P^{\alpha\beta} + g\phi^{\alpha\beta},$$

$$i\overline{\psi}\Gamma_\alpha D^\alpha\psi \rightarrow i\overline{\psi}\Gamma_\alpha\Delta^\alpha\psi = i\overline{\psi}\Gamma_\alpha J^{\alpha\beta}D_\beta\psi = i\overline{\psi}\gamma_a E^{a\beta}D_\beta\psi, \tag{8.10}$$

$$\Delta^\mu = J^{\mu\alpha}D_\alpha, \quad E^{a\alpha} = e^a_\mu J^{\mu\alpha}.$$

If one considers $E^{a\alpha}$ as an 'effective tetrad', then the relation for an 'effective metric tensor' is

$$\eta_{ab}E^{a\alpha}E^{b\beta} = \eta_{ab}e^a_\mu J^{\mu\alpha}e^b_\nu J^{\nu\beta} = P_{\mu\nu}J^{\mu\alpha}J^{\nu\beta} = G^{\alpha\beta}. \tag{8.11}$$

Such an 'effective metric tensor' $G^{\alpha\beta}$ also shows up if we consider the Lagrangian of a scalar field Φ with the same replacement of D^μ as that in (8.10),

$$\frac{1}{2}\left[P_{\mu\nu}(D^\mu\Phi)(D^\nu\Phi) - m^2\Phi^2\right] \rightarrow \frac{1}{2}\left[G^{\alpha\beta}(D_\alpha\Phi)(D_\beta\Phi) - m^2\Phi^2\right]. \tag{8.12}$$

Although it may appear as if the geometry of the space-time has changed from a pseudo-Euclidean space-time to a non-Euclidean space-time due to the presence of the tensor gauge field (or spin-2 field) $\phi_{\mu\nu}$, in the Yang–Mills approach, the presence of $E^{a\alpha}$ in (8.10) and $G^{\alpha\beta}$ is simply the manifestation of the translational gauge symmetry in flat space-time.[3]

In the literature, when one arrives at the crucial step (8.10),[4-6] most authors usually abandon the Yang–Mills approach of a truly gauge invariant theory with a quadratic gauge curvature for a Lagrangian in flat space-time, and instead follow Einstein's approach to gravity by postulating a Riemannian space-time due to the presence of $\phi_{\mu\nu}$ or $J_{\mu\nu}$ in (8.11). In other words, $E^{a\alpha}$ and $G^{\alpha\beta}$ in (8.11) are postulated to be a real tetrad and a real metric tensor of physical space-time, respectively. Such an approach, together with general coordinate invariance, inevitably leads to "the most glaring incompatibility of concepts in contemporary physics" as observed by Dyson.[7]

In the particle-physics approach described here, however, we consistently treat $G^{\mu\nu}$ in (8.11) as merely an 'effective metric tensor' for the motion of a classical object in flat space-time and in the presence of the tensor gauge field. This treatment will be substantiated in Section 8-5.

8-3. Gauge invariant action and quadratic gauge curvature

In the YMUW framework with the external translational gauge symmetry, we have the T_4 gauge curvature

$$C^{\mu\nu\alpha} = J^{\mu\lambda}(D_\lambda J^{\nu\alpha}) - J^{\nu\lambda}(D_\lambda J^{\mu\alpha}), \tag{8.13}$$

which is given by the commutation relation of the gauge covariant derivative and satisfies two simple identities,

$$[\Delta^\mu, \Delta^\nu] = C^{\mu\nu\alpha}D_\alpha, \quad \Delta^\mu = J^{\mu\nu}D_\nu,$$
$$C^{\mu\nu\alpha} = -C^{\nu\mu\alpha}, \quad C^{\mu\nu\alpha} + C^{\nu\alpha\mu} + C^{\alpha\mu\nu} = 0. \tag{8.14}$$

The translational gauge curvature $C^{\mu\nu\alpha}$ in (8.13) involves the symmetric tensor gauge field $\phi^{\mu\nu} = \phi^{\nu\mu}$. Thus, it differs from the usual Yang–Mills gauge curvature $f_{\mu\nu}^k = \partial_\nu b_\mu^k - \partial_\mu b_\nu^k - b_\mu^i b_\nu^j c_{ij}^k$, where c_{ij}^k is the structure constant of an internal gauge group whose generators have constant matrix representations. The T(4) gauge covariant derivative, Δ^ν in (8.14) satisfies the Jacobi identity, $[\Delta^\lambda, [\Delta^\mu, \Delta^\nu]] + [\Delta^\mu, [\Delta^\nu, \Delta^\lambda]] + [\Delta^\nu, [\Delta^\lambda, \Delta^\mu]] \equiv 0$, which leads to gauge-Bianchi identity in Yang–Mills gravity:

$$(\delta_\alpha^\lambda \Delta^\rho - (D^\lambda J_\alpha^\rho))C_{\ \lambda}^{\mu\nu} + (\delta_\alpha^\lambda \Delta^\mu - (D^\lambda J_\alpha^\mu))C_{\ \lambda}^{\nu\rho} + (\delta_\alpha^\lambda \Delta^\nu - (D^\lambda J_\alpha^\nu))C_{\ \lambda}^{\rho\mu} \equiv 0. \tag{8.15}$$

It turns out that there are two independent quadratic gauge-curvature scalars, $C_{\mu\alpha\beta}C^{\mu\beta\alpha}$ and $C_{\mu\alpha}^{\ \ \alpha}C^{\mu\beta}_{\ \ \beta}$. Other quadratic gauge-curvature scalars can be expressed in terms of them because of the identities in (8.14).

We postulate that, in a general frame with arbitrary coordinates and Poincaré metric tensors, the action $S_{\phi\psi}$ for fermion matter and spin-2 fields involves the linear combination of the two independent quadratic terms of the gauge-curvature and the symmetrized fermion Lagrangian:

$$S_{\phi\psi} = \int L_{\phi\psi}\sqrt{-P}d^4x, \quad P = det\, P_{\mu\nu}, \tag{8.16}$$

$$L_{\phi\psi} = \frac{1}{2g^2}\left(C_{\mu\alpha\beta}C^{\mu\beta\alpha} - C_{\mu\alpha}{}^{\alpha}C^{\mu\beta}{}_{\beta}\right)$$
$$+ \frac{i}{2}\left[\overline{\psi}\Gamma_\mu\Delta^\mu\psi - \left(\Delta^\mu\overline{\psi}\right)\Gamma_\mu\psi\right] - m\overline{\psi}\psi, \tag{8.17}$$

$$\Delta^\mu\psi = J^{\mu\nu}D_\nu\psi, \quad J^{\mu\nu} = P^{\mu\nu} + g\phi^{\mu\nu} = J^{\nu\mu}, \quad D_\lambda P_{\mu\nu} = 0. \tag{8.18}$$

Note that the quadratic gauge-curvature term in (8.17) can also be expressed as

$$L_{\phi\psi} = \frac{1}{2g^2}\left(\frac{1}{2}C_{\mu\alpha\beta}C^{\mu\alpha\beta} - C_{\mu\alpha}{}^{\alpha}C^{\mu\beta}{}_{\beta}\right)$$
$$+ \frac{i}{2}\left[\overline{\psi}\Gamma_\mu\Delta^\mu\psi - \left(\Delta^\mu\overline{\psi}\right)\Gamma_\mu\psi\right] - m\overline{\psi}\psi, \tag{8.19}$$

because $C_{\mu\alpha\beta}C^{\mu\alpha\beta} = 2C_{\mu\alpha\beta}C^{\mu\beta\alpha}$. The different relative sign of the two quadratic gauge curvatures in the Lagrangian (8.17) leads to a simple linearized equation which is mathematically the same as that in Einstein gravity, as we shall see below.

Now let us show explicitly that the action (8.16) of Yang–Mills gravity is invariant under the T_4 gauge transformations (8.3)–(8.6) in taiji symmetry framework. Based on the translational gauge transformations, we can show that

$$L_{\phi\psi} \to (L_{\phi\psi})^{\$} = L_{\phi\psi} - \Lambda^\lambda(\partial_\lambda L_{\phi\psi}). \tag{8.20}$$

Since Λ^μ is an infinitesimal gauge vector function, the gauge transformation of $P_{\mu\nu}$ can be written in the form,

$$(P_{\mu\nu})^{\$} = P_{\mu\nu} - \Lambda^\lambda\partial_\lambda P_{\mu\nu} - P_{\mu\beta}\partial_\nu\Lambda^\beta - P_{\alpha\nu}\partial_\mu\Lambda^\alpha$$
$$= \left[\left(1 - \Lambda^\sigma\partial_\sigma\right)P_{\alpha\beta}\right]\left(\delta^\alpha_\mu - \partial_\mu\Lambda^\alpha\right)\left(\delta^\beta_\nu - \partial_\nu\Lambda^\beta\right). \tag{8.21}$$

It follows from (8.21) that

$$\sqrt{-P} \to \sqrt{-P^{\$}} = \left[\left(1 - \Lambda^\sigma\partial_\sigma\right)\sqrt{-P}\right]\left(1 - \partial_\lambda\Lambda^\lambda\right), \quad P = det\, P_{\mu\nu}, \tag{8.22}$$

where we have used det $(AB) = \det A \det B$ and

$$\sqrt{(1 - \Lambda^\sigma \partial_\sigma)(-P)} = [(-P) - 2\sqrt{-P}\Lambda^\lambda \partial_\lambda \sqrt{-P}]^{1/2}$$

$$= \sqrt{-P}\left[\left(1 - \frac{\Lambda^\lambda \partial_\lambda \sqrt{-P}}{\sqrt{-P}}\right)^2\right]^{1/2}$$

$$= \sqrt{-P}\left(1 - \frac{\Lambda^\lambda \partial_\lambda \sqrt{-P}}{\sqrt{-P}}\right),$$

for infinitesimal Λ^λ. Thus, the Lagrangian $\sqrt{-P}L_{\phi\psi}$ changes only by a divergence under the gauge transformation,

$$\int \sqrt{-P}L_{\phi\psi}d^4x \rightarrow \int [\sqrt{-P}L_{\phi\psi} - \partial_\lambda(\Lambda^\lambda L_{\phi\psi}\sqrt{-P})]d^4x$$

$$= \int \sqrt{-P}L_{\phi\psi}d^4x. \tag{8.23}$$

The divergence term in (8.23) does not contribute to field equations because one can transform an integral over a 4-dimensional volume into the integral of a vector over a hypersurface on the boundaries of the volume of integration where fields and their variations vanish. Thus, we have shown that the action $S_{\phi\psi}$ is invariant under the T_4 gauge transformations (8.4), although the Lagrangian (8.17) is not gauge invariant. The gauge invariant action is sufficient to ensure that the wave equations in Yang–Mills gravity are T_4 gauge invariant.

8-4. The gravitational field equation and fermion equations in general frames of reference

In general, gauge field equations with gauge symmetry are not well defined. One usually includes a suitable gauge-fixing term in the Lagrangian to make the solutions of the gauge field equation well-defined. Yang–Mills gravity in a general frame is based on the total Lagrangian $L_{tot}\sqrt{-P}$, which is the original Lagrangian $L_{\phi\psi}$ with an additional gauge-fixing term L_{gf} involving the gauge parameter ξ

$$L_{tot}\sqrt{-P} = (L_{\phi\psi} + L_{gf})\sqrt{-P}, \tag{8.24}$$

$$L_{gf} = \frac{\xi}{2g^2}\left[\partial_\mu J_\alpha^\mu - \frac{1}{2}\partial_\alpha J_\mu^\mu\right]\left[\partial_\lambda J_\beta^\lambda - \frac{1}{2}\partial_\beta J_\lambda^\lambda\right]P^{\alpha\beta}. \tag{8.25}$$

Because the gauge fixing terms are, by definition, not gauge invariant, they involve the ordinary derivatives ∂_μ rather than the covariant derivative D_μ.[b] Here, the gauge-fixing term corresponds to the usual gauge condition $\eta^{\mu\nu}(\partial_\mu J_{\nu\alpha} - \frac{1}{2}\partial_\alpha J_{\mu\nu}) = 0$ for tensor fields. The Poincaré metric tensor $P_{\mu\nu}$ in (8.25) indicates that when one uses spherical coordinates to solve for static solutions, it denotes the metric tensor of the spherical coordinates and that when one uses inertial frames to carry out quantization, it is the Minkowski metric tensor.

The Lagrange equations for the gravitational tensor field $\phi^{\mu\nu}$ in a general frame can be derived from the action $\int L_{tot}\sqrt{-P}d^4x$. We obtain

$$H^{\mu\nu} + \xi A^{\mu\nu} = g^2 S^{\mu\nu}, \tag{8.26}$$

$$H^{\mu\nu} = D_\lambda(J_\rho^\lambda C^{\rho\mu\nu} - J_\alpha^\lambda C^{\alpha\beta}{}_\beta P^{\mu\nu} + C^{\mu\beta}{}_\beta J^{\nu\lambda})$$
$$- C^{\mu\alpha\beta}D^\nu J_{\alpha\beta} + C^{\mu\beta}{}_\beta D^\nu J_\alpha^\alpha - C^{\lambda\beta}{}_\beta D^\nu J_\lambda^\mu, \tag{8.27}$$

$$A^{\mu\nu} = \partial_\alpha\left[\left(\partial_\beta J_\rho{}^\beta - \frac{1}{2}\partial_\rho J_\lambda^\lambda\right) P^{\mu\alpha} P^{\nu\rho}\right]$$
$$+ \frac{1}{2}P^{\lambda\rho}(\partial_\alpha P_{\lambda\rho})P^{\mu\alpha} P^{\nu\rho}\left(\partial_\beta J_\rho{}^\beta - \frac{1}{2}\partial_\rho J_\gamma^\gamma\right)$$
$$- \frac{1}{2}\partial_\sigma\left[\left(\partial_\beta J_\rho{}^\beta - \frac{1}{2}\partial_\rho J_\lambda^\lambda\right) P^{\mu\nu} P^{\sigma\rho}\right]$$
$$- \frac{1}{2}P^{\lambda\rho}(\partial_\sigma P_{\lambda\rho})P^{\mu\nu} P^{\sigma\rho}\left(\partial_\beta J_\rho{}^\beta - \frac{1}{2}\partial_\rho J_\gamma^\gamma\right), \tag{8.28}$$

where μ and ν should be made symmetric. We have used the identities in (8.14) for the derivation of (8.26). The 'source tensor' $S^{\mu\nu}$ of fermion matter is given by

$$S^{\mu\nu} = \frac{1}{2}\left[\bar\psi i\Gamma^\mu D^\nu\psi - i(D^\nu\bar\psi)\Gamma^\mu\psi\right]. \tag{8.29}$$

The Dirac equation for a fermion interacting with the tensor fields $\phi^{\mu\nu}$ in a general frame can also be derived from (8.24):

$$i\Gamma_\mu(P^{\mu\nu} + g\phi^{\mu\nu})D_\nu\psi - m\psi + \frac{i}{2}\left[D_\nu(J^{\mu\nu}\Gamma_\mu)\right]\psi = 0,$$
$$i(P^{\mu\nu} + g\phi^{\mu\nu})(D_\nu\bar\psi)\Gamma_\mu + m\bar\psi + \frac{i}{2}\bar\psi\left[D_\nu(J^{\mu\nu}\Gamma_\mu)\right] = 0, \tag{8.30}$$

---

[b]If one replaces all ∂_μ by D_μ in (8.25), then the static solutions of the field Equation (8.26) do not exist beyond the second order approximation. This property was also verified by symbolic computing. We would like to thank D. W. Yang and J. Westgate for their help.

and the partial covariant derivative D_λ associated with the Poincaré metric tensor $P_{\mu\nu}$ is given by

$$D_\lambda J^{\mu\nu} = \partial_\lambda J^{\mu\nu} + \Gamma^\mu_{\lambda\rho} J^{\rho\nu} + \Gamma^\nu_{\lambda\rho} J^{\mu\rho}, \quad \text{etc.}$$

$$\Gamma^\mu_{\lambda\rho} = \frac{1}{2} P^{\mu\sigma} (\partial_\lambda P_{\sigma\rho} + \partial_\rho P_{\sigma\lambda} - \partial_\sigma P_{\lambda\rho}), \tag{8.31}$$

where $\Gamma^\mu_{\lambda\rho}$ is the Christoffel symbol and we have used the relation $(1/\sqrt{-P})\partial_\nu[Q^\nu\sqrt{-P}] = D_\nu Q^\nu$.

If one compares the fermion equation (8.30) with the Dirac equation in quantum electrodynamics (i.e., $(i\gamma^\mu\partial_\mu - e\gamma^\mu A_\mu - m)\psi = 0$) in inertial frames, one can see a distinct difference: Namely, if one takes the complex conjugate of the Dirac equations, the kinematical term $i\gamma^\mu\partial_\mu$ and the electromagnetic coupling term $e\gamma^\mu A_\mu$ have different signs. This implies the presence of both repulsive and attractive forces between two charges. However, if one takes the complex conjugate of the fermion equation for ψ in (8.30), the kinematical term and the spin-2 coupling term have the same signs. Thus, the translation gauge symmetry of gravity naturally explains why the gravitational interaction is always attractive between fermion matter and anti-fermion matter. The physical reason for the electric repulsion and the gravitational attraction between, say, electrons will be explained in terms of the exchange of virtual gauge bosons in Section 14-1. The different nature of these forces originates from the property that the electromagnetic potential transforms like a vector, while the gravitational potential $\phi_{\mu\nu}$ is a tensor.

In inertial frames with $P_{\mu\nu} = \eta_{\mu\nu}$, the gauge-field equation (8.27) can be linearized for weak fields as follows

$$\partial_\lambda\partial^\lambda\phi^{\mu\nu} - \partial^\mu\partial_\lambda\phi^{\lambda\nu} - \eta^{\mu\nu}\partial_\lambda\partial^\lambda\phi + \eta^{\mu\nu}\partial_\alpha\partial_\beta\phi^{\alpha\beta}$$
$$+ \partial^\mu\partial^\nu\phi - \partial^\nu\partial_\lambda\phi^{\lambda\mu} - gS^{\mu\nu} = 0. \tag{8.32}$$

This equation can also be written in the form:

$$\partial_\lambda\partial^\lambda\phi^{\mu\nu} - \partial^\mu\partial_\lambda\phi^{\lambda\nu} + \partial^\mu\partial^\nu\phi^\lambda_\lambda - \partial^\nu\partial_\lambda\phi^{\lambda\mu} = g\left(S^{\mu\nu} - \frac{1}{2}\eta^{\mu\nu}S^\lambda_\lambda\right), \tag{8.33}$$

where we have set $\xi = 0$ and used $J^{\mu\nu} = \eta^{\mu\nu} + g\phi^{\mu\nu}$. It is interesting that the linearized gauge-field equation (8.32) is mathematically the same as the corresponding equation in general relativity, which may be related to the fact that the transformation (8.2) is mathematically the same as that in general relativity.

8-5. Derivations of the T_4 eikonal and Einstein–Grossmann equations in the geometric-optics limit

In Yang–Mills gravity, the fundamental field equation in the geometric optics limit can be derived as follows: We first postulate the translation gauge invariant Lagrangian L_{em} for the electromagnetic potential A^μ,

$$L_{em} = -\frac{1}{4} P^{\mu\alpha} P^{\nu\beta} F_{\mu\nu} F_{\alpha\beta},$$

$$F_{\mu\nu} = \Delta_\mu A_\nu - \Delta_\nu A_\mu, \quad \Delta_\mu = J_{\mu\nu} D^\nu, \tag{8.34}$$

where we have used the same replacement as in (8.7). For simplicity, let us consider an inertial frame with $P^{\mu\nu} = \eta^{\mu\nu}$, $\Delta_\mu = J_{\mu\lambda} \partial^\lambda$, and the expression for field A^μ in the geometric optics limit,

$$A^\mu = a^\mu \exp(i\Psi),$$

where the eikonal Ψ and the wave vector $\partial_\mu \Psi$ are very large.[8]

The generalized electromagnetic wave equation in the geometric-optics limit is

$$0 = \Delta_\mu(\Delta^\mu A^\lambda - \Delta^\lambda A^\mu) + (\partial_\alpha J_\mu^\alpha)(\Delta^\mu A^\lambda - \Delta^\lambda A^\mu) \tag{8.35}$$

$$= \delta_\mu^\beta J_\lambda^\alpha J^{\lambda\sigma} \partial_\alpha \partial_\sigma A^\mu - g J^{\beta\sigma} \phi_\mu^\alpha \partial_\alpha \partial_\sigma A^\mu \equiv Z_\mu^\beta a^\mu,$$

$$Z_\mu^\beta = (\delta_\mu^\beta G^{\alpha\sigma} \partial_\alpha \Psi \partial_\sigma \Psi - g J^{\beta\sigma} \phi_\mu^\alpha \partial_\alpha \Psi \partial_\sigma \Psi), \quad G^{\alpha\sigma} = J_\lambda^\alpha J^{\lambda\sigma},$$

where the terms involving $(\partial_\alpha J_\mu^\alpha)$ are small and the terms involving $\partial_\alpha \partial_\sigma A^\mu$ or $\partial_\alpha \Psi \partial_\sigma \Psi$ are large. We have also used

$$-\Delta_\mu \Delta^\lambda A^\mu = -J_\mu^\sigma \partial_\sigma (J^{\lambda\alpha} \partial_\alpha A^\mu) = -(\delta_\mu^\sigma + g\phi_\mu^\sigma) \partial_\sigma (J^{\lambda\alpha} \partial_\alpha A^\mu)$$

$$\approx -g\phi_\mu^\sigma J^{\lambda\alpha} \partial_\alpha \partial_\mu A^\mu - g\phi_\mu^\sigma J^{\lambda\alpha} \partial_\sigma \partial_\alpha A^\mu = -g\phi_\mu^\sigma J^{\lambda\alpha} \partial_\sigma \partial_\alpha A^\mu,$$

where $\partial_\mu A^\mu = 0$. Since we are interested in the law for the propagation of light rays and the further simplification of (8.35), we have expressed the amplitude a^μ in terms of the space-like polarization vector $\epsilon^\mu(\lambda)$, i.e., $a^\mu = \epsilon^\mu(\lambda)b(x)$, $b(x) \neq 0$ in the limiting expression for A^μ. As usual, we have $\sum_\lambda \epsilon^\mu(\lambda)\epsilon^\nu(\lambda) \to -\eta^{\mu\nu}$ by summing over all polarizations.[9] Multiplying $Z_\mu^\beta a^\mu$ in (8.35) by $a^\nu \eta_{\nu\beta}$ and summing over all polarizations, we obtain

$$\frac{1}{b^2(x)} \sum_\lambda Z_\mu^\beta a^\mu a^\nu \eta_{\nu\beta} = -\delta_\beta^\mu Z_\mu^\beta = 0.$$

After some more straightforward mathematical manipulation, we obtain a new eikonal equation that involves an effective metric tensor $G_L^{\mu\nu}$,

$$G_L^{\mu\nu} \partial_\mu \Psi \partial_\nu \Psi = 0,$$

$$G_L^{\mu\nu} = G^{\mu\nu} - \frac{g}{4}\phi_\lambda^\mu J^{\lambda\nu} = \eta_{\alpha\beta}(\eta^{\mu\alpha} + g\phi^{\mu\alpha})\left(\eta^{\nu\beta} + \frac{3g}{4}\phi^{\nu\beta}\right),$$

(8.36)

in the geometric-optics (or classical) limit.

It must be stressed that we have simple $G_L^{\mu\nu}$ and eikonal Equation (8.36) only in this geometric-optics limit. Otherwise, Equation (8.35) has very complicated dependence on the wave vector $\partial_\mu \Psi$ and, hence, it is difficult to identify the effective Riemannian metric tensors (and the corresponding effective interval ds_{ei} in (8.41) below) in general.

Next, let us consider the relation between the classical equation of motion and the massive fermion wave equation. The fermion wave Equation (8.30) can be derived from the Lagrangian (8.24), i.e.,

$$i\Gamma_\mu \Delta^\mu \psi - m\psi + \frac{i}{2}\gamma_a[D_\nu(J^{\mu\nu}e_\mu^a)]\psi = 0.$$

(8.37)

Using the expression for the field $\psi = \psi_o \exp(iS)$, we can derive the equation

$$\gamma_a E^{a\mu} \partial_\mu S + m - \frac{i}{2}\gamma_a[D_\nu(J^{\mu\nu}e_\mu^a)] = 0.$$

(8.38)

In the classical limit, the momentum $\partial_\mu S$ and mass m are large quantities, and one can neglect the small gravitational interaction term involving e_μ^a. To eliminate the spin variables, we multiply the large terms in (8.38) by a factor of $(\gamma_a E^{a\mu} \partial_\mu S - m)$ so that the resultant equation can be written in the form

$$\frac{1}{2}(\gamma_b\gamma_a + \gamma_a\gamma_b)E^{a\mu}E^{b\nu}(\partial_\mu S)(\partial_\nu S) - m^2 = 0.$$

(8.39)

With the help of the anti-commutation relation for γ_a in (8.9) and the effective metric tensor (8.11), we can then simplify (8.39), obtaining the following equation for the motion of a classical object in the presence of the gravitational tensor field

$\phi^{\mu\nu}$ in flat space-time,

$$G^{\mu\nu}(\partial_\mu S)(\partial_\nu S) - m^2 = 0, \quad G^{\mu\nu} = P_{\alpha\beta}J^{\alpha\mu}J^{\beta\nu}. \tag{8.40}$$

This equation is mathematically the same as the corresponding Hamilton–Jacobi equation in general relativity. Within the framework of Yang–Mills gravity in flat space-time, we shall call (8.40) the Einstein–Grossmann equation of motion.[c] It appears that the classical equation of motion must have the form of (8.40) (or its equivalent form in terms of an 'effective geodesic equation' such as (8.44) below) in order to accurately describe the motion of macroscopic objects in a gravitational field, whether the theory is based on a curved or a flat space-time.

We note that the effective metric tensor $G_L^{\mu\nu}$ in (8.36) for light rays differs slightly from $G^{\mu\nu}$ for fermions in the geometric-optics limit. This is due to their different T_4 gauge invariant couplings to the gravitational field $\phi_{\mu\nu}$. If the second non-dominant term in $G_L^{\mu\nu}$ in (8.36) is neglected, then all wave equations for fermions and bosons lead to the same effective Riemann metric tensor in the geometric-optics limit, as one can see from (8.36) and (8.40). The physical effects of this non-dominant term in (8.36) will be discussed in Section 9-3.

8-6.   Effective action for classical objects and gauge invariance

All the observable effects of gravity are directly related to the motion of classical objects and light rays. Thus, it is important to understand the relation between the wave equations of fields and the corresponding classical equations of particles (or objects) and light rays. In Yang–Mills gravity, we have seen that the eikonal equation (8.36) for a light ray and the Einstein–Grossmann equation of motion (8.40) for classical objects are the wave equations in the geometric-optics limit.

We now demonstrate that if one postulates the following effective action S_p for classical particles in general frames with Poincaré metric tensors $P_{\mu\nu}$,

$$S_p = -\int m\, ds_{ei},$$

$$ds_{ei}^2 = I_{\mu\nu}dx^\mu dx^\nu, \quad I_{\mu\nu}G^{\nu\alpha} = \delta_\mu^\alpha, \tag{8.41}$$

$$G^{\mu\nu} = P_{\alpha\beta}J^{\alpha\mu}J^{\beta\nu} = P^{\mu\nu} + 2g\phi^{\mu\nu} + g^2\phi^{\mu\lambda}\phi^{\nu\sigma}P_{\lambda\sigma},$$

---

[c] Marcel Grossmann was associated with Einstein in elucidating the mathematical foundation of general relativity.

one can derive the Einstein–Grossmann Equation (8.40). We note that the action S_p (8.41) is not invariant under the gauge transformations (8.5) and (8.6). However, this is not surprising or problematic since S_p is only an effective action for classical particles (in the geometric optics limit) rather than the gauge invariant action (8.16) for basic tensor and fermion fields. However, one can show that the effective interval ds_{ei}^2 and hence, the action S_p are invariant under the following tensor transformations

$$Q'^{\mu_1...\mu_m}_{\alpha_1...\alpha_n}(x') = Q^{\nu_1...\nu_m}_{\beta_1...\beta_n}(x)\frac{\partial x'^{\mu_1}}{\partial x^{\nu_1}} \cdots \frac{\partial x'^{\mu_m}}{\partial x^{\nu_m}}\frac{\partial x^{\beta_1}}{\partial x'^{\alpha_1}} \cdots \frac{\partial x^{\beta_n}}{\partial x'^{\alpha_n}}. \tag{8.42}$$

If we consider the variation of the action (8.41), we have

$$\delta S_p = -m\int\left[I_{\lambda\gamma}\frac{d}{ds}\frac{dx^\lambda}{ds} + \Gamma_{\gamma,\alpha\beta}\frac{dx^\alpha}{ds}\frac{dx^\beta}{ds}\right]\delta x^\gamma ds - mI_{\mu\nu}\frac{dx^\mu}{ds}\delta x^\nu\bigg|,$$
$$\tag{8.43}$$
$$\Gamma_{\gamma,\alpha\beta} = \frac{1}{2}(\partial_\beta I_{\alpha\gamma} + \partial_\alpha I_{\beta\gamma} - \partial_\gamma I_{\alpha\beta}),$$

where $ds = ds_{ei}$. Suppose we consider only the actual path with one of its end points left unfixed,[8] i.e.,

$$I_{\lambda\gamma}\frac{d}{ds}\frac{dx^\lambda}{ds} + \Gamma_{\gamma,\alpha\beta}\frac{dx^\alpha}{ds}\frac{dx^\beta}{ds} = 0, \quad ds = ds_{ei}. \tag{8.44}$$

We then have

$$\delta S_p = -mI_{\mu\nu}\left(\frac{dx^\mu}{ds_{ei}}\right)\delta x^\nu. \tag{8.45}$$

The momentum of a classical particle is defined by

$$p_\nu = -\frac{\partial S_p}{\partial x^\nu} = -m\frac{dx^\mu}{ds_{ei}}I_{\mu\nu}, \quad I_{\mu\nu}G^{\nu\alpha} = \delta^\alpha_\mu. \tag{8.46}$$

Since $ds_{ei}^2 = I_{\mu\nu}dx^\mu dx^\nu$, we have

$$G^{\mu\nu}p_\mu p_\nu - m^2 = 0, \quad I_{\mu\nu}G^{\nu\lambda} = \delta^\lambda_\mu. \tag{8.47}$$

Equations (8.46) and (8.47) lead to the Einstein–Grossmann equation for a classical particle with mass m,

$$G^{\mu\nu}(\partial_\mu S)(\partial_\nu S) - m^2 = 0, \quad S \equiv S_p. \tag{8.48}$$

This equation is the same as (8.40), which is obtained from the Dirac wave Equation (8.37) in the presence of gravity and in the geometric-optics limit. Equations (8.47) and (8.48) are obtained in Yang–Mills gravity based on flat space-time. They are formally the same as the corresponding equations in general relativity. Thus, in the presence of gravity, classical objects in Yang–Mills gravity appear to move in a space-time with an effective metric tensor $G^{\mu\nu}$.

Since the effective action (8.41) and the effective metric tensor $G^{\mu\nu}$ are only approximate quantities in Yang–Mills gravity, it is natural for one to expect that the observable results in the classical limit may not be gauge invariant to all orders (see also discussions at the end of Section 9-1).

8-7.   Classical objects as a coherent collection of constituent particles

Equations (8.38) through (8.40) show the effect on the field equations of going to the classical limit. This limiting process is accomplished by making quantities representing the mass and momentum large (taking the limits $m \to \infty$ and $\hbar \to 0$). (In unit systems other than natural units, the phase iS_k in

$$\psi = \psi_o \exp{(iS)}$$

should be replaced by $iS_k/\hbar$). However, in the real world, the large mass and large eikonal S of classical objects have their origin in the summation of m_k and S_k over all of the large number of quantum particles k in a macroscopic system, rather than in the existence of any single particle with a large mass and momentum. In this sense, the conventional steps (8.38)–(8.40) do not directly reveal the physical connection between a macroscopic body with a large mass m and the actual physical situation of a macroscopic body consisting of a large number N of constituent particles with masses m_k, $k = 1, 2, 3, \ldots N$,

$$m_t = \sum_{k=1}^{N} m_k. \tag{8.49}$$

To explore the consequences of applying a more realistic process of taking the classical limit to Equations (8.38) to (8.40), we begin with the relation for each particle k,

$$[\gamma_\mu(\eta^{\mu\nu} + g\phi^{\mu\nu})\partial_\nu S]_{(k)} + m_{(k)} = 0, \quad k = 1, 2, \ldots N. \tag{8.50}$$

After summing over all particles, the expression (8.38) with m replaced by m_t from (8.49) is

$$\gamma_\mu\{\eta^{\mu\nu} + g\phi^{\mu\nu}(x)\}\partial_\nu S(x) + m_t = 0. \tag{8.51}$$

A further natural requirement for the classical limit is that the momentum of particles in a macroscopic object must add up 'coherently' in the following sense,

$$\sum_{k=1}^{N}[\partial_{\nu}S]_{(k)} = \partial_{\nu}S(x), \quad \sum_{k=1}^{N}\left[\phi^{\mu\nu}\partial_{\nu}S\right]_{(k)} = \phi^{\mu\nu}(x)\partial_{\nu}S(x). \tag{8.52}$$

Equations (8.51) and (8.52) are reasonable because a classical macroscopic object is, by construction, made of N bounded particles. Although the summation of constant masses in (8.49) is simple, the summation of space-time-dependent momenta in the first equation of (8.52) within the relativistic framework is not, because each particle has its own time and space coordinates in an inertial frame. However, one simplification that might be used to make this equation in (8.52) tractable is to assume that the relative motions of the quantum particles making up a single macroscopic object can be ignored in the classical limit, viewing them as being at rest relative to each other and, hence, describable using a single set of space-time coordinates. As can be seen, the passage from a system of bounded quantum particles to a single macroscopic classical object is not as simple as one might think.

Once one has the result (8.51), one can ignore the spin effect by removing γ_{μ} from the equation, as done in (8.38)–(8.40). In this way, one obtains the relativistic Einstein–Grossmann Equation (8.40) with an effective Riemannian metric tensor $G^{\mu\nu}$, whose presence in the classical limit is crucial for Yang–Mills gravity to be consistent with experiments.

8-8.   Remarks on Yang–Mills gravity and teleparallel gravity

At a first glimpse, the Lagrangian for the electromagnetic field in the presence of a gravitational field $\phi_{\mu\nu}$ appears to be quite different from the Lagrangian (8.17) of the Dirac field. However, if one follows the Yang–Mills approach with the T_4 gauge group, i.e., replacing ∂^{μ} by the T_4 gauge covariant derivative $\partial^{\mu} + g\phi^{\mu\nu}\partial_{\nu}$, one obtains a relativistic Einstein–Grossmann Equation (8.40) with $m = 0$ for light rays in the geometric optics limit.

Tensor fields with spin 2 were discussed by Fierz and Pauli in 1939.[10,11] In order to have 2 independent polarization states for a massless symmetric tensor field '$\psi_{\mu\nu}$', certain subsidiary conditions were imposed. In particular, they discussed tensor field equations which are invariant under a 'gauge transformation,'

$$\psi_{\mu\nu} \rightarrow \psi_{\mu\nu} + \partial_{\mu}\Lambda_{\nu} + \partial_{\nu}\Lambda_{\mu}.$$

Without the guidance of modern gauge symmetry, it is non-trivial to find a Lagrangian for tensor fields in such a way that wave equations and subsidiary conditions follow simultaneously from the Hamilton principle.[11] It turns out that the simplest free Lagrangian is equivalent to the free Lagrangian in (8.17), and the free field equation for a massless $\psi_{\mu\nu}$ is exactly the same as the linearized Equation (8.32) or (8.33) in Yang–Mills gravity and that in Einstein's theory of gravity.

It is generally believed that general coordinate invariance in Einstein's gravity can be considered a gauge symmetry. In particular, one can use the affine connection to construct covariant derivatives of tensors, just as one uses a gauge field to construct gauge covariant derivatives of matter fields. However, a fundamental difference between Einstein's gravity and Yang–Mills gravity is that Einstein's gravity is based on curved space-time so that the commutator of two covariant derivatives with respect to x^μ and x^ν leads to the Riemann–Christoffel curvature tensor uniquely and unambiguously. Yang–Mills gravity, however, is based on the T_4 group in flat space-time, just like the usual Yang–Mills theory and conventional field theories so that there is no Riemann–Christoffel curvature, but only a T_4 gauge curvature $C^{\mu\nu\lambda}$.

The fact that the action (8.16) in Yang–Mills theory involves a quadratic gauge curvature $C_{\mu\nu\alpha}$, while the Hilbert–Einstein action involves a linear curvature of space-time implies a basic and important breakdown of the analogy between the action (8.16) and the Hilbert–Einstein action.[12] Namely, gauge fields are fundamental and cannot be expressed in terms of any more fundamental fields, while the affine connection in Einstein's gravity is itself constructed from first derivatives of the metric tensor. There is no breakdown in the analogy between Yang–Mills gravity and the usual Yang–Mills theory, however.

There is a formulation of gravity called teleparallel gravity (TG)[13,14] that, like Yang–Mills gravity, is based on a translational gauge symmetry in a flat space-time, but unlike Yang–Mills gravity, has a torsion tensor. In TG, a flat connection with torsion makes translations local gauge symmetries and the curvature tensor is the torsion field. Furthermore, the concept of translational gauge symmetry is realized differently in Yang–Mills gravity and in TG. Although there is a one-one correspondence between equations in Yang–Mills gravity and TG, there are conceptual differences between them and the differences show up in the properties of the gauge potentials, gauge curvature and actions as follows.

(a) Gauge covariant derivative.

In Yang–Mills gravity, the $T(4)$ gauge covariant derivative $\Delta_\mu \psi$ is defined in (8.14), where the gauge potential field $\phi_{\mu\nu}$ and $J_{\mu\nu}$ are symmetric tensors in flat space-time. In contrast, the gauge covariant derivative in TG is defined by[13,14]

$$D_\mu \psi = \partial_\mu \psi + B^a_\mu \partial_a \psi = h^a_\mu \partial_a \psi, \quad h^a_\mu = \partial_\mu x^a + B^a_\mu, \tag{8.53}$$

where x^μ are the space-time coordinates with the metric tensor $g_{\mu\nu}(x) = \eta_{ab} h^a_\mu(x) h^b_\nu(x)$, and $x^a(x^\mu)$ are the coordinates in a tangent Minkowski space-time with the metric tensor $\eta_{ab} = (+,-,-,-)$. Thus, h^a_μ in TG is treated as a tetrad rather than a tensor.

(b) Gauge curvature.

In Yang–Mills gravity, the gauge curvature $C^{\mu\nu\alpha}$ is a tensor of the third rank and is defined by

$$[\Delta_\mu, \Delta_\nu] = C_{\mu\nu\lambda} \partial^\lambda, \quad C^{\mu\nu\lambda} = J^{\mu\sigma} \partial_\sigma J^{\nu\lambda} - J^{\nu\sigma} \partial_\sigma J^{\mu\lambda},$$
$$J_{\mu\nu} = \eta_{\mu\nu} + g\phi_{\mu\nu} = J_{\nu\mu}. \tag{8.54}$$

In contrast, the gauge curvature in TG is a torsion tensor $T^a_{\mu\nu}$:

$$[D_\mu, D_\nu] = T^a_{\mu\nu} \partial_a, \quad T^a_{\mu\nu} = \partial_\mu h^a_\nu - \partial_\nu h^a_\mu. \tag{8.55}$$

(c) Gravitational action.

A real physical difference between Yang–Mills gravity and TG shows up in the gravitational actions.[15,16] The gravitational Lagrangian L_ϕ in Yang–Mills gravity is given in Equations (8.16) and (8.17). In TG, the gravitational action and Lagrangian L_h are given by Hayashi and Shirafuji[13,14]

$$S_{TG} = \int L_h \sqrt{-\det g_{\mu\nu}}\, d^4 x, \tag{8.56}$$

$$L_h = a_1(t_{\lambda\mu\nu} t^{\lambda\mu\nu}) + a_2(v^\mu v_\mu) + a_3(a^\mu a_\mu), \tag{8.57}$$

$$t_{\lambda\mu\nu} = \frac{1}{2}(T_{\lambda\mu\nu} + T_{\mu\lambda\nu}) + \frac{1}{6}(g_{\nu\lambda} v_\mu + g_{\nu\mu} v_\lambda) - \frac{1}{3} g_{\lambda\mu} v_\nu, \tag{8.58}$$

$$v_\mu = T^\lambda_{\lambda\mu}, \quad a_\mu = \frac{1}{6}\epsilon_{\mu\nu\rho\sigma} T^{\nu\rho\sigma}. \tag{8.59}$$

$$T^\lambda_{\mu\nu} = h^\lambda_a(\partial_\nu h^a_\mu - \partial_\mu h^a_\nu), \quad g_{\mu\nu}(x) = \eta_{ab} h^a_\mu(x) h^b_\nu(x), \tag{8.60}$$

where $T^\lambda_{\mu\nu}$ is the torsion tensor. It appears to be impossible to adjust the three parameters a_1, a_2 and a_3 in (8.57) to make S_{TG} in (8.56) equal to the $S = \int L_\phi d^4x$ in Yang–Mills gravity for both inertial and non-inertial frames. In general, if the tetrad in a field theory involves dynamical fields, it leads to complicated vertices in the Feynman rules due to the presence of $\sqrt{-\det g_{\mu\nu}}$ in the action, if the field can be quantized.

Yang–Mills gravity is a local gauge field theory for the microscopic world in flat space-time and hence is not related to the equivalence principle. However, in the classical limit (i.e., geometric-optics limit), the fermion wave equation reduces to the relativistic Einstein–Grossmann Equation (8.40), which describes the free-fall motion of a classical object in inertial frames. In this sense, Yang–Mills gravity is approximately compatible with the equivalence principle in the geometric-optics limit.

References

1. A. A. Logunov, *The Theory of Gravity* (Translated by G. Pontecorvo, Moscow, Nauka, 2001) and references therein. See also ref. 15.

2. Jong-Ping Hsu, *Phys. Lett.* **119B**, 328 (1982).

3. J. P. Hsu and L. Hsu, *A Broader View of Relativity, General Implications of Lorentz and Poincaré Invariance* (World Scientific, 2006), Appendix D. Available online: 'google books, a broader view of relativity, Hsu'; J. P. Hsu, *Intl. J. Mod Phys. A* **21** 5119 (2006).

4. R. Utiyama and T. Fukuyama, *Prog. Theor. Phys.* **45**, 612 (1971); Y. M. Cho, *Phys. Rev. D* **14**, 2521 (1976).

5. Y. M. Cho, *Phys. Rev. D* **14**, 3341 (1976).

6. K. Hayashi and T. Shirafuji, *Phys. Rev. D* **19**, 3524 (1979).

7. F. J. Dyson, in *100 Years of Gravity and Accelerated Frames, The Deepest Insights of Einstein and Yang–Mills* (Ed. J. P. Hsu and D. Fine, World Scientific, 2005) p. 348.

8. L. Landau and E. Lifshitz, *The Classical Theory of Fields* (Trans. by M. Hamermesh, Addison-Wesley, Cambridge, Mass. 1951), pp. 29–30, pp. 136–137 and pp. 268–270.

9. J. D. Bjorken and S. D. Drell, *Relativistic Quantum Fields* (McGraw-Hill, 1965) p. 386.

10. M. Fierz, *Helv. Phys. Acta* **12**, 3 (1939).

11. M. Fierz and W. Pauli, *Proc. Roy. Soc.* **173**, 211 (1939).

12. S. Weinberg, *The Quantum Theory of Fields, vol. 2* (Cambridge Univ. Press, 1996) pp. 6–7.

13. See, for example, K. Hayashi and T. Shirafuji, *Phys. Rev. D* **19**, 3524 (1979).

14. V. C. de Andrade and J. G. Pereira, *Phys. Rev. D* **56**, 4689 (1997).

15. A. A. Logunov and M. A. Mestvirishvili, and Ning Wu discussed a theory of gravity based on flat space-time. Wu proposed interpreting the Lagrangian in the Hilbert–Einstein action for gravity in terms of a gauge field with a T_4 gauge symmetry in flat space-time. However, in their formulation, the graviton still involved an N-vertex of self-coupling, where N is an arbitrarily large number and hence, their formulation is more complicated than the maximum 4-vertex for the self-coupling of gravitons in Yang–Mills gravity. See A. A. Logunov, ref. 1; A. A. Logunov and M. A. Mestvirishvili, *Prog. Theor. Phys.* **74**, 31, (1985); Ning Wu, *Commun. Theor. Phys.* (Beijing, China) **42**, 543 (2004). All these formulations of gravity are different from Yang–Mills gravity.

16. D. Fine has pointed out that, from the viewpoint of the geometric picture, Yang–Mills gravity and teleparallel gravity are very similar. (Private correspondence, 2012).

9.

Experimental Tests of Classical Yang–Mills Gravity

9-1.   Motion in a central gravitational field

In physics, the ultimate test of any theory is its ability to explain and predict experimental phenomena. In this chapter, we test the classical approximation of Yang–Mills gravity against the results of a number of well-established experiments, such as the perihelion shift of the planet Mercury, the deflection of light in a gravitational field, and the gravitational quadruple energy radiation from systems such as binary stars. In particular, we find that a more accurate measurement of the deflection of light in a gravitational field could provide a test of Yang–Mills gravity versus Einstein gravity (see Section 9-3).

We begin by considering the motion of a classical object in a central gravitational force field. Before we can use the Einstein–Grossmann equation of motion (8.40) to determine the path of an object, however, we must first calculate the effective metric tensors $G_{\mu\nu} = P_{\alpha\beta}J^{\alpha\mu}J^{\beta\nu}$ in (8.40). Thus, we first solve the field Equation (8.26) to obtain the static potential $\phi_{\mu\nu}$ and hence, $J_{\mu\nu}$.

The only non-vanishing component $S^{00} = m\delta^3(\mathbf{r})$ from Equations (8.32) and (8.33) leads to the well-known result, $g\phi^{00} = g^2m/(8\pi r)$, in the Newtonian limit. As usual, $G^{00} = 1 + 2Gm/r$ in this limit in the Einstein–Grossman Equation (8.40), where G is the gravitational constant. Based on these results, together with $G^{\mu\nu}$ in (8.40) and $I_{\mu\nu}$ in (8.47),

$$g = \sqrt{8\pi G}, \quad \text{and} \quad g\phi^{00} = g\phi^{11} = \frac{Gm}{r}, \quad g\phi^{22} = \frac{1}{r^2}\frac{Gm}{r}, \qquad (9.1)$$

to the first order approximation in an inertial frame. These results can be obtained by solving the linearized field Equation (8.33) in spherical coordinates, $x^\mu = (w, r, \theta, \phi)$.

Because the perihelion shift of Mercury is sensitive to the coefficient of the second-order term of G^{00} or I_{00} and the coefficient of the first-order term of G^{11}, G^{22} and G^{33}, we shall also calculate the second-order terms of the components of the effective metric tensor $G^{\mu\nu}$. We solve the non-linear gauge field equations by the method of successive approximation and carry out the related post-Newtonian approximation to the second order. For the gauge field equations to be well defined, it is convenient to use the gauge-field Equation (8.26) with the gauge parameter ξ.

For simplicity, we consider an inertial frame and a static, spherically symmetric system, in which the tensor gauge fields are produced by a spherical object at rest with mass m. Based on symmetry considerations,[1] the non-vanishing components of the exterior solutions $\phi^{\mu\nu}(r)$ are $\phi^{00}(r), \phi^{11}(r), \phi^{22}(r)$ and $\phi^{33}(r) = \phi^{22}/\sin^2\theta$, where $x^\mu = (w, r, \theta, \phi)$. To solve the static gauge field, we write

$$J^{00} = J_0^0 = S(r), \quad -J^{11} = J_1^1 = R(r),$$

$$-r^2 J^{22} = J_2^2 = -r^2 \sin^2\theta J_{33} = J_3^3 = T(r). \tag{9.2}$$

The metric tensor is given by $P_{\mu\nu} = (1, -1, -r^2, -r^2 \sin^2\theta)$. In this coordinate system, the non-vanishing components of the Christoffel symbol $\Gamma^\alpha_{\mu\nu}$ are

$$\Gamma^1_{22} = -r, \quad \Gamma^1_{33} = -r \sin^2\theta, \quad \Gamma^2_{12} = 1/r,$$

$$\Gamma^2_{33} = -\sin\theta\cos\theta, \quad \Gamma^3_{13} = 1/r, \quad \Gamma^3_{23} = \cot\theta. \tag{9.3}$$

After some tedious but straightforward calculations (see Appendix B for more details), the gauge field Equation (8.26) with $(\mu, \nu) = (0, 0), (1, 1), (2, 2), (3, 3)$, can be written as

$$\frac{d}{dr}\left(R^2 \frac{dS}{dr}\right) + \frac{2}{r} R^2 \frac{dS}{dr} - \left(R\frac{d}{dr} + \frac{dR}{dr} + \frac{2R}{r}\right)$$

$$\times \left(R\left[\frac{dS}{dr} + 2\frac{dT}{dr}\right] + \frac{2T^2}{r} - \frac{2TR}{r}\right) + \xi\left[\frac{1}{4}\frac{d^2}{dr^2}(S - R + 2T)\right]$$

$$+ \frac{\xi}{r}\left[\frac{d}{dr}(S - R + 2T)\right] = 0, \tag{9.4}$$

$$R\left(\frac{dS}{dr}\right)^2 + 2r^3 \frac{d(T/r)}{dr}\left[\frac{R}{r}\frac{d(T/r)}{dr} + \frac{T^2}{r^3}\right] + \left(\frac{dS}{dr} + 2\frac{dT}{dr} - \frac{2T}{r}\right)$$

$$\times \left(-R\left[\frac{dS}{dr} + 2\frac{dT}{dr}\right] - \frac{2}{r}T^2 + \frac{2}{r}TR\right)$$

$$+ \xi\left[\frac{1}{4}\frac{d^2}{dr^2}(S - R + 2T)\right] = 0, \tag{9.5}$$

$$\left(R\frac{d}{dr} + \frac{dR}{dr} + \frac{5R}{r} - \frac{2T}{r} \right) \left[\frac{R}{r}\frac{d(T/r)}{dr} + \frac{T^2}{r^3} \right]$$

$$+ \left[\frac{1}{r^2} \left(R\frac{d}{dr} + \frac{dR}{dr} \right) + \frac{3R}{r^3} - \frac{2T}{r^3} \right]$$

$$\times \left[-R\left(\frac{dS}{dr} + 2\frac{dT}{dr} \right) - \frac{2}{r}T^2 + \frac{2}{r}TR \right]$$

$$+ \xi \left[\frac{1}{4r^2}\frac{d^2}{dr^2}(S - R + 2T) \right] + \xi \left[\frac{1}{2r^3}\frac{d}{dr}(S - R + 2T) \right] = 0, \quad (9.6)$$

respectively. The equation for $(\mu, \nu) = (3,3)$ is the same as that in (9.6).

We can solve the gauge field Equations (9.4)–(9.6) to the second order approximation by setting

$$S = 1 + a_1/r + a_2/r^2, \quad R = 1 + b_1/r + b_2/r^2, \quad T = 1 + c_1/r + c_2/r^2. \quad (9.7)$$

We find

$$a_1 = Gm, \quad a_2 = \frac{1}{2}G^2m^2, \quad b_1 = -Gm,$$

$$b_2 = G^2m^2 \left(-\frac{5}{6} + \frac{2}{3\xi} \right), \quad (9.8)$$

$$c_1 = -Gm, \quad c_2 = -G^2m^2 \left(\frac{2}{3} + \frac{2}{3\xi} \right).$$

From (9.2), (9.7), (9.8) and $J^{\mu\nu} = P^{\mu\nu} + g\phi^{\mu\nu}$, we obtain

$$g\phi^{00} = S - 1 = \frac{Gm}{r} + \frac{G^2m^2}{2r^2},$$

$$g\phi^{11} = 1 - R = \frac{Gm}{r} + \frac{G^2m^2}{r^2} \left(\frac{5}{6} - \frac{2}{3\xi} \right),$$

$$g\phi^{22} = \frac{1}{r^2}(1 - T) = \frac{1}{r^2} \left[\frac{Gm}{r} + \frac{G^2m^2}{r^2} \left(\frac{2}{3} + \frac{2}{3\xi} \right) \right],$$

$$g\phi^{33} = \frac{g\phi^{22}}{\sin^2\theta}. \quad (9.9)$$

Thus, only the second order terms in ϕ^{11}, ϕ^{22} and ϕ^{33} depend on the gauge parameter ξ. As expected, the first order terms and the second order terms in ϕ^{00}, i.e., all those crucial to the observable quantities in measuring the perihelion shift of Mercury, are independent of the gauge parameter ξ.

From the result (9.9) and the relations $G^{\mu\nu} = P_{\alpha\beta} J^{\alpha\mu} J^{\beta\nu}$ and $P_{\mu\nu} = (1, -1, -r^2, -r^2 \sin^2 \theta)$, we can now write down the effective metric tensor,

$$G^{00}(r) = 1 + \frac{2Gm}{r} + \frac{2G^2 m^2}{r^2},$$

$$G^{11}(r) = -\left[1 - \frac{2Gm}{r} + \frac{G^2 m^2}{r^2}\left(\frac{-2}{3} + \frac{4}{3\xi}\right)\right],$$

$$G^{22}(r) = -\frac{1}{r^2}\left(1 - \frac{2Gm}{r} - \frac{G^2 m^2}{r^2}\left(\frac{1}{3} + \frac{4}{3\xi}\right)\right),$$

$$G^{33}(r) = G^{22}(r)/\sin^2 \theta. \tag{9.10}$$

These results are well defined in the limit $\xi \to \infty$ and this particular choice of gauge may be called the 'static gravity gauge.' In this gauge, the effective metric tensors are

$$G^{00}(r) = 1 + \frac{2Gm}{r} + \frac{2G^2 m^2}{r^2},$$

$$G^{11}(r) = -\left[1 - \frac{2Gm}{r} - \frac{2G^2 m^2}{3}\right],$$

$$G^{22}(r) = -\frac{1}{r^2}\left(1 - \frac{2Gm}{r} - \frac{G^2 m^2}{3r^2}\right),$$

$$G^{33}(r) = G^{22}(r)/\sin^2 \theta. \tag{9.11}$$

In the calculation of the perihelion shift, we shall use the metric in its usual form in spherical coordinates $x^\mu = (w, \rho, \theta, \phi)$ with $G^{22}(\rho) = -1/\rho^2$. This can be accomplished by a change of variable

$$\rho^2 = r^2 \left(1 - \frac{2Gm}{r} - \frac{G^2 m^2}{r^2}\left[\frac{1}{3} + \frac{4}{3\xi}\right]\right)^{-1}. \tag{9.12}$$

From (9.12), we solve for r to obtain

$$r = \rho B, \quad B \equiv \left[1 - \frac{Gm}{\rho} - \frac{G^2 m^2}{\rho^2}\left(\frac{5}{3} + \frac{2}{3\xi}\right)\right],$$

$$dr = d\rho \left[1 + \frac{G^2 m^2}{\rho^2}\left(\frac{5}{3} + \frac{2}{3\xi}\right)\right]. \tag{9.13}$$

The Einstein–Grossmann equation of motion (8.40) can be expressed in terms of ρ,

$$G^{00}(\rho)\frac{\partial S}{\partial w}\frac{\partial S}{\partial w} + G^{11}(\rho)\frac{\partial S}{\partial \rho}\frac{\partial S}{\partial \rho} + G^{22}(\rho)\frac{\partial S}{\partial \mathrm{b}}\frac{\partial S}{\partial \mathrm{b}} + G^{33}(\rho)\frac{\partial S}{\partial \phi}\frac{\partial S}{\partial \phi} - m^2 = 0,$$

$$G^{00}(\rho) = G^{00}(r)_{r=\rho B}, \quad G^{11}(\rho) = G^{11}(r)\left(\frac{d\rho}{dr}\right)^2_{r=\rho B},$$

$$G^{22}(\rho) = G^{22}(r)_{r=\rho B}, \quad G^{33}(\rho) = G^{33}(r)_{r=\rho B}. \tag{9.14}$$

In spherical coordinates $x^\mu = (w, \rho, \theta, \phi)$, the effective metric tensor $G^{\mu\nu}(\rho)$ in the Einstein–Grossmann Equation (9.14) is

$$G^{00}(\rho) = 1 + \frac{2Gm}{\rho} + \frac{4G^2m^2}{\rho^2},$$

$$G^{11}(\rho) = -\left[1 - \frac{2Gm}{\rho} - \frac{6G^2m^2}{\rho^2}\right],$$

$$G^{22}(\rho) = -\frac{1}{\rho^2}, \quad G^{33}(\rho) = -\frac{1}{\rho^2\sin^2\theta}, \tag{9.15}$$

where we have used (9.13) and (9.14). The gauge parameter ξ in $G^{11}(r)$ from (9.10) and in $(d\rho/dr)^2$ from (9.13) cancel so that $G^{11}(\rho)$ in (9.15) is ξ-independent, in agreement with the gauge invariance of Yang–Mills gravity. Therefore, all components of $G^{\mu\nu}(\rho)$ in spherical coordinates are independent of the gauge parameter ξ to the second-order approximation.

This ξ-independent property of $G^{\mu\nu}(\rho)$ is notable because the complete Yang–Mills theory of gravity at the quantum level is T_4 gauge invariant. However, in the geometric-optics limit, results such as the Einstein–Grossmann equation of motion for classical objects may not, in general, be gauge invariant because the equation of motion (9.14) and the result (9.15) are only approximations. Thus, it is not unreasonable that the effective metric tensors may not be ξ-independent to all orders of Gm/ρ.

9-2.   The perihelion shift of Mercury

To compare the predictions of the classical approximation of Yang–Mills gravity to the observations of the perihelion shift of Mercury, we choose $\theta = \pi/2$ in (9.15) so that the Einstein–Grossmann Equation (9.14) for a planet with mass m_p is

$$G^{00}(\rho)\left(\frac{\partial S}{\partial w}\right)^2 + G^{11}(\rho)\left(\frac{\partial S}{\partial \rho}\right)^2 + G^{33}(\rho)\left(\frac{\partial S}{\partial \phi}\right)^2 - m_p^2 = 0. \quad (9.16)$$

To solve (9.16), we follow the general procedure for solving the Hamilton–Jacobi equation, writing the solution in the form $S = -E_o w + M\phi + f(\rho)$.[2] Solving for $f(\rho)$, we obtain

$$S = -E_o w + M\phi + \int \frac{1}{\sqrt{|G^{11}|(\rho)}}\sqrt{E_o^2 G^{00}(\rho) - m_p^2 - \frac{M^2}{\rho^2}}\,d\rho, \quad (9.17)$$

where E_o and M are the constant energy and angular momentum of the planet, respectively. The trajectory is determined by $\partial S/\partial M = constant$, so that

$$\phi = \int \frac{(M/\rho^2)\,d\rho}{\sqrt{E_o^2 G^{00}|G^{11}| - m_p^2|G^{11}| - M^2|G^{11}|/\rho^2}}. \quad (9.18)$$

To find the trajectory, it is convenient to write (9.18) as a differential equation of $\sigma = 1/\rho$. We obtain

$$\frac{d^2\sigma}{d\phi^2} = \frac{1}{P} - \sigma(1 + Q) + 3Gm\sigma^2, \quad (9.19)$$

$$P = \frac{M^2}{m_p^2 Gm}, \quad Q = \frac{6Gm}{P}\left(\frac{E_o^2 - m_p^2}{m_p^2}\right),$$

by differentiating the equation with respect to ϕ. Thus, the equation for the trajectory (9.19) in Yang–Mills gravity differs slightly from the corresponding result in general relativity by a correction term Q. This correction term Q is of the order of $(Gm/P)\beta^2$ which is undetectable by present observations because the velocity β of the planet is very small compared to the speed of light, $\beta \ll 1$.

By the usual method of successive approximations,[2] we obtain the solution

$$\sigma = \frac{1}{P(1 + Q)}\left[1 + e\cos\left(\phi\left(1 - \frac{3Gm}{P} + \frac{Q}{2}\right)\right)\right]. \quad (9.20)$$

The advance of the perihelion for one revolution of the planet is given by

$$\delta\phi \approx \frac{6\pi Gm}{P}\left(1 - \frac{(E_o^2 - m_p^2)}{4m_p^2} - \frac{2Gm}{3P}\right). \quad (9.21)$$

The second and the third terms on the right-hand side of (9.21) differentiate Yang–Mills gravity from Einstein's general relativity. However, these correction terms are only on the order of 10^{-7} for Mercury and the predictions of both theories fall well within the range of experimental observations (the observational accuracy of the perihelion shift of Mercury is about 1%,).[3] Thus, the predicted perihelion shift in Yang–Mills gravity is independent of the gauge parameter ξ that appears in the second order approximation of the solution of $g\phi^{\mu\nu}$.

9-3. Deflection of light in a gravitational field

The deflection of light can be derived from the equation for the propagation of a light ray in an inertial frame in the geometric-optics limit. In the presence of a gravitational field, the path of a light ray is determined by the eikonal Equation (8.36),

$$G_L{}^{\mu\nu}\partial_\mu\Psi\partial_\nu\Psi = 0,$$

$$G_L{}^{\mu\nu} = \eta_{\alpha\beta}\left(\eta^{\mu\alpha} + g\phi^{\mu\alpha}\right)\left(\eta^{\nu\beta} + \frac{3g}{4}\phi^{\nu\beta}\right), \tag{9.22}$$

in the geometric-optics limit. Following the usual convention, we assume that the motion of the light ray is in the plane defined by $\theta = \pi/2$ in spherical coordinates (w, r, θ, ϕ) (the $x - y$ plane in Cartesian coordinates). Using $G_L^{\mu\nu}$ from (9.22), the static solution (9.9), and $x^\mu = (w, r, \theta, \phi)$ and going through the steps (9.10) to (9.15), the eikonal Equation (9.22) can be written as

$$G_L^{00}(\rho)\left(\frac{\partial\Psi}{\partial w}\right)^2 + G_L^{11}(\rho)\left(\frac{\partial\Psi}{\partial\rho}\right)^2 - \frac{1}{\rho^2}\left(\frac{\partial\Psi}{\partial\phi}\right)^2 = 0,$$

$$G_L^{00}(\rho) = 1 + \frac{7Gm}{4\rho}, \quad G_L^{11}(\rho) = -1 + \frac{7Gm}{4\rho}, \quad G_L^{22} = \frac{-1}{\rho^2}. \tag{9.23}$$

Similar to the general procedure for solving (9.16) in a spherical symmetric tensor field, we look for the eikonal Ψ in the form[2]

$$\Psi = -E_0 w + M\phi + f(\rho). \tag{9.24}$$

One can then determine $f(\rho)$ and solve for the trajectory of the ray, which is similar to (9.19) with $m_p \to 0$ and E_0 replaced by $\omega_o = -\partial\Psi/\partial w$ ($c = 1$). We have

$$\frac{d^2\sigma}{d\phi^2} = -\sigma(1 + Q_o) + \frac{21}{8}Gm\sigma^2, \quad \sigma = \frac{1}{\rho} \tag{9.25}$$

$$Q_o = \frac{49G^2m^2\omega_o^2}{M^2}, \tag{9.26}$$

where the new correction term Q_o that differentiates Yang–Mills gravity from general relativity is extremely small and negligible. Following the usual procedure for calculating the deflection of light due to the sun,[2] we find the following result,

$$\Delta\phi \approx \frac{7Gm}{2R} \approx 1.53'', \quad \frac{\omega_o}{M} = \frac{1}{R}, \tag{9.27}$$

for the deflection of a light ray passing through the spherically symmetric tensor field generated by the sun at a distance R from the center of the sun to the first order approximation. Because the accuracy of observations from experiments to measure the light deflection from the sun is no better than $10 - 20\%$, result (9.27) is consistent with experimental observation s of $\Delta\phi_{exp} \approx 1.75''$.[3]

Although the Yang–Mills result is smaller than the value of $1.75''$ predicted by general relativity by about 12%,[2] this comparison should be made with caution because the calculations in Yang–Mills gravity are carried out in an inertial frame, while the corresponding result in Einstein gravity is not calculated in the inertial frame.[4a]

If the accuracy of the observations of the deflection of light rays by the sun can be improved to the level of a few percent (perhaps using space-based telescopes), then one could distinguish between the predictions of Yang–Mills gravity and general relativity. The significance of such an experiment cannot be over-emphasized since it would determine whether the Yang–Mills idea of gauge symmetry for all interactions or the Einstein idea of general coordinate invariance for gravity (or perhaps neither) stands the test of experimental consistency. The burning questions are (i) whether the Yang–Mills 'gauge curvature' of the local translational group or the general relativistic 'space-time curvature' dictates the gravitational interaction and (ii) whether or not the curvature of space-time revealed by classical tests of gravity is simply a manifestation of the flat space-time translational gauge symmetry for the motion of particles in the classical limit, as suggested by Yang–Mills gravity.

It is important to note that the result (9.27) is valid only for electromagnetic waves in the (high frequency) geometric-optics limit. The eikonal Equation (9.22)

---

[a]Wigner said: "Evidently, the usual statements about future positions of particles, as specified by their coordinates, are not meaningful statements in general relativity. This is a point which cannot be emphasized strongly enough ... Expressing our results in terms of the values of coordinates became a habit with us to such a degree that we adhere to this habit also in general relativity, where values of coordinates are not per se meaningful." (See ref. 4).

has the simple effective metric tensor $G_L^{\mu\nu}$ only for high-frequency waves such as visible light. The results are not valid for experiments involving electromagnetic radiation at radio frequencies. Radar echo experiments to test time dilation due to a gravitational field have used radiation with frequencies in the thousands of megahertz, roughly 10^9 (1/sec),[1] which are too small in comparison with the frequencies of visible light, $\approx 10^{14}$ (1/sec.) and so are not suitable as tests of Yang–Mills gravity. However, if those experiments can be repeated using radiation such as laser light, then they could also be used to test Yang–Mills gravity versus general relativity.

Finally, we note that the gravitational red shift in Yang–Mills gravity is identical to the conventional result. A photon in a gravitational field has kinetic energy $\hbar\omega$ and potential energy $\hbar\omega g\phi^{00}$.[5] In Yang–Mills gravity, the conservation of energy, $\hbar\omega + \hbar\omega g\phi^{00} = $ constant, and the static potentials in (9.9) lead to the usual gravitational red-shift,

$$\omega_2/\omega_1 = \left(1 + g\phi_1^{00}\right) / \left(1 + g\phi_2^{00}\right), \tag{9.28}$$

which has been experimentally confirmed to 1% accuracy.[3]

9-4. Gravitational quadrupole radiation

Although Yang–Mills gravity is formulated for both inertial and non-inertial frames,[6,7] for simplicity, we choose inertial frames, where $P_{\mu\nu} = \eta_{\mu\nu} = (+,-,-,-)$ and $D_\mu = \partial_\mu$, in which to discuss the gravitational quadrupole radiation. We note that the calculations of gravitational quadrupole radiation performed using Einstein gravity are not based in an inertial frame because the framework of Einstein gravity does not allow for inertial frames.

Ordinary matter consists of fermions such as protons and neutrons (or up- and down-quarks). The Lagrangian L involving the gravitational tensor field $\phi_{\mu\nu}$ and a fermion field ψ,

$$L = \frac{1}{2g^2} \left(C_{\mu\alpha\beta} C^{\mu\beta\alpha} - C_{\mu\alpha}{}^{\alpha} C^{\mu\beta}{}_{\beta} \right)$$

$$+ \frac{i}{2} \left[\overline{\psi}\gamma^\mu (\Delta_\mu \psi) - (\Delta_\mu \overline{\psi})\gamma^\mu \psi \right] - m\overline{\psi}\psi, \tag{9.29}$$

$$C^{\mu\nu\alpha} = J^{\mu\lambda}(\partial_\lambda J^{\nu\alpha}) - J^{\nu\lambda}(\partial_\lambda J^{\mu\alpha}), \quad C_{\mu\alpha\beta} C^{\mu\beta\alpha} = (1/2)C_{\mu\alpha\beta} C^{\mu\alpha\beta},$$

$$\Delta_\mu \psi = J_{\mu\nu}\partial^\nu \psi, \quad J_{\mu\nu} = \eta_{\mu\nu} + g\phi_{\mu\nu} = J_{\nu\mu}, \quad c = \hbar = 1. \tag{9.30}$$

For the gravitational quadrupole radiation, we consider only the tensor field $\phi_{\mu\nu}$. Moreover, it suffices to calculate the gravitation radiation to just the second order in $g\phi_{\mu\nu}$. As usual, we impose the gauge condition

$$\partial^\mu \phi_{\mu\nu} = \frac{1}{2}\partial_\nu \phi_\lambda^\lambda, \tag{9.31}$$

which is always possible.[1] The gauge invariant action with the Lagrangian (9.29) leads to the gravitational tensor field equation in inertial frames,[7]

$$H^{\mu\nu} = g^2 S^{\mu\nu}, \tag{9.32}$$

$$H^{\mu\nu} \equiv \partial_\lambda \left(J_\rho^\lambda C^{\rho\mu\nu} - J_\alpha^\lambda C^{\alpha\beta}{}_\beta \eta^{\mu\nu} + C^{\mu\beta}{}_\beta J^{\nu\lambda} \right)$$
$$- C^{\mu\alpha\beta} \partial^\nu J_{\alpha\beta} + C^{\mu\beta}{}_\beta \partial^\nu J_\alpha^\alpha - C^{\lambda\beta}{}_\beta \partial^\nu J_\lambda^\mu, \tag{9.33}$$

where μ and ν should be made symmetric and we have used the identities in (8.14). It is not necessary to write this symmetry of μ and ν in (9.32) explicitly in the following discussions of gravitational radiation. The source tensor $S^{\mu\nu}$ in the gravitational field equation (9.32) is given by

$$S^{\mu\nu} = \frac{1}{2}\left[\overline{\psi}i\gamma^\mu \partial^\nu \psi - i(\partial^\nu \overline{\psi})\gamma^\mu \psi \right]. \tag{9.34}$$

For weak fields in inertial frames with the gauge condition (9.31), the field equation can be linearized as follows:

$$\partial_\lambda \partial^\lambda \phi^{\mu\nu} - \partial^\mu \partial_\lambda \phi^{\lambda\nu} + \partial^\mu \partial^\nu \phi_\lambda^\lambda - \partial^\nu \partial_\lambda \phi^{\lambda\mu} = g\left(S^{\mu\nu} - \frac{1}{2}\eta^{\mu\nu}S_\lambda^\lambda \right), \tag{9.35}$$

where we have used $J^{\mu\nu} = \eta^{\mu\nu} + g\phi^{\mu\nu}$. With the help of the gauge condition (9.31), (9.35) can be written as

$$\partial_\lambda \partial^\lambda \phi_{\mu\nu} = g\left(S_{\mu\nu} - \frac{1}{2}\eta_{\mu\nu}S_\lambda^\lambda \right) \equiv gS_{\mu\nu}^o, \quad g = \sqrt{8\pi G}, \tag{9.36}$$

where G is the Newtonian gravitational constant. To the first-order approximation, we have the static solution,[5]

$$g\phi_{00} = g\phi_{11} = \frac{Gm}{r}, \quad \text{etc.} \tag{9.37}$$

Note that the source tensor $S^{\mu\nu}$ given by (9.34) in Yang–Mills gravity is independent of $\phi^{\mu\nu}$. It satisfies the conservation law,

$$\partial_\mu S^{\mu\nu} = 0, \tag{9.38}$$

in the weak field approximation.

From Equation (9.36), one has the usual retarded potential

$$\phi_{\mu\nu}(\mathbf{x}, t) = \frac{g}{4\pi} \int d^3x' \frac{S^o_{\mu\nu}(\mathbf{x}', t - |\mathbf{x} - \mathbf{x}'|)}{|\mathbf{x} - \mathbf{x}'|},$$

$$\mathbf{x} \equiv \mathbf{r}, \quad \mathbf{x}' \equiv \mathbf{r}', \quad x^\mu = (w, x, y, z), \quad w = ct = t, \tag{9.39}$$

which is generated by the source $S_{\mu\nu}$ in (9.36). This equation is usually used to discuss the gravitational radiation. When discussing the radiation in the wave zone at a distance much larger than the dimension of the source, the solution can be approximated by a plane wave,[1]

$$\phi_{\mu\nu}(x) = e_{\mu\nu} \exp\left(-ik_\lambda x^\lambda\right) + e^*_{\mu\nu} \exp\left(ik_\lambda x^\lambda\right), \tag{9.40}$$

where $e_{\mu\nu}$ is the polarization tensor. The plane wave property and the usual gauge condition (9.31) lead to

$$k_\mu k^\mu = 0, \quad k_\mu e^\mu_\nu = \frac{1}{2}k_\nu e^\mu_\mu, \quad k^\mu = \eta^{\mu\nu}k_\nu. \tag{9.41}$$

For the symmetric polarization tensor, $e_{\mu\nu} = e_{\nu\mu}$, of the massless tensor field in flat space-time, there are only two physical states with helicity ± 2 that are invariant under the Lorentz transformation.

Let us write $S_{\mu\nu}(\mathbf{x}, t)$ in terms of a Fourier integral,[1]

$$S_{\mu\nu}(\mathbf{x}, t) = \int_0^\infty S_{\mu\nu}(\mathbf{x}, \omega)e^{-i\omega t}\,d\omega + c.c. \tag{9.42}$$

The retarded field emitted by a single Fourier component $S_{\mu\nu}(\mathbf{x}, t) = [S_{\mu\nu}(\mathbf{x}, \omega)e^{-i\omega t} + c.c.]$ is given by

$$\phi_{\mu\nu}(\mathbf{x}, t) = \frac{g}{4\pi} \int \frac{d^3x'}{\mathbf{x} - \mathbf{x}'} S^o_{\mu\nu}(\mathbf{x}', \omega) \exp\left(-i\omega t + i\omega|\mathbf{x} - \mathbf{x}'|\right) + c.c.$$

$$S^o_{\mu\nu}(\mathbf{x}, \omega) = S_{\mu\nu}(\mathbf{x}, \omega) - \frac{1}{2}\eta_{\mu\nu}S^\lambda{}_\lambda(\mathbf{x}, \omega). \tag{9.43}$$

The source tensor $s^{\mu\nu}$ of gravitation is defined by the exact field Equation (9.32) written in the following form,

$$\partial^\lambda \partial_\lambda \phi^{\mu\nu} = g(S^{\mu\nu} - s^{\mu\nu}). \tag{9.44}$$

Thus we have

$$s^{\mu\nu} = \frac{1}{g^2}\Big[C^{\rho\mu\nu}\partial_\lambda J^\lambda_\rho + \partial_\rho\left(g\phi^{\rho\lambda}\partial_\lambda J^{\mu\nu}\right) + g\phi^\lambda_\rho\partial_\lambda\left(J^{\rho\alpha}\partial_\alpha J^{\mu\nu}\right)$$

$$- J^\lambda_{\ \rho}\partial_\lambda\left(J^{\mu\alpha}\partial_\alpha J^{\rho\nu}\right) - C^{\mu\beta\alpha}\partial^\nu J_{\alpha\beta} - \eta^{\mu\nu}\partial_\lambda\left(J^\lambda_{\ \rho}C^{\rho\beta}_{\ \ \beta}\right)$$

$$+ \partial_\lambda\left(C^{\mu\beta}_{\ \ \beta}J^{\nu\lambda}\right) + C^{\mu\beta}_{\ \ \beta}\partial^\nu J^\lambda_{\ \lambda} - C^{\lambda\beta}_{\ \ \beta}\partial^\nu J^\mu_{\ \lambda}\Big]. \tag{9.45}$$

To the second order then, we obtain the 'source tensors' of the gravitational field,

$$s^{\mu\nu} = s_1^{\mu\nu} + s_2^{\mu\nu}, \tag{9.46}$$

$$s_1^{\mu\nu} = (\partial_\lambda\phi)\partial^\lambda\phi^{\mu\nu} - \frac{1}{2}(\partial_\lambda\phi)\partial^\mu\phi^{\lambda\nu} + 2\phi^{\lambda\sigma}\partial_\lambda\partial_\sigma\phi^{\mu\nu}$$

$$- (\partial_\lambda\phi^{\mu\sigma})\partial_\sigma\phi^{\lambda\nu} - \phi^\lambda_\rho\partial_\lambda\partial^\mu\phi^{\rho\nu} - \frac{1}{2}\phi^{\mu\sigma}\partial_\sigma\partial^\nu\phi$$

$$- (\partial^\mu\phi^{\beta\alpha})\partial^\nu\phi_{\alpha\beta} + (\partial^\beta\phi^{\mu\alpha})\partial^\nu\phi_{\alpha\beta}, \tag{9.47}$$

$$s_2^{\mu\nu} = -\frac{3}{4}(\partial_\lambda\phi)\partial^\lambda\phi\eta^{\mu\nu} - \phi^{\lambda\sigma}\partial_\lambda\partial_\sigma\phi\eta^{\mu\nu} + (\partial_\lambda\phi^{\beta\sigma})\partial_\sigma\phi^\lambda_{\ \beta}\eta^{\mu\nu}$$

$$+ \frac{1}{2}(\partial^\nu\phi^{\mu\lambda})\partial_\lambda\phi + \frac{1}{2}\phi^{\nu\lambda}\partial^\mu\partial_\lambda\phi + \phi^{\mu\sigma}\partial_\sigma\partial^\nu\phi$$

$$- (\partial^\nu\phi^{\beta\sigma})\partial_\sigma\phi^\mu_{\ \beta} - \phi^{\beta\sigma}\partial_\sigma\partial^\nu\phi^\mu_{\ \beta} + \frac{3}{4}(\partial^\nu\phi)\partial^\mu\phi, \tag{9.48}$$

where $\phi \equiv \phi^\lambda_\lambda$, and we have used the gauge condition (9.31). The gravitational source tensors $s_1^{\mu\nu}$ and $s_2^{\mu\nu}$ arise from the first and the second quadratic gauge-curvatures, $C_{\mu\alpha\beta}C^{\mu\beta\alpha}$ and $-C_{\mu\alpha}^{\ \ \alpha}C^{\mu\beta}_{\ \ \beta}$ respectively, in the Lagrangian (9.29). One can then use the plane wave solution (9.40) and the gauge condition (9.41) to calculate the gravitational source tensors (9.46) in an inertial frame. This complicated result can be simplified by taking the average of $s^{\mu\nu}$ over a region of space and time much larger than the wavelengths of the radiated waves.[1] After such an averaging process, one obtains the following results:

$$\langle s_1^{\mu\nu}\rangle = -2k^\mu k^\nu e^{\lambda\rho}e^*_{\lambda\rho} + \frac{1}{2}k^\mu k^\nu e^\lambda_\lambda e^{*\alpha}_\alpha,$$

$$\langle s_2^{\mu\nu}\rangle = \frac{1}{2}k^\mu k^\nu e^\lambda_\lambda e^{*\alpha}_\alpha. \tag{9.49}$$

In the wave zone approximation, one can write the polarization tensor in terms of the Fourier transform of $S_{\mu\nu}$:

$$e_{\mu\nu}(\mathbf{x},\omega) = \frac{g}{4\pi r}\Big[S_{\mu\nu}(\mathbf{k},\omega) - \frac{1}{2}\eta_{\mu\nu}S^\lambda_\lambda(\mathbf{k},\omega)\Big], \tag{9.50}$$

$$S_{\mu\nu}(\mathbf{k}, \omega) \equiv \int d^3\mathbf{x}' [S_{\mu\nu}(\mathbf{x}', \omega)] \exp(-i\mathbf{k} \cdot \mathbf{x}'), \tag{9.51}$$

$$\phi_{\mu\nu}(\mathbf{x}, t) \approx e_{\mu\nu}(\mathbf{x}, \omega) \exp(-ik_\lambda x^\lambda) + c.c., \tag{9.52}$$

where we have used (9.41) and (9.42) with the wave zone approximation $|\mathbf{x} - \mathbf{x}'| \approx r - \mathbf{x}' \cdot \mathbf{x}/|\mathbf{x}|$ and $\mathbf{k} = \omega\mathbf{x}/|\mathbf{x}|$. Thus, the average source tensor of a gravitational plane wave can be written as

$$\langle s^{\mu\nu} \rangle = -\frac{G}{\pi r^2} k^\mu k^\nu \left(S^{\lambda\rho}(\mathbf{k}, \omega) S^*_{\lambda\rho}(\mathbf{k}, \omega) - \frac{1}{2} S^\lambda_\lambda(\mathbf{k}, \omega) S^{*\sigma}_\sigma(\mathbf{k}, \omega) \right). \tag{9.53}$$

The power P_o emitted per unit solid angle in the direction $\mathbf{x}/|\mathbf{x}|$ is[1]

$$\frac{dP_o}{d\Omega} = r^2 \frac{x^i \langle s_{i0} \rangle}{|\mathbf{x}|}, \tag{9.54}$$

which can be written in terms of $S_{\mu\nu}(\mathbf{k}, \omega)$ in (9.51),

$$\frac{dP_o}{d\Omega} = \frac{G\omega^2}{\pi} \left(S^{\lambda\rho}(\mathbf{k}, \omega) S^*_{\lambda\rho}(\mathbf{k}, \omega) - \frac{1}{2} S^\lambda_\lambda(\mathbf{k}, \omega) S^{*\sigma}_\sigma(\mathbf{k}, \omega) \right). \tag{9.55}$$

Although the energy-momentum tensor of the gravitational field (9.46) in Yang–Mills gravity is quite different from that in general relativity, the result (9.55) for the power emitted per solid angle turns out to be the same as that obtained in general relativity and consistent with experimental observations.[1,8] Following the usual method and using the standard approximations, one can calculate the power radiated by a body rotating around one of the principal axes of the ellipsoid of inertia.[1] At twice the rotating frequency Ω, i.e., $\omega = 2\Omega$, the total power $P_o(\omega)$ emitted by a rotating body is

$$P_o(2\Omega) = \left[\frac{32G\Omega^6 I^2 e_q^2}{5} \right], \quad c = 1, \tag{9.56}$$

where I and e_q are the moment of inertia and equatorial ellipticity, respectively. Thus, to the second order approximation, the gravitational quadrupole radiation (9.56) predicted by Yang–Mills gravity is also the same as that predicted by general relativity.[1]

References

1. S. Weinberg, *Gravitation and Cosmology* (John Wiley and Sons, 1972), pp. 79–80, pp. 175–210 and pp. 251–272.

2. For the procedure of calculations, see L. Landau and E. Lifshitz, *The Classical Theory of Fields* (Trans. by M. Hamermesh, Addison-Wesley, Cambridge, Mass. 1951), pp. 276–277 and pp. 312–316. J. P. Hsu, *Eur. Phys. J. Plus*, (2013) **128**, 31, DOI 10.1140/epjp/i2013-1303-3.

3. Wei-Tou Ni, in *100 Years of Gravity and Accelerated Frames, The Deepest Insights of Einstein and Yang–Mills* (Ed. J. P. Hsu and D. Fine, World Scientific, 2005) p. 478 and p. 481. For a recent light deflection experiment, see E. B. Fomalont and S. M. Kopeikin, *Astrophys. J.* **598**, 704 (2003).

4. E. P. Wigner, *Symmetries and Reflections, Scientific Essays* (The MIT Press, 1967), pp. 52–53.

5. S. Weinberg, ref. 1, p. 85.

6. J. P. Hsu and L. Hsu, *A Broader View of Relativity, General Implications of Lorentz and Poincaré Invariance* (World Scientific, 2006), Appendix D. Available online: 'google books, a broader view of relativity, Hsu.'

7. J. P. Hsu, *Int. J. Mod. Phys. A* **21**, 5119 (2006).

8. J. H. Taylor, *Rev. Mod. Phys.* **66**, 771 (1994).

10.

The S-Matrix in Yang–Mills Gravity

10-1.   The gauge-invariant action and gauge conditions

Beginning with this chapter, we discuss a quantum theory of gravity under the Yang–Mills–Utiyama–Weyl framework. Although Yang–Mills gravity can be formulated in both inertial and non-inertial frames and in the presence of both fermion and scalar fields,[1] it is difficult to discuss quantum field theory and particle physics even in the simplest non-inertial frame.[2-4] For that reason, we consider quantum Yang–Mills gravity in inertial frames with the Minkowski metric tensor $\eta_{\mu\nu} = (1, -1, -1, -1)$ and without involving other fields. In this chapter, we demonstrate some basic physical properties of quantum Yang–Mills gravity, namely, that the S-matrix satisfies both the conditions of unitarity and gauge invariance. The action S_{pg} for pure gravity (i.e., involving only the space-time gauge field $\phi_{\mu\nu}$ with the gauge-fixing terms $L_{\xi\zeta}$) is assumed to be[1]

$$S_{pg} = \int (L_\phi + L_{\xi\zeta}) d^4 x, \tag{10.1}$$

$$L_\phi = \frac{1}{4g^2} \left(C_{\mu\nu\alpha} C^{\mu\nu\alpha} - 2 C^\alpha_{\mu\alpha} C^{\mu\beta}_\beta \right), \tag{10.2}$$

$$L_{\xi\zeta} = \frac{\xi}{2g^2} \left[\partial^\mu J_{\mu\alpha} - \frac{\zeta}{2} \partial_\alpha J \right] \left[\partial^\nu J_{\nu\beta} - \frac{\zeta}{2} \partial_\beta J \right] \eta^{\alpha\beta}, \tag{10.3}$$

$$C^{\mu\nu\alpha} = J^{\mu\sigma} \partial_\sigma J^{\nu\alpha} - J^{\nu\sigma} \partial_\sigma J^{\mu\alpha}, \quad J_{\mu\nu} = \eta_{\mu\nu} + g\phi_{\mu\nu} = J_{\nu\mu}, \tag{10.4}$$

$$J = J^\lambda_\lambda = \delta^\lambda_\lambda - g\phi, \quad \phi = \phi^\lambda_\lambda, \quad c = \hbar = 1,$$

where $C^{\mu\alpha\beta}$ is the T_4 gauge curvature for inertial frames. In quantum Yang–Mills gravity, it is necessary to include the gauge fixing terms $L_{\xi\zeta}$ in the action functional (10.1), where $L_{\xi\zeta}$ takes the form (10.3) and may involve two arbitrary gauge parameters ξ and ζ in general. We note that the gauge-fixing term $L_{\xi\zeta}$ in (10.3) is the same (i.e., involves ordinary partial derivatives) in both inertial and non-inertial frames,V so that $L_{\xi\zeta}$ is not gauge invariant.[5-7] However, in a non-inertial frame, the constant metric tensor $\eta_{\mu\nu}$ and ∂_μ in the Lagrangian L_ϕ given by (10.2) are replaced by a space-time-dependent Poincaré metric tensor $P_{\mu\nu}$ and the associated covariant derivative D_μ, respectively. In this way, the Lagrangian $L_\phi\sqrt{-P}$ in a general frame changes only by a divergence under T_4 gauge transformations and hence, the action $S_\phi = \int L_\phi\sqrt{-P}d^4x$ is T_4 gauge invariant.[1]

The Lagrangian $L_{\xi\zeta}$ in (10.3) corresponds to a class of gauge conditions of the following form,

$$\partial^\lambda J_{\rho\lambda} - \frac{\zeta}{2}\partial_\rho J = Y_\rho, \tag{10.5}$$

where $Y_\rho = Y_\rho(x)$ is independent of the fields and the gauge function Λ^μ. For simplicity, we concentrate on two cases, $\zeta = 1$ and $\zeta = 0$ (with ξ arbitrary) in the following discussion. The Lagrangian for pure gravitational field, $L_{pg} = L_\phi + L_{\xi\zeta}$, can be expressed in terms of space-time gauge fields $\phi_{\mu\nu}$,

$$L_{pg} = L_2 + L_3 + L_4 + L_{\xi\zeta}, \tag{10.6}$$

where

$$L_2 = \frac{1}{2}(\partial_\lambda\phi_{\alpha\beta}\partial^\lambda\phi^{\alpha\beta} - \partial_\lambda\phi_{\alpha\beta}\partial^\alpha\phi^{\lambda\beta} - \partial_\lambda\phi\partial^\lambda\phi$$
$$+ 2\partial_\lambda\phi\partial^\beta\phi^\lambda_\beta - \partial_\lambda\phi^\lambda_\mu\partial^\beta\phi_{\mu\beta}), \tag{10.7}$$

$$L_{\xi\zeta} = \frac{\xi}{2g^2}\left(\partial_\lambda J^{\lambda\alpha} - \frac{\zeta}{2}\partial^\alpha J\right)\left(\partial^\rho J_{\rho\alpha} - \frac{\zeta}{2}\partial_\alpha J\right)$$
$$= \frac{\xi}{2}\left[(\partial_\lambda\phi^{\lambda\alpha})\partial^\rho\phi_{\rho\alpha} - \zeta(\partial_\lambda\phi^{\lambda\alpha})\partial_\alpha\phi + \frac{\zeta^2}{4}(\partial^\alpha\phi)\partial_\alpha\phi\right]. \tag{10.8}$$

The Lagrangians L_2 and $L_{\xi\zeta}$ involve terms quadratic in $\phi_{\mu\nu}$ and determine the propagator of the graviton in Yang–Mills gravity. The Lagrangians L_3 and L_4 correspond to the interactions of 3 and 4 gravitons respectively and will be discussed in Chapter 11.

10-2. Feynman–DeWitt–Mandelstam (FDM) ghost fields in Yang–Mills gravity

Because Yang–Mills gravity has nonlinear field equations, the gauge condition (10.5) cannot be imposed for all time and the theory will have gauge non-invariant amplitudes. One must introduce 'ghost fields' to cancel these unwanted gauge non-invariant amplitudes, so that the theory, and in particular the S-matrix, is gauge invariant. The ghost fields and their interactions with the graviton in Yang–Mills gravity can be obtained by the Faddeev–Popov method.[8,9] Let us choose the gauge condition (10.5) for the following discussions. It is important to symmetrize the indices of the symmetric tensor field. Thus, we write the gauge condition (10.5) in the following form where the indices μ and ν of $J_{\mu\nu}$ have been explicitly symmetrized:

$$Y_\lambda = \frac{1}{2}(\delta^\mu_\lambda \partial^\nu + \delta^\nu_\lambda \partial^\mu - \zeta\eta^{\mu\nu}\partial_\lambda)J_{\mu\nu}. \tag{10.9}$$

The vacuum-to-vacuum amplitude is defined as the probability amplitude that the physical system will be in the vacuum state at time $w = \infty$, when it is known to be in the vacuum state at time $w = -\infty$. This is similar to that in the usual gauge theory.[9,10] It is convenient to use the method of path integrals to calculate the vacuum-to-vacuum amplitude, which completely determines the interactions of a physical system.[10] The vacuum-to-vacuum amplitude (with an external source $j^{\mu\nu}$) of quantum Yang–Mills gravity is given by

$$W_Y[j] \equiv \int d[J_{\rho\sigma}] \left(\exp\left[i \int d^4x \, (L_\phi + J_{\mu\nu}j^{\mu\nu}) \right] \right.$$

$$\left. \times (\det U) \prod_{\alpha,x} \delta\left(\partial^\lambda J_{\lambda\alpha} - \frac{1}{2}\partial_\alpha J^\lambda_\lambda - Y_\alpha \right) \right). \tag{10.10}$$

The functional determinant $\det U$ in (10.10) is determined by

$$\frac{1}{\det U} = \int d\left[\Lambda^\lambda\right] \Pi_{x,\nu} \delta\left(\frac{1}{2}\left(\delta^\mu_\lambda \partial^\nu + \delta^\nu_\lambda \partial^\mu - \zeta\eta^{\mu\nu}\partial_\lambda\right) J^\$_{\mu\nu} - Y_\lambda \right), \tag{10.11}$$

$$J^\$_{\mu\nu} = J_{\mu\nu} - \Lambda^\lambda\partial_\lambda J_{\mu\nu} - \left(\partial_\mu\Lambda^\lambda\right)J_{\lambda\nu} - \left(\partial_\nu\Lambda^\lambda\right)J_{\mu\lambda}.$$

It follows from (10.11) that

$$U_{\mu\nu} = (\partial^\lambda E_{\mu\nu\lambda}) + E_{\mu\nu\lambda}\partial^\lambda, \tag{10.12}$$

where

$$E_{\mu\nu\lambda} = J_{\mu\nu}\partial_\lambda + J_{\nu\lambda}\partial_\mu + \left(\partial_\nu J_{\mu\lambda}\right) - \zeta\eta_{\mu\lambda}J_{\nu\sigma}\partial^\sigma - \frac{\zeta}{2}\eta_{\mu\lambda}(\partial_\nu J).$$

Since the vacuum-to-vacuum amplitude $W_Y[j]$ in (10.10) is invariant under an infinitesimal change of $Y^\alpha(x)$ for all $Y^\alpha(x)$, we may write (10.10) in the form

$$W[j] = \int d[Y^v] W_Y[j] \exp\left[i \int d^4x \left(\frac{\xi}{2g^2} Y^\mu Y_\mu\right)\right]$$

$$= \int d[J_{\alpha\beta}](\det U) \exp\left[i \int d^4x (L_\phi + L_{\xi\zeta} + J_{\mu\nu} j^{\mu\nu})\right], \quad (10.13)$$

to within an unimportant multiplicative factor.

As usual, the functional determinant $(\det U)$ in (10.11) can be expressed in terms of a Lagrangian L_{gho} for ghost fields V^μ and $\overline{V}^\mu$,

$$\det(U_{\mu\nu}) = \int \exp\left(i \int L_{gho}\, d^4x\right) d[V(x)^\lambda, \overline{V}(x)^\sigma], \quad (10.14)$$

$$\int L_{gho}\, d^4x = \int \overline{V}^\mu U_{\mu\nu} V^\nu\, d^4x = -\int (\partial_\lambda \overline{V}^\mu) E_{\mu\nu\lambda} V^\nu\, d^4x$$

$$= \int \left[\overline{V}^\mu \left[\partial^2 \eta_{\mu\nu} + (1-\zeta)\partial_\mu \partial_\nu\right] V^\nu\right.$$

$$- g\left(\partial^\lambda \overline{V}^\mu\right) \phi_{\mu\nu}\partial_\lambda V^\nu - g\left(\partial^\lambda \overline{V}^\mu\right) \phi_{\lambda\nu}\partial_\mu V^\nu$$

$$- g\left(\partial^\lambda \overline{V}^\mu\right) (\partial_\nu \phi_{\mu\lambda}) V^\nu + \zeta g\left(\partial_\mu \overline{V}^\mu\right) \phi_{\sigma\nu}\partial^\sigma V^\nu$$

$$\left. + \frac{\zeta}{2} g\left(\partial_\mu \overline{V}^\mu\right) (\partial_\nu \phi_\lambda^\lambda) V^\nu\right] d^4x, \quad (10.15)$$

where $U_{\mu\nu}$ is given by (10.12) and $\overline{V}^\nu$ is considered to be an independent field.

The quanta of the ghost fields V^μ and $\overline{V}^\nu$ in the Lagrangian L_{gho} are the ghost (or fictitious) particles in Yang–Mills gravity. These ghost fields are vector fields and are quantized in the same way as fermion fields. Such 'ghost vector-fermions' obey Fermi–Dirac statics, and were discussed in quantum Einstein gravity by Feynman, DeWitt and Mandelstam.[11] By definition, they do not exist in initial and final states of a physical process. They interact with themselves and other physical particles and can only appear in the intermediate states of a process. Their interactions contribute extra amplitudes to remove the unwanted amplitudes due to the interaction of unphysical components of gauge fields.[a] Therefore, they are

---

[a]In general, the unphysical components of a gauge field depend on gauge conditions or gauge parameters. Their interactions generate amplitudes that depend on gauge parameters and also upset the unitarity of the S-matrix.

crucial for unitarity and gauge invariance of the S-matrix in any quantum theory of gravity and may be called Feynman–DeWitt–Mandelstam (FDM) ghosts in quantum Yang–Mills gravity.

Thus, the vacuum-to-vacuum amplitude (10.13) with (det U) given by (10.14) indicates that Yang–Mills gravity, including the gravitational and the ghost fields, is completely determined by the Lagrangian L_{tot},

$$L_{tot} = L_\phi + L_{\xi\zeta} + \overline{V}^\mu U_{\mu\nu} V^\nu. \tag{10.16}$$

The Feynman–Dyson rules in Yang–Mills gravity for Feynman diagrams can be derived from this total Lagrangian. The last term $\overline{V}^\mu U_{\mu\nu} V^\nu$ is the FDM ghost Lagrangian L_{gho}, which is obtained by considering the T_4 gauge transformation of the gauge condition as shown in (11.12).

10-3.   Unitarity and gauge invariance of the S-matrix and FDM ghost particles

We now give some arguments and a proof for the unitarity of the S-matrix in Yang–Mills gravity based on the total Lagrangian (10.16), similar to those of Fradkin and Tyutin for the S-matrix in Einstein gravity.[12] With the gauge condition in (10.9), one can write the FDM ghost field $V^\mu(x)$ in the following form,[12]

$$V^\mu(x) = \int d^4y \, \mathbf{D}^\mu_\nu(x, y, \phi_{\alpha\beta}) \, \hat{V}^\nu(y). \tag{10.17}$$

The total Lagrangian (10.16) implies that the FDM ghost field V^μ satisfies the equation

$$U_{\mu\nu} V^\nu = 0, \tag{10.18}$$

where $U_{\mu\nu}$ is given in (10.12). Clearly, in the limit of zero coupling strength, $g \to 0$, one has $J_{\mu\nu} \to \eta_{\mu\nu}$. Thus, the operator $U_{\mu\nu}$ reduces to a non-singular differential operator in this limit,

$$U_{\mu\nu} \to \eta_{\mu\nu} \partial_\lambda \partial^\lambda \equiv U^0_{\mu\lambda}, \quad \zeta = 1. \tag{10.19}$$

This limiting property can be seen from (10.12). One can choose the function $\mathbf{D}^\mu_\nu(x, y, \phi_{\alpha\beta})$ in Equation (10.17) to have the specific form

$$\mathbf{D}^\mu_\nu = \left[U^{-1} \right]^{\mu\lambda} \overleftarrow{U}^0_{\lambda\nu}, \tag{10.20}$$

so that $\hat{V}^\mu$ satisfies the free field equation,

$$U^0_{\lambda\mu}\hat{V}^\mu = [\partial^\sigma\partial_\sigma\eta_{\lambda\mu}]\hat{V}^\mu = 0. \tag{10.21}$$

In gauge invariant gravity, the generating functional for Green's functions can be defined after the gauge condition is specified. Similarly, in Yang–Mills gravity the generating functional for the Green's functions (or the vacuum-to-vacuum amplitude) (10.13) can be written as

$$W[j] = \int d[J_{\alpha\beta}] \exp\left[i\int d^4x \left(L_\phi + \frac{\xi}{2g^2}F_\mu F_\nu \eta^{\mu\nu} + J_{\mu\nu}j^{\mu\nu}\right) + \text{Tr}\ln U\right], \tag{10.22}$$

$$F_\alpha \equiv \eta^{\rho\lambda}\partial_\rho J_{\lambda\alpha} - \frac{1}{2}\eta^{\lambda\rho}\partial_\alpha J_{\lambda\rho}, \quad \det U = \exp(\text{Tr}(\ln U)),$$

where the external sources $j^{\mu\nu}$ are arbitrary functions. It follows from (10.21) and (10.22) that the S-matrix corresponding to the generating functional (10.22) is unitary and gauge invariant.[12,13] There is no physical difference between $\text{Tr}(\ln U)$ and $\text{Tr}(\ln U)(\overleftarrow{U}^0)^{-1}$ because U^0 is a free operator. The presence of U^0 in, for example, (10.22) contributes only an unimportant multiplicative factor to the generating functional $W[j]$.

The last term in (10.22) can be written in terms of the FDM ghost fields $V^\alpha(x)$ and $\overline{V}^\beta(x)$,[13,14]

$$\exp(\text{Tr}(\ln U)) = \int d\left[V^\alpha, \overline{V}^\beta\right] \exp\left(i\int L_{gho}d^4x\right), \tag{10.23}$$

where the ghost Lagrangian L_{gho} (given in (10.15)) describes the FDM ghosts associated with the gauge specified in (10.9). Note that $\overline{V}^\mu$ is considered to be an independent field. The quanta of the fields V^μ and $\overline{V}^\mu$ in the Lagrangian L_{gho} are the FDM ghost particles, which are vector-fermions. By definition of the physical state subspace for the S-matrix, these FDM ghost particles do not exist in the external states. They can only appear in the intermediate steps of a physical process.[13,14] The T(4) gauge field equations, $H^{\mu\nu} = 0$, for gravitational fields also hold in the physical subspace.

10-4.   Discussion

For a more intuitive and physical understanding of the process by which ghost fields can be used to restore the unitarity and gauge invariance of the action, let us consider quantum electrodynamics (QED) with a more complicated non-linear gauge condition involving an artificial 'new' photon coupling,[14]

$$\partial_\mu A^\mu - \beta' A_\mu A^\mu = a(x), \quad a(x) : \text{a suitable function.} \tag{10.24}$$

Roughly speaking, the unitarity of the S-matrix in QED hinges on whether the unphysical component, $\chi = \partial_\mu A^\mu$, of the gauge field A^μ satisfies the free equation

$$\partial_\mu \partial^\mu \chi = 0. \tag{10.25}$$

This is indeed the case if one imposes the usual linear gauge condition in QED for quantization. However, suppose one instead imposes the non-linear gauge condition (10.24), which one is free to do. Now, $\chi = \partial_\mu A^\mu$ no longer satisfies the free equation. Instead,

$$(\partial_\mu - 2\beta' A_\mu)\partial^\mu a(x) \equiv M a(x) = 0. \tag{10.26}$$

This equation implies that the unphysical components of A^μ have a new interaction with A_μ. In other words, the non-linear gauge condition (10.24) cannot be imposed for all times.[14] Such an unwanted new interaction produces extra unphysical amplitudes that violate gauge invariance and hence, the unitarity of the S matrix. These unphysical amplitudes must be removed from the theory to restore the gauge invariance and unitarity of the S-matrix in QED. The removal of these unphysical amplitudes is accomplished by the presence of

$$\det M = \exp[\text{Tr}\,(\ln M)] \tag{10.27}$$

in the generating functional of Green's function. In essence, the non-free operator M is 'subtracted' from QED with the non-linear gauge condition (10.24). As a result, there are no more amplitudes that violate unitarity and gauge invariance in QED. Effectively, the unphysical components of the photon satisfy the free equation, so that the S-matrix in QED with a non-linear gauge condition becomes unitary and gauge invariant again.

In quantum Yang–Mills gravity, the delta function in (10.10) may be understood as a way to impose the gauge condition (10.9) at all times. This is equivalent to the presence of det U, given by (10.11), in the theory to remove gauge non-invariant amplitudes and to restore the unitarity of the S-matrix.

For the quantization of gauge fields with internal gauge groups and the unitarity of the S-matrix, it is convenient to use the method of Lagrange multipliers, which may be understood as the unphysical components of gauge fields.[12,14] If a

Lagrange multiplier satisfies a free equation, then the unphysical components of gauge fields do not interact and the S-matrix is unitary. Equivalently, the chosen gauge condition can be imposed for all times. In this case, the ghost particles in the theory satisfy the free equations and can be ignored. Quantum electrodynamics with a linear gauge condition is a well-known gauge theory in which ghost particles can be ignored.

If, on the other hand, the Lagrange multipliers do not obey free equations, then the unphysical components produce extra unphysical amplitudes to violate unitarity. In this case, gauge invariance will also be violated because the unphysical components of the gauge fields are gauge dependent. Consequently, one needs to use FDM ghosts to restore unitarity and gauge invariance.[15]

References

1. J. P. Hsu and L. Hsu, *A Broader View of Relativity, General Implications of Lorentz and Poincaré Invariance* (World Scientific, 2006), Appendix D. Available online: 'google books, a broader view of relativity, Hsu'; J. P. Hsu, *Intl. J. Mod Phys. A* **21** 5119 (2006).

2. J. P. Hsu, *Chin. J. Phys.* **40**, 265 (2002).

3. D. Schmidt and J. P. Hsu, *Int. J. Mod. Phys. A* **20**, 5989 (2005).

4. J. P. Hsu and D. Fine, *Int. J. Mod. Phys. A* **20**, 7485 (2005).

5. The gauge-fixing terms are not covariant because they are introduced to fix the gauge, i.e., to remove the ambiguity in the tensor field equation due to the T_4 gauge symmetry.

6. Without this gauge fixing term (10.3), one cannot find a well-defined static solution, as one can see from equation (9.9). The solution (9.9) diverges as $\xi \to 0$, i.e., in the absence of the gauge-fixing terms (10.3).

7. In electrodynamics, where the Maxwell equations have U_1 gauge symmetry, one can remove the ambiguity in their solution by choosing a particular gauge, e.g., the Coulomb gauge.

8. V. N. Popov, *Functional Integrals in Quantum Field Theory and Statistical Physics* (trans. from Russian by J. Niederle and L. Hlavaty, D. Reidel Publishing Comp., Boston, 1983) Chapters 2 and 3.

9. L. D. Faddeev and V. N. Popov, in *100 Years of Gravity and Accelerated Frames, The Deepest Insights of Einstein and Yang–Mills* (Ed. J. P. Hsu and D. Fine, World Scientific, 2005) p. 325.

10. K. Huang, *Quarks, Leptons & Gauge Fields* (World Scientific, 1982), pp. 127–147.

11. R. P. Feynman, in *100 Years of Gravity and Accelerated Frames, The Deepest Insights of Einstein and Yang–Mills* (Ed. J. P. Hsu and D. Fine, World Scientific, 2005) pp. 272–297; B. S. DeWitt, *ibid.*, pp. 298–324; S. Mandelstam, *Phys. Rev.* **175**, 1604 (1968).

12. F. S. Fradkin and I. V. Tyutin, *Phys. Rev. D* **2**, 2841 (1970).

13. J. P. Hsu and J. A. Underwood, *Phys. Rev. D* **12**, 620 (1975) and references therein. Since the vacuum-to-vacuum amplitude (10.22) is unitary, it can only be a phase factor, which is the generating functional for connected Green's functions.

14. J. P. Hsu, *Phys. Rev. D* **8**, 2609 (1973); J. P. Hsu and E. C. G. Sudarshan, *Nucl. Phys. B* **91** 477 (1975).

15. In Yang–Mills gravity, the direct calculations of the equation of motion for the Lagrange multiplier are much more complicated, making it more difficult to obtain the corresponding FDM ghost Lagrangian. The Faddeev and Popov method used in sections 10-2 and 10-3 is easier.

11.

Quantization of Yang–Mills Gravity and Feynman–Dyson Rules

11-1. A gauge invariant action for gravitons and fermions and their field equations

In this chapter, we derive rules for Feynman diagrams for quantum Yang–Mills gravity. In the canonical quantization of fields, one uses the Schroedinger field operators and assumes equal-time commutation relations for them. Suppose the eigenstates of a field operator $\phi_{op}(\mathbf{r}) \equiv \phi_{op}^{\mu\nu}(\mathbf{r})$ are denoted by $|\phi\rangle$. One can use the eigenstates to define the transition amplitude, $\langle \phi_2, w_2|\phi_1, w_1\rangle$, which determines completely the dynamical interactions of the physical system and hence, one has a quantum field theory. Alternatively, one could also use the method of path integrals to express the transition amplitude $\langle \phi_2, w_2|\phi_1, w_1\rangle$ in terms of the classical Lagrangian density without using the Schroedinger field operators and eigenstates of field operators.[1] For the S-matrix of gravitational fields, it suffices to find the vacuum-to-vacuum amplitude, which we have done in the previous chapter. We now generalize this amplitude to include gravitons, scalar bosons and fermions. As with previous discussions, although Yang–Mills gravity can be formulated in both inertial and non-inertial frames due to the generality of the taiji symmetry framework, we limit our discussion to inertial frames because of the complexity of its formulation in even the simplest accelerated frame.

The gauge invariant action $S_{\phi\psi}$ for spin 2 field $\phi_{\mu\nu} = \phi_{\nu\mu}$, fermion field ψ and scalar field Φ is given by[2]

$$S_{\phi\psi} = \int (L_\phi + L_\psi + L_\Phi + L_{\xi\zeta})d^4x, \quad c = \hbar = 1,$$

$$L_\phi = \frac{1}{2g^2}\left(\frac{1}{2}C_{\mu\nu\alpha}C^{\mu\nu\alpha} - C_{\mu\alpha}{}^\alpha C^{\mu\beta}{}_\beta\right), \tag{11.1}$$

$$L_\psi = \frac{i}{2} \left[\overline{\psi} \gamma_\mu \left(\Delta^\mu \psi \right) - \left(\Delta^\mu \overline{\psi} \right) \gamma_\mu \psi \right] - m \overline{\psi} \psi, \tag{11.2}$$

$$L_\Phi = \frac{1}{2} \left[\eta^{\mu\nu} (\Delta_\mu \Phi)(\Delta_\nu \Phi) - m_\Phi^2 \Phi^2 \right], \tag{11.3}$$

$$L_{\xi\zeta} = \frac{\xi}{2g^2} \left[\partial^\mu J_{\mu\alpha} - \frac{\zeta}{2} \partial_\alpha J_\lambda^\lambda \right] \left[\partial^\nu J_{\nu\beta} - \frac{\zeta}{2} \partial_\beta J_\lambda^\lambda \right] \eta^{\alpha\beta}, \tag{11.4}$$

$$\Delta_\mu \psi = J_{\mu\nu} \partial^\nu \psi, \quad J_{\mu\nu} = \eta_{\mu\nu} + g\phi_{\mu\nu} = J_{\nu\mu},$$

where the T_4 gauge curvature $C^{\mu\alpha\beta}$ is given by

$$C^{\mu\nu\alpha} = J^{\mu\sigma} \partial_\sigma J^{\nu\alpha} - J^{\nu\sigma} \partial_\sigma J^{\mu\alpha}. \tag{11.5}$$

The gauge-fixing term $L_{\xi\zeta}$, involving up to two arbitrary gauge parameters ξ and ζ, is necessary for the quantization of fields with gauge symmetry. We choose the usual gauge condition of the form

$$\partial^\mu J_{\mu\nu} - \frac{\zeta}{2} \partial_\nu J = Y_\nu, \quad J = J_\lambda^\lambda = \delta_\mu^\mu - g\phi, \quad \phi = \phi_\lambda^\lambda, \tag{11.6}$$

where Y_ν is a suitable function of space-time.

The Lagrangian for just a gravitational field (or 'pure gravity') $L_{pg} = L_\phi + L_{\xi\zeta}$ in terms of the tensor field $\phi_{\mu\nu}$ can be written

$$L_{pg} = L_2 + L_3 + L_4 + L_{\xi\zeta}, \tag{11.7}$$

where

$$L_2 = \frac{1}{2} \left(\partial_\lambda \phi_{\alpha\beta} \partial^\lambda \phi^{\alpha\beta} - \partial_\lambda \phi_{\alpha\beta} \partial^\alpha \phi^{\lambda\beta} - \partial_\lambda \phi \partial^\lambda \phi \right.$$
$$\left. + 2\partial_\lambda \phi \partial^\beta \phi_\beta^\lambda - \partial_\lambda \phi_\mu^\lambda \partial_\beta \phi^{\mu\beta} \right), \tag{11.8}$$

$$L_3 = g \left[\left(\partial_\lambda \phi_{\alpha\beta} \partial_\rho \phi^{\alpha\beta} \right) \phi^{\lambda\rho} - \left(\partial_\lambda \phi_{\alpha\beta} \partial_\rho \phi^{\lambda\beta} \right) \phi^{\alpha\rho} - \left(\partial_\lambda \phi \partial_\rho \phi \right) \phi^{\lambda\rho} \right.$$
$$\left. + \left(\partial^\lambda \phi \partial_\rho \phi_{\lambda\beta} \right) \phi^{\beta\rho} + \left(\partial_\lambda \phi \partial^\beta \phi_\beta^\mu \right) \phi_\mu^\lambda - \left(\partial^\lambda \phi_{\lambda\mu} \partial_\rho \phi_\beta^\mu \right) \phi^{\beta\rho} \right], \tag{11.9}$$

$$L_4 = \frac{g^2}{2} \left[\left(\partial^\lambda \phi_{\alpha\beta} \partial_\rho \phi^{\alpha\beta} \right) \phi_{\lambda\mu} \phi^{\mu\rho} - \left(\partial_\lambda \phi_{\alpha\beta} \partial_\rho \phi^{\mu\beta} \right) \phi_\mu^\lambda \phi^{\alpha\rho} \right.$$
$$+ \left(\partial_\lambda \phi \partial_\rho \phi_\beta^\mu \right) \phi_\mu^\lambda \phi^{\beta\rho} + \left(\partial_\lambda \phi \partial_\rho \phi_\beta^\mu \right) \phi_\mu^\rho \phi^{\beta\lambda}$$
$$\left. - \left(\partial_\lambda \phi \partial_\rho \phi \right) \phi_\mu^\lambda \phi^{\mu\rho} - \left(\partial_\lambda \phi_{\alpha\mu} \partial_\rho \phi_\beta^\mu \right) \phi^{\lambda\alpha} \phi^{\beta\rho} \right], \tag{11.10}$$

$$L_{\xi\zeta} = \frac{\xi}{2} \left[\left(\partial_\lambda \phi^{\lambda\alpha} \right) \partial^\rho \phi_{\rho\alpha} - \zeta \left(\partial_\lambda \phi^{\lambda\alpha} \right) \partial_\alpha \phi + \frac{\zeta^2}{4} \left(\partial^\alpha \right) \partial_\alpha \phi \right]. \tag{11.11}$$

The action (11.1) with the Lagrangians given in (11.2), (11.3) and (11.4) leads to the T_4 field equation in inertial frames,

$$H^{\mu\nu} + \xi A^{\mu\nu} = g^2 S^{\mu\nu}, \tag{11.12}$$

$$H^{\mu\nu} \equiv Sym\left[\partial_\lambda \left(J_\rho^\lambda C^{\rho\mu\nu} - J_\alpha^\lambda C^{\alpha\beta}{}_\beta \eta^{\mu\nu} + C^{\mu\beta}{}_\beta J^{\nu\lambda} \right) \right.$$
$$\left. - C^{\mu\alpha\beta} \partial^\nu J_{\alpha\beta} + C^{\mu\beta}{}_\beta \partial^\nu J_\alpha^\alpha - C^{\lambda\beta}{}_\beta \partial^\nu J_\lambda^\mu \right],$$

$$A^{\mu\nu} = Sym\left[\partial^\mu \partial_\sigma J^{\sigma\nu} - \frac{\zeta}{2} \partial^\mu \partial^\nu J - \frac{\zeta}{2} \eta^{\mu\nu} \partial^\lambda \partial^\sigma J_{\lambda\sigma} + \frac{\zeta^2}{4} \eta^{\mu\nu} \partial^2 J \right],$$

where "*Sym*" denotes that μ and ν should be made symmetric, e.g., $L_{\mu\nu} = (1/2)$ $[L_{\mu\nu} + L_{\nu\mu}]$.

The identities for the T_4 gauge curvature are

$$C_{\mu\nu\alpha} = -C_{\nu\mu\alpha}, \quad C_{\mu\nu\alpha} + C_{\nu\alpha\mu} + C_{\alpha\mu\nu} = 0. \tag{11.13}$$

For simplicity, we ignore the scalar field Φ in the following discussion of the energy-momentum tensor. The symmetric 'source tensor' $S^{\mu\nu}$ in Equation (11.12) is given by

$$S^{\mu\nu} = \frac{1}{4}\left[\overline{\psi} i\gamma^\mu \partial^\nu \psi - i\left(\partial^\nu \overline{\psi}\right)\gamma^\mu \psi + \overline{\psi} i\gamma^\nu \partial^\mu \psi - i\left(\partial^\mu \overline{\psi}\right)\gamma^\nu \psi \right], \tag{11.14}$$

which acts as the fermion source for generating the gravitational field $\phi_{\mu\nu}$. For weak fields in inertial frames, the field Equation (11.12) can be linearized as follows:

$$\partial_\lambda \partial^\lambda \phi^{\mu\nu} - \partial^\mu \partial_\lambda \phi^{\lambda\nu} - \eta^{\mu\nu} \partial^2 \phi + \eta^{\mu\nu} \partial_\alpha \partial_\beta \phi^{\alpha\beta}$$
$$+ \partial^\mu \partial^\nu \phi - \partial^\nu \partial_\lambda \phi^{\lambda\mu} + \xi \partial^\mu \partial_\lambda \phi^{\lambda\nu} - \frac{\xi\zeta}{2} \partial^\mu \partial^\nu \phi$$
$$- \frac{\xi\zeta}{2} \eta^{\mu\nu} \partial_\alpha \partial_\beta \phi^{\alpha\beta} + \frac{\xi\zeta^2}{4} \eta^{\mu\nu} \partial^2 \phi = gS^{\mu\nu}, \tag{11.15}$$

where g is related to Newtonian constant G by $g = \sqrt{8\pi G}$. In this weak field limit, the linearized Equation (11.15) is the same as that in Einstein gravity. In momentum space, (11.15) can be written as

$$A^{\mu\nu\rho\sigma}(k)\phi_{\rho\sigma}(k) = 0,$$

$$A^{\mu\nu\rho\sigma}(k) = \frac{k^2}{2}(\eta^{\mu\rho}\eta^{\nu\sigma} + \eta^{\mu\sigma}\eta^{\nu\rho}) - \left(1 - \frac{\xi}{2}\right)k^\nu k^\sigma \eta^{\mu\rho} - \left(1 - \frac{\xi}{2}\right)k^\mu k^\sigma \eta^{\nu\rho}$$

$$-\left(1 - \frac{\xi\zeta^2}{4}\right) k^2 \eta^{\mu\nu} \eta^{\rho\sigma} + \left(1 - \frac{\xi\zeta}{2}\right) k^\rho k^\sigma \eta^{\mu\nu}$$

$$+\left(1 - \frac{\xi\zeta}{2}\right) k^\mu k^\nu \eta^{\rho\sigma}, \tag{11.16}$$

when $g = 0$. Equation (11.16) will be useful for deriving the Feynman propagator for gravitons in the next section.

In the presence of a space-time gauge field, the Dirac equation of a fermion can be derived from the fermion Lagrangian (11.2),

$$i\left(\gamma^\nu + g\gamma_\mu \phi^{\mu\nu}\right) \partial_\nu \psi + \frac{ig}{2} \left(\partial_\nu \phi^{\mu\nu}\right) \gamma_\mu \psi - m\psi = 0,$$

$$i\left(\partial^\mu + g\phi^{\mu\nu}\partial_\nu\right) \overline{\psi}\gamma_\mu + \frac{ig}{2}\overline{\psi}\gamma_\mu \left(\partial_\nu \phi^{\mu\nu}\right) + m\overline{\psi} = 0. \tag{11.17}$$

It follows from these two equations that the modified fermion current j^ν in the presence of the gravitational field is conserved,

$$\partial_\nu j^\nu = 0, \tag{11.18}$$

where

$$j^\nu = \overline{\psi}\gamma^\nu\psi + g\phi^{\mu\nu}\overline{\psi}\gamma_\mu\psi = J^{\mu\nu}\overline{\psi}\gamma_\mu\psi.$$

The conserved total energy-momentum tensor $T^{\mu\nu}(\phi\psi)$ of the graviton-fermion system is given by the Lagrangian $L_{\phi\psi} = L_\phi + L_\psi$,

$$T^{\mu\nu}(\phi\psi) = T(\phi)^{\mu\nu} + T(\psi)^{\mu\nu}, \tag{11.19}$$

$$T(\phi)^{\mu\nu} = \frac{\partial L_\phi}{\partial(\partial_\mu \phi_{\alpha\beta})} \partial^\nu \phi_{\alpha\beta} - \eta^{\mu\nu} L_\phi$$

$$T(\psi)^{\mu\nu} = \frac{\partial L_\psi}{\partial(\partial_\mu \psi)} \partial^\nu \psi + \partial^\nu \overline{\psi} \frac{\partial L_\psi}{\partial(\partial_\mu \overline{\psi})} - \eta^{\mu\nu} L_\psi,$$

according to Noether's theorem. Note that L_ψ does not contain $\partial_\mu \phi_{\alpha\beta}$ and L_ϕ does not involve $\partial_\mu \psi$ and $\partial_\mu \overline{\psi}$. From equation (11.19), we obtain

$$T(\phi)^{\mu\nu} = \frac{1}{g}(J_\lambda^\mu C^{\lambda\alpha\beta} - \eta^{\alpha\beta} J_\lambda^\mu C^{\lambda\sigma}{}_\sigma + J^{\beta\mu} C^{\alpha\sigma}{}_\beta)\partial^\nu \phi_{\alpha\beta}$$

$$- \eta^{\mu\nu} \frac{1}{4g^2}(C_{\mu\alpha\beta} C^{\mu\alpha\beta} - 2C_{\mu\alpha}{}^\alpha C^{\mu\beta}{}_\beta), \tag{11.20}$$

$$T(\psi)^{\mu\nu} = \frac{i}{2}\left[\overline{\psi}\gamma_\alpha J^{\alpha\mu} \partial^\nu \psi - \left(\partial^\nu \overline{\psi}\right) J^{\alpha\mu} \gamma_\alpha \psi\right] - \eta^{\mu\nu} L_\psi. \tag{11.21}$$

These conserved energy-momentum tensors for the fermion and gauge fields are not inherently symmetric, but they can be symmetrized.[3,4]

In Yang–Mills gravity, it is the symmetric source tensor $S^{\mu\nu}$ in (11.12) and (11.14) that produces the gravitational field $\phi_{\alpha\beta}$, rather than the non-symmetric energy-momentum tensor $T(\psi)^{\mu\nu}$ for fermions. Thus, the source tensor $S^{\mu\nu}$ differs from the energy-momentum tensor in Einstein gravity.

From the linearized Equation (11.15) for weak fields, one can derive the conservation law

$$\partial_\mu S^{\mu\nu} = 0.$$

It is gratifying that the source tensor $S^{\mu\nu}$ in the field equation (11.12) does not involve the tensor gauge field, as shown in (11.14), in contrast to the Noether tensor $T^{\mu\nu}(\psi)$ in (11.21). This makes the coupling of fields in Yang–Mills gravity simpler than in Einstein gravity.

11-2.   The Feynman–Dyson rules for quantum Yang–Mills gravity

We now use the results from the previous section to write down the Feynman–Dyson rules for Yang–Mills gravity. Although these are conventionally referred to as the Feynman rules, we choose to refer to them as the Feynman–Dyson rules in this case because the intuitive Feynman rules for calculating the invariant amplitude (a Lorentz scalar) are incomplete. It was Dyson's rigorous derivation and elaboration of the rules that allowed their application to higher-order processes, such as Kinoshita's higher-order corrections to the magnetic moment of the electron.[5]

Let us consider the general propagator for the graviton corresponding to the general gauge-fixing terms in (11.4) in Yang–Mills gravity. This general Feynman propagator involves two arbitrary parameters, ξ and ζ, and can be derived from the free Equation (11.16) in momentum space. The graviton propagator $G_{\alpha\beta\rho\sigma}(k)$ is defined as the inverse of $A^{\mu\nu\rho\sigma}(k)$ in (11.16) as follows,

$$G_{\alpha\beta\rho\sigma}(k)A^{\mu\nu\rho\sigma}(k) = \frac{i}{2}(\delta_\alpha^\mu \delta_\beta^\nu + \delta_\alpha^\nu \delta_\beta^\mu), \qquad (11.22)$$

with a factor i. Its general structure can be inferred from Lorentz and Poincaré covariance. We can write the graviton propagator $G_{\alpha\beta\rho\sigma}(k)$ in the following form,

$$G_{\alpha\beta\rho\sigma} = A(\eta_{\alpha\beta}\eta_{\rho\sigma} - \eta_{\rho\alpha}\eta_{\sigma\beta} - \eta_{\rho\beta}\eta_{\sigma\alpha}) - B(k_\alpha k_\beta \eta_{\rho\sigma} + k_\rho k_\sigma \eta_{\alpha\beta})$$
$$+ C(k_\sigma k_\beta \eta_{\rho\alpha} + k_\alpha k_\sigma \eta_{\rho\beta} + k_\rho k_\beta \eta_{\sigma\alpha} + k_\rho k_\alpha \eta_{\sigma\beta}) - D k_\alpha k_\beta k_\rho k_\sigma,$$

where the unknown functions $A(k^2), B(k^2), C(k^2), D(k^2)$ are to be determined by the equation (11.22). It follows from (11.16), (11.22) and the above expression for $G_{\alpha\beta\rho\sigma}$ that

$$G_{\alpha\beta\rho\sigma} = \frac{-i}{2k^2} \left(\left[\eta_{\alpha\beta}\eta_{\rho\sigma} - \eta_{\rho\alpha}\eta_{\sigma\beta} - \eta_{\rho\beta}\eta_{\sigma\alpha}\right] - \frac{1}{k^2}\frac{(2\zeta - 2)}{(\zeta - 2)} \right.$$
$$\times \left(k_\alpha k_\beta \eta_{\rho\sigma} + k_\rho k_\sigma \eta_{\alpha\beta} \right) + \frac{1}{k^2}\frac{(\xi - 2)}{\xi}(k_\sigma k_\beta \eta_{\rho\alpha} + k_\alpha k_\sigma \eta_{\rho\beta} + k_\rho k_\beta \eta_{\sigma\alpha}$$
$$\left. + k_\rho k_\alpha \eta_{\sigma\beta}) - \frac{1}{k^4}\frac{(2\zeta - 2)}{\xi(\zeta - 2)}\left[\frac{4 + 2\xi - 4\xi\zeta}{\zeta - 2} - 4 + 2\xi\right] k_\alpha k_\beta k_\rho k_\sigma \right),$$

$$(11.23)$$

where the usual prescription of $i\epsilon$ for the Feynman propagator (11.23) is understood. For example, $k^2 = k_0^2 - \mathbf{k}^2$ in (11.23) is understood as $k^2 + i\epsilon$. This is related to the property that the integral $\int d^4x(L_\phi + J_{\mu\nu}j^{\mu\nu})$ in the vacuum-to-vacuum amplitude (10.10) is defined by a passage to Euclidean space-time and then a continuation back to Minkowski space-time. This procedure gives the correct $i\epsilon$ prescription in the Feynman propagator (11.23). In this book, it should always be understood as being done, even when the $i\epsilon$ is not explicitly included in the Feynman propagators.

In the calculation of higher-order Feynman diagrams such as the self-energy of the graviton or gravitational vertex corrections, the general propagator (11.23) with two arbitrary parameters leads to a complicated result because the divergent terms are gauge dependent. It is very time consuming to calculate, even using symbolic computing. Thus, it is useful to limit oneself to a special case such as $\zeta = 1$ and arbitrary ξ for an explicit calculation. In this case, the propagator (11.23) reduces to the simpler form

$$G_{\alpha\beta\rho\sigma} = \frac{-i}{2k^2} \left[(\eta_{\alpha\beta}\eta_{\rho\sigma} - \eta_{\rho\alpha}\eta_{\sigma\beta} - \eta_{\rho\beta}\eta_{\sigma\alpha}) \right.$$
$$\left. + \frac{1}{k^2}\frac{(\xi - 2)}{\xi}(k_\sigma k_\beta \eta_{\rho\alpha} + k_\alpha k_\sigma \eta_{\rho\beta} + k_\rho k_\beta \eta_{\sigma\alpha} + k_\rho k_\alpha \eta_{\sigma\beta}) \right]. \quad (11.24)$$

In the special case of $\zeta = 0$ and arbitrary ξ, the graviton propagator (11.23) reduces to the form

$$
\begin{aligned}
G_{\alpha\beta\rho\sigma} = \frac{-i}{2k^2} \Bigg(& [\eta_{\alpha\beta}\eta_{\rho\sigma} - \eta_{\rho\alpha}\eta_{\sigma\beta} - \eta_{\rho\beta}\eta_{\sigma\alpha}] \\
& - \frac{1}{k^2}[k_\alpha k_\beta \eta_{\rho\sigma} + k_\rho k_\sigma \eta_{\alpha\beta}] + \frac{1}{k^2}\left[\frac{\xi - 2}{\xi}\right](k_\sigma k_\beta \eta_{\rho\alpha} + k_\alpha k_\sigma \eta_{\rho\beta} \\
& + k_\rho k_\beta \eta_{\sigma\alpha} + k_\rho k_\alpha \eta_{\sigma\beta}) - \frac{1}{k^4}\left[\frac{\xi - 6}{\xi}\right]k_\alpha k_\beta k_\rho k_\sigma \Bigg).
\end{aligned}
\tag{11.25}
$$

In Yang–Mills gravity, the simple gauge specified by $\xi = 2$ and $\zeta = 1$ has been discussed by DeWitt and may be called the DeWitt gauge.[5] In this gauge, the covariant graviton propagator takes the simplest form, $G_{\alpha\beta\rho\sigma} = (-i/2k^2)(\eta_{\alpha\beta}\eta_{\rho\sigma} - \eta_{\rho\alpha}\eta_{\sigma\beta} - \eta_{\rho\beta}\eta_{\sigma\alpha})$ and is consistent with that obtained in previous works by Leibbrandt, Capper and others in general relativity.[6–8] As an aside, the overall factor of $(1/2)$ in the graviton propagators (11.23) is necessary for Yang–Mills gravity to be consistent with its T_4 gauge identity—a generalized Ward-Takahasi identity for the Abelian group T_4 with ghosts. (Cf. Sec. 13-2 in ch. 13.) The free Lagrangian (11.8) in Yang–Mills gravity is mathematically the same as that in the Hilbert–Einstein Lagrangian. In quantum Einstein gravity as formulated and discussed by Feynman and DeWitt,[5] this factor of $(1/2)$ for the covariant graviton propagator is necessary in order for it to be consistent with the Slavnov–Taylor identity, as shown in References 5–8.

We now consider the Feynman–Dyson rules for graviton interactions corresponding to iL_3. The momenta in the Feynman–Dyson rules in this chapter are in-coming to the vertices. Suppose the tensor-indices and momenta of the 3-vertex are denoted by $[\phi^{\mu\nu}(p)\phi^{\sigma\tau}(q)\phi^{\lambda\rho}(k)]$. The graviton 3-vertex is

$$
\begin{aligned}
ig\, Sym\, P_6 \big(& -p^\lambda q^\rho \eta^{\sigma\mu}\eta^{\tau\nu} + p^\sigma q^\rho \eta^{\lambda\mu}\eta^{\tau\nu} + p^\lambda q^\rho \eta^{\sigma\tau}\eta^{\mu\nu} \\
& - p^\sigma q^\rho \eta^{\mu\nu}\eta^{\tau\lambda} - p^\rho q^\sigma \eta^{\mu\nu}\eta^{\tau\lambda} + p^\mu q^\rho \eta^{\nu\sigma}\eta^{\lambda\tau} \big).
\end{aligned}
\tag{11.26}
$$

As before, we use the symbol *Sym* to denote that a symmetrization is to be performed on each index pair $(\mu\nu)$, $(\sigma\tau)$ and $(\lambda\rho)$ in $[\phi^{\mu\nu}(p)\phi^{\sigma\tau}(q)\phi^{\lambda\rho}(k)]$. The symbol P_n denotes a summation to be carried out over permutations of the momentum-index triplets, and the subscript gives the number of permutations in each case. For example,

$$P_6(p^\lambda q^\rho \eta^{\sigma\mu} \eta^{\tau\upsilon}) = (p^\lambda q^\rho \eta^{\sigma\mu} \eta^{\tau\upsilon} + p^\sigma k^\tau \eta^{\lambda\mu} \eta^{\rho\upsilon}$$
$$+ p^\rho q^\lambda \eta^{\mu\sigma} \eta^{\upsilon\tau} + p^\tau k^\sigma \eta^{\mu\lambda} \eta^{\upsilon\rho} + q^\mu k^\upsilon \eta^{\lambda\sigma} \eta^{\rho\tau} + k^\mu q^\upsilon \eta^{\sigma\lambda} \eta^{\tau\rho}),$$

which is the result that one obtains directly from the first term in the Lagrangian L_3 in (11.9) by following the usual method for obtaining the vertex using the Feynman–Dyson method.[9–11]

The 4-vertex (i.e., iL_4 with $\phi^{\mu\nu}(p)\phi^{\sigma\tau}(q)\phi^{\lambda\rho}(k)\phi^{\alpha\beta}(l)$) is given by

$$\frac{ig^2}{2} Sym\, P_{24}(-p^\rho q^\beta \eta^{\mu\sigma} \eta^{\upsilon\tau} \eta^{\lambda\alpha} + p^\rho q^\beta \eta^{\lambda\sigma} \eta^{\upsilon\tau} \eta^{\mu\alpha}$$
$$- p^\rho q^\beta \eta^{\lambda\sigma} \eta^{\tau\alpha} \eta^{\mu\nu} - p^\beta q^\rho \eta^{\lambda\sigma} \eta^{\tau\alpha} \eta^{\mu\nu}$$
$$+ p^\rho q^\beta \eta^{\mu\nu} \eta^{\sigma\tau} \eta^{\lambda\alpha} + p^\rho q^\beta \eta^{\lambda\nu} \eta^{\tau\alpha} \eta^{\mu\sigma}). \tag{11.27}$$

The fermion propagator has the usual form

$$\frac{i}{\gamma^\mu p_\mu - m}. \tag{11.28}$$

The fermion–graviton 3-vertex (i.e., iL_ψ^{int} with $\overline{\psi}(q)\psi(p)\phi_{\mu\nu}(k)$) takes the form

$$Sym\, \frac{i}{2} g\gamma_\mu(p_\nu + q_\nu), \tag{11.29}$$

where $\psi(p)$ is treated as an annihilation operator of a fermion with momentum p_ν, and $\overline{\psi}(q)$ is a creation operator of a fermion with the momentum q_ν.

To obtain the Feynman–Dyson rules for the general propagator (involving an arbitrary gauge parameter ζ) for the ghost particle and the interaction vertices of the ghost particle, it is convenient to write the Lagrangian L_{eff} in Equation (10.15) in the following form

$$L_{eff} = \overline{V}^\mu \left[\eta_{\mu\nu} \partial_\lambda \partial^\lambda + (1 - \zeta)\partial_\mu \partial_\nu \right] V^\nu$$
$$- g \left[\left(\partial^\alpha \overline{V}^\beta \right) V^\nu \partial_\nu \phi_{\alpha\beta} + \left(\partial^\lambda \overline{V}^\mu \right) \left(\partial_\lambda V^\nu \right) \phi_{\mu\nu} + \left(\partial^\lambda \overline{V}^\mu \right) \left(\partial_\mu V^\nu \right) \phi_{\lambda\nu} \right]$$
$$+ g\zeta(\partial^\lambda \overline{V}_\lambda)(\partial^\mu V^\nu)\phi_{\mu\nu} + g(\zeta/2) \left(\partial_\mu \overline{V}^\mu \right) V^\nu \left(\partial_\nu \phi_{\alpha\beta} \right) \eta^{\alpha\beta}. \tag{11.30}$$

The propagator of the FDM ghosts can be derived from the Lagrangian L_{eff} in (11.30) for ghost fields,

$$G^{\mu\nu} = \frac{-i}{k^2} \left(\eta^{\mu\nu} - \frac{k^\mu k^\nu}{k^2} \frac{(1 - \zeta)}{(2 - \zeta)} \right). \tag{11.31}$$

Similarly, the ghost–ghost–graviton vertex (denoted by $\overline{V}^{\mu}(p)V^{\nu}(q)\phi^{\alpha\beta}(k))$ is found to be

$$\frac{ig}{2}\left[p^{\alpha}k^{\nu}\eta^{\mu\beta}+p^{\beta}k^{\nu}\eta^{\mu\alpha}+p\cdot q\left(\eta^{\mu\alpha}\eta^{\nu\beta}+\eta^{\mu\beta}\eta^{\nu\alpha}\right)+p^{\alpha}q^{\mu}\eta^{\nu\beta}\right.$$

$$\left.+p^{\beta}q^{\mu}\eta^{\nu\alpha}-\zeta p^{\mu}k^{\nu}\eta^{\alpha\beta}-\zeta p^{\mu}q^{\beta}\eta^{\nu\alpha}-\zeta p^{\mu}q^{\alpha}\eta^{\nu\beta}\right]. \qquad (11.32)$$

We have used the convention that all momenta are incoming to the vertices. All ghost-particle vertices are bilinear in the FDM ghost,[12] as shown in the effective Lagrangian (11.30). Thus the FDM ghosts appear, by definition of the physical subspace, only in closed loops in the intermediate states of a physical process, and there is a factor of -1 for each vector-fermion loop in Yang–Mills gravity.

The Feynman–Dyson rules for the coupling between the graviton and a scalar particle can be obtained from the Lagrangian L_{Φ} in (11.3). As usual, the propagator $G(k)$ for the scalar particle is given by

$$G(k)=\frac{i}{k^2-m_{\Phi}^2}, \qquad (11.33)$$

where the $i\epsilon$ prescription is understood. The 3-vertex coupling between graviton and the scalar particle, $\phi_{\mu\nu}(k)\Phi(p)\Phi(q)$ is

$$-ig(p_{\mu}q_{\nu}+q_{\mu}p_{\nu}). \qquad (11.34)$$

The 4-vertex, $\phi_{\alpha\beta}(l)\phi_{\mu\nu}(k)\Phi(p)\Phi(q)$, is given by

$$\frac{-ig^2}{4}\left[\eta_{\alpha\mu}(p_{\beta}q_{\nu}+p_{\nu}q_{\beta})+\eta_{\beta\mu}(p_{\alpha}q_{\nu}+p_{\nu}q_{\alpha})\right.$$

$$\left.+\eta_{\alpha\nu}(p_{\beta}q_{\mu}+p_{\mu}q_{\beta})+\eta_{\beta\nu}(p_{\alpha}q_{\mu}+p_{\mu}q_{\alpha})\right], \qquad (11.35)$$

where the symmetrization for each index pair $(\mu\nu)$, $(\alpha\beta)$ has been carried out explicitly. Again, all momenta are, by definition, incoming into the vertices. There is a factor $(1/2)$ for each closed loop containing only two graviton lines. For example, this factor $(1/2)$ applies to the process $\Phi\Phi \rightarrow \phi_{\alpha\beta}\phi_{\mu\nu} \rightarrow \Phi\Phi$.

References

1. K. Huang, *Quarks, Leptons & Gauge Fields* (World Scientific, 1982), pp. 122–135.

2. J. P. Hsu, *Int. J. Mod. Phys. A* **21**, 5119 (2006); J. P. Hsu and L. Hsu, *A Broader View of Relativity, General Implications of Lorentz and Poincaré Invariance* (World Scientific, 2006), Appendix D. Available online: 'google books, a broader view of relativity, Hsu.'

3. G. Wentzel, *Quantum Theory of Fields* (translated from German by C. Houtermans and J. M. Jauch, Interscience, 1943). Appendix I, pp. 203–211.

4. V. Fock, *The Theory of Space, Time and Gravitation* (translated N. Kemmer, Pergamon, 1959) pp. 158–166; L. Landau and E. Lifshitz, *The Classical Theory of Fields* (Trans. M. Hamermesh, Addison-Wesley, 1951) pp. 293–296 and pp. 316–318.

5. See, for example, R. P. Feynman, *Theory of Fundamental Processes* (W. A. Benjamin, 1962) p. 168; F. J. Dyson, in *Lorentz and Poincaré Invariance* (J. P. Hsu and Y. Z. Zhang, World Scientific, 2001) p. 316; T. Kinoshita, talk at Nishina Hall, RIKEN (2010). (google search: Kinoshita, Magnetic moment.) B. S. DeWitt, *Phys. Rev.* **162**, 1239 (1967). See also Ref. 12.

6. D. M. Capper and M. Ramon Medrano, *Phys. Rev.* **D 9**, 1641 (1974); G. Leibbrandt, *Rev. Mod. Phys.* **47**, 849 (1975).

7. D. M. Capper and M. A. Namazie, *Nucl. Phys.* **B 142**, 535 (1978).

8. N. Grillo, *Class. Quantum Grav.* **18**, 141 (2001).

9. J. J. Sakurai, *Advanced Quantum Mechanics* (Addison-Wesley, 1967) p. 313.

10. N. N. Bogoliubov and D. V. Shirkov, *Introduction to the Theory of Quantized Fields* (tr. by G. M. Volkov, Interscience, 1959) pp. 206–226.

11. J. P. Hsu and J. A. Underwood, *Phys. Rev.* **D 12**, 620 (1975).

12. R. P. Feynman and B. S. DeWitt in *100 Years of Gravity and Accelerated Frames* (Ed. J. P. Hsu and D. Fine, World Scientific, 2005) pp. 272–324; S. Mandelstam, *Phys. Rev.* **175**, 1604 (1968).

12.

Gravitational Self-Energy of the Graviton

12-1.   Graviton self-energy in the DeWitt gauge

Yang–Mills gravity with the space-time translational gauge symmetry has some interesting properties. On one hand, the graviton self-coupling has a maximum of four gravitons at a vertex, which resembles the situation in renormalizable gauge theories such as QCD and electroweak theory. On the other hand, the total Lagrangian (10.16) appears not to be renormalizable by the criterion of power counting because the coupling strength g has the dimension of length, in contrast to usual gauge theories. In order to shed some light on how Yang–Mills gravity might be made renormalizable despite this difference from the usual gauge theories, we now consider the graviton self-energy at the one-loop level and compare the result in Yang–Mills gravity with other known renormalizable theories. In the process, we will employ the D-dimensional regularization to preserve the space-time translational gauge symmetry of physical amplitudes and to define the divergent quantities.

The divergent self-energy of the graviton has contributions from several Feynman diagrams. The most important of these corresponds to the following process,

$$\phi_{\mu\nu}(p) \to \phi_{\rho\sigma}(q)\phi_{\gamma\eta}(p-q) \to \phi_{\alpha\beta}(p). \tag{12.1}$$

Its amplitude, denoted by $S_1^{\mu\nu\alpha\beta}(p)$, can be calculated using the Feynman rules with $\xi = 2$ and $\zeta = 1$ in the gauge-fixing Lagrangian (11.4), for simplicity. Process (12.1) leads to the following amplitude

$$S_1^{\mu\nu\alpha\beta} = \int \frac{d^D q}{(2\pi)^D} V^{\mu\nu\rho\sigma\gamma\lambda}(p,q,k) G_{\rho\sigma\rho'\sigma'}(q) G_{\tau\eta\gamma\lambda}(k_1) V^{\alpha\beta\rho'\sigma'\tau\eta}(p_1,q_1,k_1),$$

$$\tag{12.2}$$

where $D \to 4$ at the end of calculations. Since all momenta are, by definition, incoming to the vertex, we have $p + q + k = 0$, $q_1 = -q$, $k_1 = -k = p + q$, and

$p_1 = -p$. The graviton 3-vertex is given by (11.26), while the graviton propagator in the DeWitt gauge is given by (11.23) with $\xi = 2$ and $\zeta = 1$, i.e.,

$$G_{\rho\sigma,\rho'\sigma'}(q) = \frac{-i}{2q^2}(\eta_{\rho\sigma}\eta_{\rho'\sigma'} - \eta_{\rho\rho'}\eta_{\sigma\sigma'} - \eta_{\rho\sigma'}\eta_{\sigma\rho'}). \tag{12.3}$$

Using FeynCalc[1] to evaluate the integral (12.2) in the limit, $D \to 4$, we obtain

$$\begin{aligned}
S_1^{\mu\nu\alpha\beta} = \frac{ig^2}{24(4\pi)^2}\Gamma(0)\,[&p^4(8\eta^{\alpha\nu}\eta^{\beta\mu} + 8\eta^{\alpha\mu}\eta^{\beta\nu} + 31\eta^{\alpha\beta}\eta^{\mu\nu}) \\
&+ 56p^\alpha p^\beta p^\mu p^\nu - 30p^2(p^\alpha p^\beta\eta^{\mu\nu} + p^\mu p^\nu\eta^{\alpha\beta}) \\
&- 7p^2(p^\alpha p^\mu\eta^{\beta\nu} + p^\beta p^\mu\eta^{\alpha\nu} + p^\alpha p^\nu\eta^{\beta\mu} + p^\beta p^\nu\eta^{\alpha\mu})], \quad (12.4)
\end{aligned}$$

where $p^4 = (p^2)^2 = (p_\mu p^\mu)^2$, and $\Gamma(0)$ is the Euler function $\Gamma(n)$ with $n = 0$. The factor of $1/2$ for the Feynman diagram for the process (12.1) is included in (12.4), which is consistent with that of the symbolic computation using xAct.[1]

A second contribution to the graviton self-energy comes from a diagram involving a ghost-loop, which corresponds to the process

$$\phi_{\mu\nu}(p) \to V_\rho(q)V'_\lambda(p-q) \to \phi_{\alpha\beta}(p), \tag{12.5}$$

where the vector ghost-loop is associated with a factor of (-1) due to the fermion property of the Feynman–DeWitt–Mandelstam (FDM) ghost. The virtual process (12.5) leads to the self-energy amplitude,

$$S_2^{\mu\nu\alpha\beta}(p) = \int \frac{d^D q}{(2\pi)^D} V^{\mu\nu\rho\lambda}(p,q,k)G_{\rho\rho'}(q)G_{\lambda\lambda'}(k)V^{\alpha\beta\rho'\lambda'}(p_1,q_1,k_1). \tag{12.6}$$

The FDM ghost propagator $G^{\mu\nu}(q)$ and the graviton–ghost–ghost vertex $\phi^{\mu\nu}V'^\rho(k)V^\lambda(q)$ are given by (11.31) and (11.32), respectively, with $\zeta = 1$. With the help of FeynCalc, (12.6) gives the result

$$\begin{aligned}
S_2^{\mu\nu\alpha\beta} = \frac{-i}{48}\frac{g^2}{(4\pi)^2}\Gamma(0)p^2[&7p^4\eta^{\alpha\beta}\eta^{\mu\nu} + 2p^4(\eta^{\alpha\mu}\eta^{\beta\nu} + \eta^{\alpha\nu}\eta^{\beta\mu}) \\
&- p^2(p^\alpha p^\mu\eta^{\beta\nu} + p^\beta p^\mu\eta^{\alpha\nu} + p^\alpha p^\nu\eta^{\beta\mu} + p^\beta p^\nu\eta^{\alpha\mu}) \\
&- 6p^2(p^\alpha p^\beta\eta^{\mu\nu} + p^\mu p^\nu\eta^{\alpha\beta}) + 8p^\mu p^\nu p^\alpha p^\beta]. \quad (12.7)
\end{aligned}$$

At the one-loop level, there are three more tadpole diagrams involving (i) a graviton-loop and a graviton 4-vertex, (ii) a graviton-loop and two graviton 3-vertices, and (iii) a ghost-loop with one graviton 3-vertex and a graviton–ghost–ghost vertex. When we use dimensional regularization[2] to calculate the diagram corresponding to (i) with a factor of $1/2$ for this Feynman diagram, we obtain

$$S_3^{\mu\nu\alpha\beta} = \frac{1}{2} \int \frac{d^D q}{(2\pi)^D} V^{\mu\nu\sigma\rho\lambda\alpha\beta}(p, q, k, l) G_{\sigma\tau\lambda\rho}(q), \qquad (12.8)$$

where the 4-vertex $V^{\mu\nu\sigma\rho\lambda\alpha\beta}(p, q, k, l)$ involves terms of the form

$$p^\rho q^\beta \eta^{\mu\sigma} \eta^{\nu\tau} \eta^{\lambda\alpha}, \text{ etc.} \qquad (12.9)$$

Each term has one momentum q^β and a D-dimensional integration, so that it vanishes,

$$\int \frac{d^D q}{(2\pi)^D} q^\beta f(q^2) = 0, \qquad (12.10)$$

because it is an odd function in q^α.

The other two tadpole diagrams (ii) and (iii) also have vanishing loop-integrals under dimensional regularizations due to (12.10) and

$$\int \frac{d^D q}{(2\pi)^D} (q^2)^{\beta-1} = 0, \quad \beta = 0, 1, 2, 3, \dots. \qquad (12.11)$$

These two tadpole diagrams have an additional complication because the loop and the external graviton line $\phi_{\mu\nu}(p) \rightarrow \phi_{\alpha\beta}(p')$ are connected by a graviton propagator involving a factor $1/(p - p')^2$, which diverges in the limit $p \rightarrow p'$. Thus, these two tadpole amplitudes have the form $0/0$, which is undefined. To overcome this complication, we adopt the dimensional regularization method with the prescription that one first calculates the D-dimensional integrals before carrying out the other limiting procedures such as $p \rightarrow p'$. In this way, the gravitational tadpoles do not contribute. In gauge theories with internal gauge groups, one might include counter terms to remove the contributions from such tadpole diagrams.[3]

The total divergent graviton self-energy amplitude $S^{\mu\nu\alpha\beta}(p)$ is thus the sum of (12.4) and (12.7),

$$S^{\mu\nu\alpha\beta} = S_1^{\mu\nu\alpha\beta} + S_2^{\mu\nu\alpha\beta} = \frac{ig^2}{48(4\pi)^2} \Gamma(0) \left[p^4(14\eta^{\alpha\nu}\eta^{\beta\mu} + 14\eta^{\alpha\mu}\eta^{\beta\nu} + 55\eta^{\alpha\beta}\eta^{\mu\nu}) \right.$$
$$+ 104 p^\alpha p^\beta p^\mu p^\nu - 54 p^2 (p^\alpha p^\beta \eta^{\mu\nu} + p^\mu p^\nu \eta^{\alpha\beta})$$
$$\left. - 13 p^2 (p^\alpha p^\mu \eta^{\beta\nu} + p^\beta p^\mu \eta^{\alpha\nu} + p^\alpha p^\nu \eta^{\beta\mu} + p^\beta p^\nu \eta^{\alpha\mu}) \right]. \qquad (12.12)$$

This self-energy amplitude in the DeWitt gauge is particularly simple. The calculation of this amplitude is done first for a general consistency check of symbolic computing. This calculation also paves the way to much more complicated calculations of self-energy in general gauge involving two gauge parameters ξ and ζ. The physical implications of this result will be discussed in section 12-3.

12-2.   Graviton self-energy in a general gauge

In general, the self-energy of a particle in gauge theories depends on gauge conditions or the gauge parameters. To wit, let us consider a general gauge-fixing Lagrangian involving two parameters, ξ and ζ:

$$L_{\xi\zeta} = \frac{\xi}{2g^2}\left[\partial^\mu J_{\mu\alpha} - \frac{\zeta}{2}\partial_\alpha J^\lambda_\lambda\right]\left[\partial_\nu J^{\nu\alpha} - \frac{\zeta}{2}\partial^\alpha J^\lambda_\lambda\right]. \tag{12.13}$$

In this case, the corresponding graviton propagator also involves two parameters,

$$\begin{aligned}
G_{\alpha\beta\rho\sigma}(k) = \frac{-i}{2}\Bigg[&\frac{1}{k^2}(\eta_{\alpha\beta}\eta_{\rho\sigma} - \eta_{\rho\alpha}\eta_{\sigma\beta} - \eta_{\rho\beta}\eta_{\sigma\alpha}) \\
&+ \frac{1}{k^4}\frac{\xi - 2}{\xi}(k_\sigma k_\beta\eta_{\rho\alpha} + k_\alpha k_\sigma\eta_{\rho\beta} + k_\rho k_\beta\eta_{\sigma\alpha} + k_\rho k_\alpha\eta_{\sigma\beta}) \\
&- \frac{2(\zeta - 1)}{k^4(\zeta - 2)}(k_\alpha k_\beta\eta_{\rho\sigma} + k_\rho k_\sigma\eta_{\alpha\beta}) \\
&- \frac{4(\zeta - 1)}{k^6\xi(\zeta - 2)^2}(6 - \xi - 2\xi\zeta - 2\zeta)k_\alpha k_\beta k_\rho k_\sigma\Bigg],
\end{aligned} \tag{12.14}$$

where again, the $i\epsilon$ prescription for the Feynman propagators is understood. In the limit, $D \to 4$, FeynCalc[1] gives the self-energy $S_1^{\mu\nu\alpha\beta}$ due to the graviton loop as

$$\begin{aligned}
S_1^{\mu\nu\alpha\beta} = \frac{ig^2}{(4\pi)^2}\frac{\Gamma(0)}{240\xi^2(\zeta - 2)^4}\Big[&S_a p^4\left(\eta^{\alpha\nu}\eta^{\beta\mu} + \eta^{\alpha\mu}\eta^{\beta\nu}\right) \\
&+ S_b p^4\eta^{\alpha\beta}\eta^{\mu\nu} + S_c p^\alpha p^\beta p^\mu p^\nu + S_d p^2\left(p^\alpha p^\beta\eta^{\mu\nu} + p^\mu p^\nu\eta^{\alpha\beta}\right) \\
&+ S_e p^2\left(p^\alpha p^\mu\eta^{\beta\nu} + p^\beta p^\mu\eta^{\alpha\nu} + p^\alpha p^\nu\eta^{\beta\mu} + p^\beta p^\nu\eta^{\alpha\mu}\right)\Big],
\end{aligned} \tag{12.15}$$

where

$$\begin{aligned}
S_a = \Big[&\left(69\xi^2 + 80\xi + 92\right)\zeta^4 - 8\left(54\xi^2 + 47\xi + 92\right)\zeta^3 \\
&+ 8\left(165\xi^2 + 52\xi + 252\right)\zeta^2 - 32\left(60\xi^2 - \xi + 68\right)\zeta \\
&+ 4\left(273\xi^2 - 60\xi + 196\right)\Big],
\end{aligned} \tag{12.16}$$

$$S_b = 8\left[\left(3\xi^2 + 20\xi + 19\right)\zeta^4 + \left(101\xi^2 - 62\xi - 152\right)\zeta^3 \right.$$
$$+ \left(-425\xi^2 + 112\xi + 312\right)\zeta^2 + 8\left(85\xi^2 - 67\xi - 4\right)\zeta$$
$$\left. - 381\xi^2 + 620\xi - 212\right], \tag{12.17}$$

$$S_c = 16\left[\left(17\xi^2 + 10\xi + 36\right)\zeta^4 - 3\left(17\xi^2 + 6\xi + 96\right)\zeta^3 \right.$$
$$+ 2\left(75\xi^2 - 56\xi + 404\right)\zeta^2 - 8\left(35\xi^2 - 22\xi + 116\right)\zeta$$
$$\left. + 186\xi^2 - 40\xi + 392\right], \tag{12.18}$$

$$S_d = -2\left[\left(17\xi^2 + 60\xi + 76\right)\zeta^4 + 8\left(58\xi^2 - 11\xi - 76\right)\zeta^3 \right.$$
$$- 8\left(225\xi^2 + 34\xi - 156\right)\zeta^2 + 16\left(155\xi^2 - 44\xi - 8\right)\zeta$$
$$\left. - 4\left(301\xi^2 - 380\xi + 212\right)\right], \tag{12.19}$$

$$S_e = \left[\left(21\xi^2 - 120\xi - 92\right)\zeta^4 + \left(-68\xi^2 + 776\xi + 736\right)\zeta^3 \right.$$
$$- 32\left(10\xi^2 + 53\xi + 68\right)\zeta^2 + 16\left(65\xi^2 + 98\xi + 176\right)\zeta$$
$$\left. - 4\left(193\xi^2 + 100\xi + 356\right)\right], \tag{12.20}$$

where we have used (12.2), (12.3) and (12.14).

For the general gauge-fixing Lagrangian (12.13), the graviton–ghost–ghost vertex $\overline{V}^\mu(p)V^\nu(q)\phi^{\alpha\beta}(k) \equiv U^{\mu\nu\alpha\beta}(p, q, k)$ and the ghost propagator $G^{\mu\nu}(q)$ are both dependent on the gauge parameter ζ and are given by

$$U^{\mu\nu\alpha\beta}(p, q, k) = \frac{ig}{2}\left[p^\alpha k^\nu \eta^{\mu\beta} + p^\beta k^\nu \eta^{\alpha\mu} + p \cdot q\left(\eta^{\mu\alpha}\eta^{\nu\beta} + \eta^{\mu\beta}\eta^{\nu\alpha}\right)\right.$$
$$+ p^\alpha q^\mu \eta^{\beta\nu} + p^\beta q^\mu \eta^{\alpha\nu} - \zeta p^\mu k^\nu \eta^{\alpha\beta}$$
$$\left. - \zeta p^\mu q^\beta \eta^{\alpha\nu} - \zeta p^\mu q^\alpha \eta^{\beta\nu}\right], \tag{12.21}$$

$$G^{\mu\nu}(q) = \frac{-i}{q^2}\left(\eta^{\mu\nu} - \frac{q^\mu q^\nu}{q^2}\frac{(1-\zeta)}{(2-\zeta)}\right), \quad \zeta \neq 2, \tag{12.22}$$

respectively. FeynCalc gives the following contribution of the ghost-loop to the graviton self-energy,

$$S_2^{\mu\nu\alpha\beta} = -\frac{ig^2}{(4\pi)^2}\frac{\Gamma(0)}{480\,(\zeta - 2)^2}\left[\left(15\zeta^2 - 46\zeta + 51\right)p^4\left(\eta^{\alpha\nu}\eta^{\beta\mu} + \eta^{\alpha\mu}\eta^{\beta\nu}\right)\right.$$
$$+ \left(5\zeta^2 + 14\zeta + 51\right)p^4\eta^{\alpha\beta}\eta^{\mu\nu} + 8\left(15\zeta^2 - 46\zeta + 41\right)p^\alpha p^\beta p^\mu p^\nu$$
$$+ 2(18\zeta - 23)p^2\left(p^\alpha p^\mu \eta^{\beta\nu} + p^\beta p^\mu \eta^{\alpha\nu} + p^\alpha p^\nu \eta^{\beta\mu} + p^\beta p^\nu \eta^{\alpha\mu}\right)$$
$$\left. - 2\left(5\zeta^2 + 22\zeta + 3\right)p^2\left(p^\alpha p^\beta \eta^{\mu\nu} + p^\mu p^\nu \eta^{\alpha\beta}\right)\right], \tag{12.23}$$

where the ghost-loop is associated with a factor (-1) due to the fermion nature of FDM ghost. Finally, we can now obtain the total self-energy amplitude $T^{\mu\nu\alpha\beta}$ for the graviton:[4]

$$T^{\mu\nu\alpha\beta} = S_1^{\mu\nu\alpha\beta} + S_2^{\mu\nu\alpha\beta}$$

$$= \frac{ig^2}{(4\pi)^2} \frac{\Gamma(0)}{480\,\xi^2(\zeta-2)^4} \Big[T_1\, p^4 (\eta^{\alpha\nu}\eta^{\beta\mu} + \eta^{\alpha\mu}\eta^{\beta\nu}) + T_2\, p^4 \eta^{\alpha\beta}\eta^{\mu\nu}$$

$$+ T_3\, p^\alpha p^\beta p^\mu p^\nu + T_4\, p^2 (p^\alpha p^\beta \eta^{\mu\nu} + p^\mu p^\nu \eta^{\alpha\beta})$$

$$+ T_5\, p^2 \left(p^\alpha p^\mu \eta^{\beta\nu} + p^\beta p^\mu \eta^{\alpha\nu} + p^\alpha p^\nu \eta^{\beta\mu} + p^\beta p^\nu \eta^{\alpha\mu} \right) \Big], \qquad (12.24)$$

where the coefficients $T_1, T_2 \ldots T_5$ are given by

$$T_1 = \left(123\xi^2 + 160\xi + 184\right)\zeta^4 - 2\left(379\xi^2 + 376\xi + 736\right)\zeta^3$$

$$+ \left(2345\xi^2 + 832\xi + 4032\right)\zeta^2 - 4\left(863\xi^2 - 16\xi + 1088\right)\zeta$$

$$+ 4\left(495\xi^2 - 120\xi + 392\right), \qquad (12.25)$$

$$T_2 = \left(43\xi^2 + 320\xi + 304\right)\zeta^4 + 2\left(811\xi^2 - 496\xi - 1216\right)\zeta^3$$

$$- \left(6815\xi^2 - 1792\xi - 4992\right)\zeta^2 + 4\left(2757\xi^2 - 2144\xi - 128\right)\zeta$$

$$- 6300\xi^2 + 9920\xi - 3392, \qquad (12.26)$$

$$T_3 = 8\Big[\left(53\xi^2 + 40\xi + 144\right)\zeta^4 - 2\left(49\xi^2 + 36\xi + 576\right)\zeta^3$$

$$+ \left(315\xi^2 - 448\xi + 3232\right)\zeta^2 - 4\left(193\xi^2 - 176\xi + 928\right)\zeta$$

$$+ 4\left(145\xi^2 - 40\xi + 392\right)\Big], \qquad (12.27)$$

$$T_4 = -2\Big[\left(29\xi^2 + 120\xi + 152\right)\zeta^4 + \left(926\xi^2 - 176\xi - 1216\right)\zeta^3$$

$$- \left(3535\xi^2 + 544\xi - 2496\right)\zeta^2 + \left(4884\xi^2 - 1408\xi - 256\right)\zeta$$

$$- 4\left(605\xi^2 - 760\xi + 424\right)\Big], \qquad (12.28)$$

$$T_5 = 2\Big[\left(21\xi^2 - 120\xi - 92\right)\zeta^4 - \left(86\xi^2 - 776\xi - 736\right)\zeta^3$$

$$- \left(225\xi^2 + 1696\xi + 2176\right)\zeta^2 + 4\left(219\xi^2 + 392\xi + 704\right)\zeta$$

$$- 8\left(85\xi^2 + 50\xi + 178\right)\Big]. \qquad (12.29)$$

The self-energy has a very complicated dependence on the gauge parameters ξ and ζ. In the special case $\xi = 2$ and $\zeta = 1$, the general result (12.24) reduces to the graviton self-energy in the DeWitt gauge given by (12.12), as expected.

12-3.   Discussion

The physical graviton propagator is different from the free propagator (12.3) in a general gauge because a graviton can disintegrate virtually into two gravitons and a FDM ghost pair for a small interval of time, as indicated in (12.1) and (12.5) respectively. A similar situation also occurs in quantum Einstein gravity.[5] After dimensional regularization, the ultraviolet divergence of the graviton self-energy contains only the simple pole terms in (12.24). In this sense, the ultraviolet divergence in Yang–Mills gravity is no worse than that in QED at the one-loop level. Yang–Mills gravity and Einstein gravity also have another type of divergence at low energies, i.e., the infrared divergence, which is found in other massless gauge theories, and which will not be discussed here.

In QED, it is easy to see how such a self-interaction affects the photon self-energy (or mass) by summing up the free photon propagator and its one-loop correction.[6] The situation is much more complicated for the graviton, however. For simplicity, let us consider the graviton propagator in the DeWitt gauge, i.e., (12.3). We have

$$G_{\alpha\beta\rho\sigma}(p) = -i\left[\frac{1}{2p^2}(\eta_{\alpha\beta}\eta_{\rho\sigma} - \eta_{\rho\alpha}\eta_{\sigma\beta} - \eta_{\rho\beta}\eta_{\sigma\alpha})\right], \qquad (12.30)$$

which is independent of the gauge parameters and corresponds to the Feynman gauge in QED. The one-loop correction to $G_{\mu\nu\alpha\beta}$ is

$$G^{(1)}_{\mu\nu\alpha\beta} = G_{\mu\nu\mu'\nu'}S^{\mu'\nu'\alpha'\beta'}G_{\alpha'\beta'\alpha\beta}, \qquad (12.31)$$

using (12.12) and (12.30), so that

$$G^{(1)}_{\mu\nu\alpha\beta} = \frac{-ig^2\Gamma(0)}{48(4\pi)^2 p^4}\left[14p^4\left(\eta^{\alpha\nu}\eta^{\beta\mu} + \eta^{\alpha\mu}\eta^{\beta\nu} + \eta^{\alpha\beta}\eta^{\mu\nu}\right)\right.$$

$$+ 104p^\alpha p^\beta p^\mu p^\nu + 28p^2\left(p^\alpha p^\beta \eta^{\mu\nu} + p^\mu p^\nu \eta^{\alpha\beta}\right)$$

$$\left. - 13p^2\left(p^\alpha p^\mu \eta^{\beta\nu} + p^\beta p^\mu \eta^{\alpha\nu} + p^\alpha p^\nu \eta^{\beta\mu} + p^\beta p^\nu \eta^{\alpha\mu}\right)\right]. \qquad (12.32)$$

Unlike the equivalent situation for photons in QED, the form of (12.32) differs from that of a free graviton propagator (12.30). This difference appears to suggest that the quantum correction of the graviton propagator does not affect the graviton mass.

We have thus seen that the divergences in the graviton self-energy at the one-loop level resembles those in QED, i.e., they all involve simple poles $\Gamma(0)$, corresponding to a logarithmic divergence. Thus, it is indeed possible to remove those divergences using gauge-dependent terms, even if the structure of the counter terms necessary to remove the divergences in the graviton self-energy is more complicated than the corresponding terms for the photon self-energy in QED. Presumably, these differences in the structure of the counter terms reflect the difference in the gauge symmetry groups of Yang–Mills gravity and QED: the external space-time translational (non-compact) group T_4 and the internal (compact) group U_1. To see more dramatic differences in the divergences, it is likely necessary to calculate the graviton self-energy at the 2-loop level.

The motivation for investigating the high-energy behavior of quantum gravity is to have a complete understanding of the whole of field theory, rather than just its experimental verifications. Because of the weakness of the coupling strength of the gravitational interaction ($\approx 10^{-39}$), the experimental effects of quantum gravity may never be detected in Earth-bound laboratories. However, we would like to make sure that observable results from QCD, the electroweak theory, and all other calculations at relatively low energies[7,8] will not be upset by higher order corrections due to the gravitational interaction. This is important theoretically because all physical particles in nature participate in gravitational interactions.[9]

References

1. We have employed R. Mertig's 'FeynCalc' and J. M. Martin-Garcia's package 'xAct' to carry out symbolic computations and to check the consistency of those computations. The symbolic computations were carried out in collaboration with S. H. Kim.

2. C. G. Bollini and J. J. Giambiagi, *Phys Lett. B* **40**, 566 (1972); *Nuovo Cimento B* **12**, 20 (1972); G. 't Hooft and M. T. Veltman, *Nucl. Phys. B* **44**, 189 (1972); J. F. Ashmore, *Lett. Nuovo Cimento* **4**, 289 (1972); G. Cicuta and E. Montaldi, *Lett. Nuovo Cimento* **4**, 329 (1972). For a brief review of different regularizations, see T. Muta, *Foundations of Quantum Chromodynamics* (World Scientific, 1987) pp. 103–109.

3. J. P. Hsu and J. A. Underwood, *Phys. Rev. D* **15**, 1668 (1977) and references therein.

4. J. P. Hsu and S. H. Kim, "The S-matrix and graviton self-energy in quantum Yang–Mills gravity", *Eur. Phys. J. Plus* (2012) **127**:146, DOI 10.1140/epjp/i2012-12146-3. Online: arXiv:1210.4503v1.

5. D. M. Capper and M. Ramon Medrano, *Phys. Rev. D* **9**, 1641 (1974); G. Leibbrandt, *Rev. Mod. Phys.* **47**, 849 (1975); D. M. Capper and M. A. Namazie, *Nucl. Phys. B* **142**, 535 (1978).

6. J. J. Sakurai, *Advanced Quantum Mechanics* (Addison-Wesley, 1967) pp. 273–275.

7. Q. Wang and Y. P. Kuang, in *Symposium on the Frontiers of Physics at Millennium.* (eds. Y. L. Wu and J. P. Hsu, World Scientific, 2001) pp. 51–59 and pp. 344–354; Y. P. Kuang and B. Fu, *Phys. Rev. D* **42**, 2300 (1990).

8. K. C. Chou and Y. L. Wu, *Phys. Rev. D* **53**, R3492 (1996), (Rapid Communication).

9. Thus, strictly speaking, unless one has a finite quantum theory of gravity, one cannot say that the observed physical mass of say, an electron, is finite and independent of the gauge condition.

13.

Space-Time Gauge Identities and Finite-Loop Renormalization

13-1.   Space-time (T_4) gauge identities

We continue our discussion of renormalizability in Yang–Mills gravity. In quantum electrodynamics, the Abelian U_1 gauge symmetry implies that a particular linear gauge condition can be imposed for all times.[1] This local gauge symmetry leads to the conservation of the electric charge and the Ward–Takahasi–Fradkin (WTF) identities,[2] which imply relations between different renormalization constants. However, in non-Abelian gauge theories with internal gauge groups, such as the Yang–Mills theory, a particular gauge condition cannot be imposed for all times. If one imposes a gauge condition at a certain time, it does not hold at a later time. Faddeev and Popov introduced a δ-function involving the gauge condition to the vacuum-to-vacuum amplitude in gauge theory, as shown in equation (10.10). As a result, there emerges a new Lagrangian involving additional couplings of gauge fields with anti-commuting scalar fields (or ghost scalar-fermions). These ghosts are called Faddeev–Popov ghosts, and appear only in the intermediate steps of a physical process. The non-Abelian gauge symmetry leads to Slavnov–Taylor identities, which are a generalization of the WTF identities.[3]

In Yang–Mills gravity, the situation is different from either of the cases described. It has both the Abelian T_4 gauge symmetry and a non-linear gauge field equation, so that a particular gauge condition for the Abelian T_4 gauge field cannot be imposed for all times. In order to restore gauge invariance and unitarity of the S matrix, there must be additional couplings between the T_4 gauge fields $\phi_{\mu\nu}$ and the anti-commuting vector fields $V_\mu(x)$ and $V'_\nu(x)$, which are Feynman–DeWitt–Mandelstam (FDM) ghosts in quantum Yang–Mills gravity.[4–6]

In this first part of this chapter, we derive generalized WTF identities in Yang–Mills gravity for a class of gauge conditions involving two gauge parameters ξ and ζ, as shown in the gauge-fixing Lagrangian (11.4). By explicit calculation in the lowest order (i.e., at the lowest tree-level), we show that the T_4 gauge identities are satisfied by a general propagator of the graviton involving ξ and ζ, and by FDM ghosts involving the parameter ζ.

Let us derive the general T_4 gauge identities for the class of gauge conditions

$$F_\lambda(J) = Y_\lambda(x), \quad J \equiv J_{\mu\nu} = \eta_{\mu\nu} + g\phi_{\mu\nu}, \tag{13.1}$$

$$F_\lambda(J) = \frac{1}{2}\left(\delta_\lambda^\mu \partial^\nu + \delta_\lambda^\nu \partial^\mu - \zeta\eta^{\mu\nu}\partial_\lambda\right)J_{\mu\nu} \equiv h_\lambda^{\mu\nu}J_{\mu\nu}. \tag{13.2}$$

In Yang–Mills gravity, we shall consistently use the covariant tensor $J_{\mu\nu}$ in (13.1) to express the T_4 gauge curvature $C_{\mu\nu\alpha} = J_{\mu\sigma}\partial^\sigma J_{\nu\alpha} - J_{\nu\sigma}\partial^\sigma J_{\mu\alpha}$, the gravitational action $S_\phi = \int d^4x L_\phi$, the gauge conditions, etc., and to perform variations and calculations.

First, we consider the vacuum-to-vacuum amplitude with the gauge conditions (13.1). It is given by

$$W = \int d[J]\,(\det U)\exp\left[i\int d^4x\left(L_\phi + L_{\xi\zeta}\right)\right] \tag{13.3}$$

$$= \int d\left[J, \overline{V}, V\right]\exp\left[i\int d^4x\left(L_\phi + L_{\xi\zeta} + \overline{V}^\mu U_{\mu\nu}V^\nu\right)\right], \tag{13.4}$$

$$L_{\xi\zeta} = \frac{\xi}{2g^2}F_\alpha F_\beta \eta^{\alpha\beta}, \quad F_\alpha = h_\alpha^{\mu\nu}J_{\mu\nu}, \tag{13.5}$$

where L_ϕ and $U_{\mu\nu}$ are given by (10.2) and (10.12), respectively. To generate Green's functions,[7,8] we introduce an external source term $j_{\mu\nu}$ for the T_4 gauge fields in the vacuum-to-vacuum amplitudes (13.3),

$$W[j] \equiv \int d[J]\,(\det U)\exp\left[i\int d^4x\left(L_\phi + L_{\xi\zeta} + J_{\mu\nu}j^{\mu\nu}\right)\right], \tag{13.6}$$

where $W[j]$, $\{d[J]\det U\}^a$ and $\int d^4x L_\phi$ are invariant under the infinitesimal T_4 gauge transformations (8.5)–(8.6),

$$T_{\mu\nu} \to (T_{\mu\nu})^\$ = T_{\mu\nu} - \Lambda^\lambda \partial_\lambda T_{\mu\nu} - T_{\mu\alpha}\partial_\nu\Lambda^\alpha - T_{\alpha\nu}\partial_\mu\Lambda^\alpha, \tag{13.7}$$

$$Q^{\mu\nu} \to (Q^{\mu\nu})^\$ = Q^{\mu\nu} - \Lambda^\lambda \partial_\lambda Q^{\mu\nu} + Q^{\lambda\nu}\partial_\lambda\Lambda^\mu + Q^{\mu\lambda}\partial_\lambda\Lambda^\nu. \tag{13.8}$$

In the derivation of the T_4 gauge identities, we impose the following constraint for the gauge function $\Lambda_\lambda(x)$

$$\Lambda_\lambda = \left[U^{-1}(J)\right]_{\lambda\sigma}\rho^\sigma \equiv \int\left[U^{-1}(x,y)\right]_{\lambda\sigma}\rho^\sigma(y)d^4y, \tag{13.9}$$

---

[a]The invariance of $d[J]\det U$ has been well-established for all compact Lie groups. However, for non-compact and infinite continuous groups, the formal proof of its invariance in the literature is not completely satisfactory. Cf. eqs. (13.56)–(13.57) and discussions below.

where ρ^σ is an arbitrarily infinitesimal function independent of $\phi_{\mu\nu}(x)$.

Under the infinitesimal T_4 gauge transformation (13.7)–(13.8), we have

$$J_{\mu\nu}^\$ = J_{\mu\nu} + \delta J_{\mu\nu}, \quad F_\nu^\$ = F_\nu + \delta F_\nu, \quad W[j]^\$ = W[j], \tag{13.10}$$

where

$$\delta J_{\mu\nu} = f_{\mu\nu\lambda}\Lambda^\lambda = f_{\mu\nu\lambda}\left[U^{-1}(J)\right]^{\lambda\alpha}\rho_\alpha, \tag{13.11}$$

$$\delta F_\lambda(J) = h_{\mu\nu\lambda}f^{\mu\nu\alpha}\Lambda_\alpha = h_{\mu\nu\lambda}f^{\mu\nu\alpha}[U^{-1}(J)]_{\alpha\beta}\rho^\beta = \rho_\lambda, \tag{13.12}$$

where we have used (13.9) and

$$\begin{aligned} f_{\mu\nu\lambda} &= -\left[(\partial_\lambda J_{\mu\nu}) + J_{\lambda\nu}\partial_\mu + J_{\mu\lambda}\partial_\nu\right], \\ h_\lambda{}^{\mu\nu} &= \frac{1}{2}\left(\delta_\lambda^\mu\partial^\nu + \delta_\lambda^\nu\partial^\mu - \zeta\eta^{\mu\nu}\partial_\lambda\right). \end{aligned} \tag{13.13}$$

The operator $U_{\rho\lambda}$ is given by

$$U_{\rho\lambda}(J) = -h_\rho{}^{\mu\nu}f_{\mu\nu\lambda}, \tag{13.14}$$

which is the same as that in equation (10.12) and consistent with (10.15) in chapter 10.

From (13.6) and (13.10), we have

$$\frac{\delta W[j]}{\delta\rho_\alpha} = \int d[\phi](\det U)\exp\left[i\int d^4x\left(L_\phi + L_{\xi\zeta} + J_{\mu\nu}j^{\mu\nu}\right)\right]Z^\alpha = 0, \tag{13.15}$$

where

$$Z^\alpha = \frac{\delta}{\delta\rho_\alpha}\int d^4x\left(\frac{\xi}{g^2}F_\alpha\delta F_\beta\eta^{\alpha\beta} + j^{\mu\nu}\delta J_{\mu\nu}\right). \tag{13.16}$$

Eq. (13.16) implies

$$Z^\alpha = \frac{\xi}{g^2}F^\alpha(J) + j^{\mu\nu}f_{\mu\nu\lambda}\left[U^{-1}(J)\right]^{\lambda\alpha}. \tag{13.17}$$

Thus, the T_4 identities (13.15) can also be written in the form

$$\left[\left(\frac{\xi}{g^2}F^\alpha(\Delta)\right)_y + \int d^4z\, j^{\mu\nu}(z)(f_{\mu\nu\lambda})_z\left[U^{-1}(\Delta)\right]^{\lambda\alpha}_{zy}\right]W[j] = 0,$$

$$\Delta \equiv \frac{1}{i}\frac{\delta}{\delta j}. \tag{13.18}$$

For those identities to be useful, we must know $[U^{-1}]W[j]$. Using (13.4) with an external source, let us consider[7]

$$W^{\mu\nu}[j] = \int d[\phi, \overline{V}, V] V^\mu \overline{V}^\nu \exp\left[i \int d^4x \left(L_{tot} + j^{\mu\nu} J_{\mu\nu}\right)\right], \quad (13.19)$$

$$L_{tot} = L_\phi + L_{\xi\zeta} + \overline{V}^\mu U_{\mu\nu} V^\nu, \quad (13.20)$$

where det U in (13.6) and (13.15) can be described as loops generated by FDM ghost fields $\overline{V}^\mu$ and V^ν, as shown in (10.14) and (10.15).

This Green's function for FDM ghosts V^μ and $\overline{V}^\nu$ in the presence of external source satisfies

$$[U(\Delta)]_{\mu\nu} W^{\nu\lambda}[j] = \delta_\mu^\lambda W[j]. \quad (13.21)$$

For the operator $U(\Delta)$, we have

$$\left[U^{-1}(\Delta)\right]^{\alpha\mu} [U(\Delta)]_{\mu\nu} = \delta_\nu^\alpha. \quad (13.22)$$

It follows from (13.18), (13.21) and (13.22) that

$$\int \exp\left[i \int d^4x (L_{tot} + J_{\mu\nu}(x) j^{\mu\nu}(x))\right]$$

$$\times \left[\frac{\xi}{g^2} \left(h_{\mu\nu\lambda} J^{\mu\nu}\right)_y + \int d^4z\, \overline{V}^\nu(y) j^{\mu\nu}(z) \left(f_{\mu\nu\lambda} V^\lambda\right)_z\right]$$

$$\times d[\phi, V, \overline{V}] = 0, \quad (13.23)$$

where

$$\left(f_{\mu\nu\lambda} V^\lambda\right)_z = \left[-\left(\frac{\partial}{\partial z^\lambda} J_{\mu\nu}(z)\right) - J_{\mu\lambda}(z)\frac{\partial}{\partial z^\nu} - J_{\lambda\nu}(z)\frac{\partial}{\partial z^\mu}\right] V^\lambda(z), \quad etc.$$

These are the gravitational space-time gauge identities for the generating functional $W[j]$ for the wide class of gauge conditions specified by (13.1) and (13.2).

13-2.   Space-time gauge identities and a general graviton propagator

In general, the graviton and ghost propagators in the Feynman–Dyson rules depend on the specific form of the gauge condition and gauge parameters. For example, if one chooses a gauge condition specified by the gauge-fixing Lagrangian (13.5) and (13.2) involving two arbitrary parameters ξ and ζ, the graviton propagator will depend on the two gauge parameters,[9]

$$G_{\alpha\beta\rho\sigma} = \frac{-i}{2k^2} \left([\eta_{\alpha\beta}\eta_{\rho\sigma} - \eta_{\rho\alpha}\eta_{\sigma\beta} - \eta_{\rho\beta}\eta_{\sigma\alpha}] - \frac{1}{k^2} \frac{(2\zeta - 2)}{(\zeta - 2)} \right.$$

$$\times (k_\alpha k_\beta \eta_{\rho\sigma} + k_\rho k_\sigma \eta_{\alpha\beta}) + \frac{1}{k^2} \frac{(\xi - 2)}{\xi}$$

$$\times (k_\sigma k_\beta \eta_{\rho\alpha} + k_\alpha k_\sigma \eta_{\rho\beta} + k_\rho k_\beta \eta_{\sigma\alpha} + k_\rho k_\alpha \eta_{\sigma\beta})$$

$$\left. - \frac{1}{k^4} \frac{(2\zeta - 2)}{\xi(\zeta - 2)} \left[\frac{4 + 2\xi - 4\xi\zeta}{\zeta - 2} - 4 + 2\xi \right] k_\alpha k_\beta k_\rho k_\sigma \right), \quad (13.24)$$

where the $i\epsilon$ prescription for the Feynman propagator is understood.

The propagator $G^{\mu\nu}$ of the FDM ghost particle can be derived from the effective Lagrangian (13.20) and the relation (13.14) for $U_{\mu\nu}$. We obtain

$$G^{\mu\nu} = \frac{-i}{k^2} \left(\eta^{\mu\nu} - \frac{k^\mu k^\nu}{k^2} \frac{(1 - \zeta)}{(2 - \zeta)} \right), \quad (13.25)$$

where ζ is not equal to 2.

Differentiating the gauge identities (13.23) with respect to $j_{\mu\nu}$ and letting all external sources vanish, we obtain the simplest identity

$$\xi \langle T (h^{\lambda\mu\nu}\phi_{\mu\nu})_x \phi^{\alpha\beta}(y) \rangle = -\langle T \overline{V}^\lambda(x)(f^{\alpha\beta\sigma} V_\sigma)_y \rangle, \quad (13.26)$$

where T denotes chronological ordering of the fields. This is a generalized Ward–Takahasi–Fradkin identity for an Abelian T_4 gauge group because (13.26) contains FDM ghosts.

In the lowest order, the identity (13.26), together with (13.2) and (13.13), leads to the identity,

$$\xi k^\beta G_{\alpha\beta\rho\sigma}(k) - \frac{\xi\zeta}{2} k_\alpha \eta^{\mu\nu} G_{\mu\nu\rho\sigma}(k) = -k_\sigma G_{\alpha\rho}(k) - k_\rho G_{\alpha\sigma}, \quad (13.27)$$

where $G_{\alpha\beta\rho\sigma}(k)$ and $G_{\alpha\rho}(k)$ are general propagators of the graviton (13.24) and the FDM ghost (13.25), respectively.

Direct calculations give the following results,

$$A_{\alpha\rho\sigma} \equiv \xi k^\beta G_{\alpha\beta\rho\sigma}(k) = \frac{-i}{2k^2} \left[-\xi k_\sigma \eta_{\rho\alpha} \frac{\zeta}{\zeta - 2} - 2(k_\sigma \eta_{\rho\alpha} + k_\rho \eta_{\sigma\alpha}) \right.$$

$$\left. + \frac{k_\alpha k_\rho k_\sigma}{k^2} \left(2\xi - 4 + \frac{2\zeta - 2}{\zeta - 2} \left[-3\xi + 4 + \frac{4\xi\zeta - 4 - 2\xi}{\zeta - 2} \right] \right) \right], \quad (13.28)$$

$$B_{\alpha\rho\sigma} \equiv -\frac{\xi\zeta}{2} k^\alpha \eta^{\mu\nu} G_{\mu\nu\rho\sigma}(k) = \frac{-i}{2k^2} \left[-\xi k_\alpha \eta_{\rho\sigma} \frac{\zeta}{\zeta - 2} + \frac{k_\alpha k_\rho k_\sigma}{k^2} \right.$$

$$\left. \times \left(-2\xi\zeta + 4\zeta + \frac{\zeta(2\zeta - 2)}{\zeta - 2} \left[-3\xi - 2 - \frac{2\xi\zeta - 2 - \xi}{\zeta - 2} \right] \right) \right]. \quad (13.29)$$

From (13.28) and (13.29), we obtain

$$A_{\alpha\rho\sigma} + B_{\alpha\rho\sigma} = \frac{-i}{k^2} \left[-(k_\sigma \eta_{\rho\alpha} + k_\rho \eta_{\sigma\alpha}) + \frac{2\zeta - 2}{\zeta - 2} \left(\frac{k_\alpha k_\rho k_\sigma}{k^2} \right) \right]. \quad (13.30)$$

Using (13.25), the right-hand-side of (13.26) gives the result,

$$C_{\alpha\rho\sigma} \equiv -k_\sigma G_{\alpha\rho}(k) - k_\rho G_{\alpha\sigma}$$

$$= \frac{-i}{k^2} \left[-k_\sigma \eta_{\alpha\rho} - k_\rho \eta_{\alpha\sigma} + \frac{2k_\sigma k_\alpha k_\rho (1 - \zeta)}{k^2(2 - \zeta)} \right]. \quad (13.31)$$

The left-hand-side of (13.27) given by (13.30) is the same as the right-hand-side of (13.27) given by (13.31). Thus, we have verified the space-time gauge identity (13.27) for general gauge conditions by explicit calculation using a very general graviton propagator involving two arbitrary gauge parameters, ξ and ζ. The calculation of identities for the next order with a one-loop correction is formidable and requires the aid of a computer.

13-3.   Gauge identities in quantum electrodynamics with a non-linear gauge condition

From the explicit calculations of the space-time gauge identities (13.27)–(13.31), one can see cancellations of a large number of terms due to the space-time gauge symmetry. The situation is similar to that in non-Abelian gauge field theory: When one calculates the imaginary part of the amplitudes coming from the interactions of say, the unphysical component of a vector boson and scalar ghost to verify unitarity of physical amplitudes, one sees many cancellations of unphysical amplitudes involving gauge parameters, so that unitarity of the S matrix is preserved. This is the power of gauge symmetry of internal groups.[8]

One might think that quantum chromodynamics has ghost particles solely because its gauge group $[SU_3]_{color}$ is non-Abelian and that QED, with the Abelian group U_1, does not have ghosts. However, this is true if and only if one imposes a linear gauge condition in QED. As a quick illustration of the power of gauge symmetry as instantiated by the generalized Ward–Takahasi–Fradkin (WTF) identities, we consider the case of QED with a class of non-linear gauge condition[1]

$$F(A) \equiv \partial_\mu A^\mu - \beta' A_\mu A^\mu = a(x), \quad \beta' \neq 0, \tag{13.32}$$

where $a(x)$ is a real function independent of A_μ and $\Lambda(x)$.

In the absence of electrons, the gauge invariant Lagrangian L_A for photons implies that photons do not couple with themselves and are free particles. However, when one chooses a non-linear gauge condition (13.32), there will be new 3-vertex and 4-vertex interactions of photons. These new photon–photon interactions upset the gauge invariance and unitarity of the theory. However, the U_1 gauge symmetry assures that the extra unwanted amplitudes produced by these new vertices will be cancelled by the interactions of new electromagnetic ghosts, so that gauge invariance and unitarity of the physical amplitudes are preserved.

The generating functional of the Green's function in the non-linear gauge can be written as[1]

$$W_\alpha\left[\eta, \overline{\eta}, \eta_\mu\right] = \int d[\Omega]\,(\det M) \exp\left[i \int \left(L + L_s - \frac{1}{2\alpha} F^2\right) d^4 x\right], \tag{13.33}$$

$$L = -\frac{1}{4} F_{\mu\nu} F^{\mu\nu} + L_\psi, \quad L_s = \eta_\mu A^\mu + \overline{\eta}\psi + \overline{\psi}\eta, \quad [\Omega] \equiv \left[A_\mu, \overline{\psi}, \psi\right], \tag{13.34}$$

$$M = (\partial_\mu - 2\beta' A_\mu)\partial^\mu, \quad \det M = \int d[\overline{c}, c] \exp\left(i \int d^4 x\, \overline{c} M c\right), \tag{13.35}$$

where $F_{\mu\nu} = \partial_\mu A_\nu - \partial_\nu A_\mu$, and L_ψ is the usual fermion Lagrangian in QED. The generalized WTF identities are obtained in the form

$$\left[-\frac{1}{\alpha}\left(\partial_\mu \frac{\delta}{i\delta\eta_\mu} + \beta' \frac{\delta^2}{\delta\eta^\mu \delta\eta_\mu}\right) + \eta_\mu \partial^\mu M^{-1}\right.$$
$$\left. - e\frac{\delta}{\delta\eta}\eta M^{-1} - e\overline{\eta}\frac{\delta}{\delta\overline{\eta}} M^{-1}\right] W_\alpha\left[\eta, \overline{\eta}, \eta_\mu\right] = 0. \tag{13.36}$$

Here, we have used the gauge transformation $A'_\mu(x) = A_\mu(x) + \partial_\mu \Lambda(x)$ with the constraint $\Lambda(x) = M^{-1}\rho(x)$, where $\rho(x)$ is an arbitrary infinitesimal real function, independent of A_μ. Following similar steps from (13.18) to (13.23), we can express the generalized WTF identities (13.36) in the form

$$\int \exp\left[i \int d^4x \left(L_{eff} + A_\mu \eta^\mu + \bar{\eta}\psi + \bar{\psi}\eta\right)\right]$$

$$\times \left[\frac{1}{\alpha}\left[(\partial_\mu - \beta' A_\mu) A^\mu\right]_y + \int d^4z\, \bar{c}(y)\eta^\mu(z)\left[\partial_\mu c\right]_z\right] d\left[\Omega, c, \bar{c}\right] = 0, \quad (13.37)$$

where

$$L_{eff} = L + \frac{1}{2\alpha}(\partial_\mu A^\mu - \beta' A_\mu A^\mu)^2 + \bar{c}Mc, \quad [\partial_\mu c]_z = \frac{\partial}{\partial z^\mu} c(z), \quad etc.$$

Thus, we have derived the generalized WTF identities involving electromagnetic ghosts $\bar{c}(x)$ and $c(x)$ for the generating functional $W_\alpha[\eta, \bar{\eta}, \eta_\mu]$ in the non-linear gauge specified by (13.32).

To the lowest order, the simplest identity obtained from (13.37) is given by

$$\frac{1}{\alpha}k^\mu G_{\mu\nu}(k) + k_\nu G(k) = 0, \tag{13.38}$$

which corresponds to (13.37). The propagators for the photon and the electromagnetic ghost are given by

$$G_{\mu\nu}(k) = \frac{-i}{k^2}\left(\eta_{\mu\nu} - (1-\alpha)\frac{k_\mu k_\nu}{k^2}\right) \quad \text{and} \quad G(k) = \frac{i}{k^2}, \tag{13.39}$$

respectively, where the $i\epsilon$ prescription for the propagator is understood. One can verify that the identity (13.38) is indeed satisfied.

The arbitrary gauge parameter β' does not appear in the lowest-order identity (13.38). It does appear in the higher order identities and also in the additional ghost–photon coupling in the effective Lagrangian L_{eff} in (13.37). However, when one calculates physical amplitudes such as e^+e^- scattering, the terms involving the gauge parameter β' are all cancelled,[1] so that the observable results are, as expected, unitary and independent of the parameter β'.

13-4. The infinite continuous group of general coordinate transformations in flat space-time

As discussed in section 7-5, both local translations and the general coordinate transformations in flat space-time can be expressed by

$$x^\mu \to x'^\mu = x^\mu + \Lambda^\mu(x), \tag{13.40}$$

where $\Lambda^\mu(x)$ is an arbitrary infinitesimal vector function. In chapter 8, we used the generators of space-time translations to formulate Yang–Mills gravity in flat space-time. Now for a detailed discussion of space-time gauge identities, we must also consider the generators and group properties of the general coordinate transformations in flat space-time.

According to the general transformation formula (8.3) for tensors such as scalars, vectors, etc., we have

$$Q(x) \to Q^\$(x) = Q(x) - \Lambda^\lambda(x)\partial_\lambda Q(x), \quad Q = \psi, \overline{\psi}, \Phi, \qquad (13.41)$$

$$V_\mu(x) \to V_\mu^\$(x) = V_\mu(x) - \Lambda^\lambda(x)\partial_\lambda V_\mu(x) - V_\lambda(x)\partial_\mu\Lambda^\lambda(x), \quad (13.42)$$

$$J_{\mu\nu}(x) \to J_{\mu\nu}^\$(x) = J_{\mu\nu}(x) - \Lambda^\lambda(x)\partial_\lambda J_{\mu\nu}(x) - J_{\mu\alpha}(x)\partial_\nu\Lambda^\alpha(x)$$
$$- J_{\alpha\nu}(x)\partial_\mu\Lambda^\alpha(x). \qquad (13.43)$$

The second term involving $-\Lambda^\lambda(x)\partial_\lambda$ on the RHS of (13.41)–(13.43) is the displacement term, which is related to the local space-time displacement. To see the structure of generators associated with the infinite continuous group of general coordinate transformations, we write (13.41)–(13.43) in the following form,

$$\delta Q(x) = Q^\$(x) - Q(x) = -\left[Q(x)\overleftarrow{\partial}_\lambda\right]\Lambda^\lambda(x)$$

$$= -\int \partial_\lambda\delta(x - x')Q(x')\Lambda^\lambda(x)dx'$$

$$= -\int \partial_\lambda\delta(x-x')\delta(x-x'')Q(x')\Lambda^\lambda(x'')dx'\,dx'', \qquad (13.44)$$

$$\delta V_\mu(x) = V_\mu^\$(x) - V_\mu(x) = -V_\alpha(x)\left[\overleftarrow{\partial}_\lambda\delta_\mu^\alpha + \delta_\lambda^\alpha\partial_\mu\right]\Lambda^\lambda(x)$$

$$= -\int \left[\partial_\lambda\delta(x - x')\delta(x - x'')\delta_\mu^\alpha + \delta_\lambda^\alpha\delta(x - x')\partial_\mu\delta(x - x'')\right]$$

$$\times V_\alpha(x')\Lambda^\lambda(x'')dx'\,dx'' \equiv H_\mu^{\alpha'}{}^{\lambda''}V_{\alpha'}\Lambda^{\lambda''}, \qquad (13.45)$$

$$\delta J_{\mu\nu}(x) = -J_{\alpha\beta}(x)\left[\delta_{\mu\nu}^{\alpha\beta}\overleftarrow{\partial}_\lambda + \delta_{\mu\lambda}^{\alpha\beta}\partial_\nu + \delta_{\lambda\nu}^{\alpha\beta}\partial_\mu\right]\Lambda^\lambda(x)$$

$$= -\int \left[\delta_{\mu\nu}^{\alpha\beta}\partial_\lambda\delta(x - x')\delta(x - x'') + \delta_{\mu\lambda}^{\alpha\beta}\delta(x - x')\partial_\nu\delta(x - x'')\right.$$

$$\left. + \delta_{\lambda\nu}^{\alpha\beta}\delta(x - x')\partial_\mu\delta(x - x'')\right]J_{\alpha\beta}(x')\Lambda^\lambda(x'')dx'\,dx''$$

$$\equiv H_{\mu\nu}^{\alpha'\beta'}{}^{\lambda''}J_{\alpha'\beta'}\Lambda^{\lambda''}, \qquad (13.46)$$

$$\delta(x) \equiv \delta^4(x), \quad dx' \equiv d^4x', \text{ etc.}$$

where summation and integration over repeated indices (and space-time variables) are understood. Let us compare (13.45) with the SU_2 gauge transformation of the Yang–Mills field b_μ^a,

$$\delta b_\mu^a(x) = \epsilon^{abc}\omega^b(x)b_\mu^c(x) + \frac{1}{f}\partial_\mu\omega_a(x). \tag{13.47}$$

We see that $H_\mu^{\alpha'}{}_{\lambda''}$ in (13.45) corresponds to the structure constant ϵ^{abc} in (13.47). For SU_2, we have $-i\epsilon_{abc} = (\tau_a/2)_{bc}$, where $\tau_a/2$ are the SU_2 generators. Similarly, one may consider $H_\mu^{\alpha'}{}_{\lambda''}$ as a matrix element of the generators $H_{\lambda''}$ of the group, which satisfy the commutation relation,[10]

$$[H_\mu, H_{\nu'}] = H_{\lambda''}C^{\lambda''}{}_{\mu\nu'}, \tag{13.48}$$

where the structure constants of the group are given by

$$C^{\lambda''}{}_{\mu\nu'} = \delta_\mu^\sigma\partial_\nu\delta(x'' - x)\delta(x'' - x') - \delta_\nu^\sigma\partial_\mu\delta(x'' - x')\delta(x'' - x). \tag{13.49}$$

The generators of the infinite continuous group are quite different from the internal groups in non-Abelian gauge theories, in which the generators have constant matrix representations.

There is a subtlety related to gauge transformations of the metric $d[J_{\mu\nu}]$ det U and their relation to the space-time gauge identities. To see it, let us examine the gauge transformation of $W[j^{\mu\nu}]$. When we calculate the Jacobian J_{co} of the gauge transformations (13.46) to the first order in $\Lambda^\lambda(x)$, we have

$$W[j]^\$ = \int d\left[J_{\alpha\beta}^\$\right]\left(\det U^\$\right)\exp\left[i\int d^4x(L_\phi + \frac{\xi}{2g^2}F_\mu^\$ F_\nu^\$ \eta^{\mu\nu} + J_{\mu\nu}^\$ j^{\mu\nu})\right].$$

$$= \int d[J_{\alpha\beta}]J_{co}\det(U + \delta U)$$

$$\times \exp\left[i\int d^4x\left(L_\phi + \frac{\xi}{2g^2}F_\mu F_\nu\eta^{\mu\nu} + J_{\mu\nu}j^{\mu\nu}\right.\right.$$

$$\left.\left. + \frac{\xi}{g^2}F_\mu\delta F_\nu\eta^{\mu\nu} + j^{\mu\nu}\delta J_{\mu\nu}\right)\right], \tag{13.50}$$

$$J_{co} = \det\left(\frac{\delta J_{\mu\nu}^\$}{\delta J_{\alpha\beta}}\right) = \det\frac{\delta}{\delta J_{\alpha\beta}}(J^{\mu\nu} + \delta J_{\mu\nu})$$

$$= \exp\left[\text{Tr}\frac{\delta}{\delta J_{\alpha\beta}}\delta J_{\mu\nu}\right] = \exp\left[\int d^4x\frac{\delta}{\delta J_{\alpha\beta}}\delta J_{\mu\nu}\right]_{\alpha\beta=\mu\nu}$$

$$= 1 + \left[\int d^4x (Q_1 + Q_2) \right], \tag{13.51}$$

$$Q_1 = \left(\frac{\delta f_{\mu\nu\lambda}}{\delta J_{\mu\nu}} \right) \Lambda^\lambda, \tag{13.52}$$

$$Q_2 = f_{\mu\nu\lambda} \left(\frac{\delta \Lambda^\lambda}{\delta J_{\mu\nu}} \right), \tag{13.53}$$

where $J^\$_{\mu\nu} = J_{\mu\nu} + \delta J_{\mu\nu}$ and $\delta J_{\mu\nu}$ are given by (13.43) and (13.46), respectively, and Λ^λ satisfies, by definition, the condition (13.9). The trace Tr in (13.51) includes both group indices and space-time labels (summation and integration over repeated indices and labels are always understood). We have also used the relations

$$\det (1 + Q) = \exp[\mathrm{Tr}\, \ln (1 + Q)] = \exp[\mathrm{Tr}\, Q], \quad \text{for small } Q. \tag{13.54}$$

The factor $\det (U + \delta U)$ in (13.50) can be evaluated as follows,

$$\begin{aligned}
\det (U + \delta U) &= \det U \det (1 + U^{-1}\delta U) \\
&= \det U \exp[\mathrm{Tr}\, \ln (1 + U^{-1}\delta U)] \\
&= (\det U)[1 + Q_3], \quad Q_3 = \mathrm{Tr}\, U^{-1}\delta U.
\end{aligned} \tag{13.55}$$

Similar to (13.15), we have the T_4 gauge identity

$$\int d[\phi] \, (\det U) \exp \left[i \int d^4x (L_\phi + L_{\xi\zeta} + J_{\mu\nu}j^{\mu\nu}) \right]$$

$$\times \left[\frac{\xi}{g^2} F_\mu \delta F_\nu \eta^{\mu\nu} + j^{\mu\nu}\delta J_{\mu\nu} + Q_1 + Q_2 + Q_3 \right] = 0, \tag{13.56}$$

which is obtained from (13.50) with the help of (13.51)–(13.55) and T_4 gauge invariance, $W[j]^\$ = W[j]$. Comparing (13.56) with (13.16), one can see that the last three terms in (13.56) are new terms arising from the gauge transformations of $d[J] (\det U)$. For any compact Lie group, one can show that the new terms corresponding to $Q_1 + Q_2 + Q_3$ in (13.56) do not contribute to the gauge identities.[7] The T_4 gauge group in Yang–Mills gravity is not a compact Lie group. It appears that one cannot evaluate $Q_1 + Q_2 + Q_3$ unambiguously because they involve delta functions and their space-time derivatives. If one uses (13.46) for $\delta J_{\mu\nu}$ to calculate Q_1, one obtains

$$\mathrm{Tr}\, H_{\mu\nu}{}^{\alpha'\beta'} \lambda''\frac{\delta J_{\alpha'\beta'}}{\delta J_{\mu\nu}} \Lambda^{\lambda''} = H_{\mu\nu}{}^{\mu\nu} \lambda'' \Lambda^{\lambda''}, \tag{13.57}$$

which is not well-defined.[11]

In Einstein gravity, the gauge transformation for tensor fields in curved space-time is similar to (13.46). The derivation of gauge identities also leads to a result similar to (13.56) in curved space-time. It has been argued that the trace $G_{\mu\nu}{}^{\mu\nu}\lambda''$ in Einstein gravity is formally zero,[12] however, such a conclusion based on a formal proof[12] is not completely satisfactory because of the presence of terms such as $0 \times \infty$.

Based on calculations of Feynman diagrams, it appears that one has well-defined T_4 gauge identities (13.23) (or (13.26)) at the tree-level, but the situation at the 1-loop level is not clear and needs further study.

13-5. Remarks on ultraviolet divergences and finite-loop renormalization for gravity

Based on the discussions of Feynman–Dyson rules for quantum Yang–Mills gravity and calculations of higher order Feynman diagrams in chapters 11 and 12, we attempt to explore a new view of ultraviolet divergences and possible 'renormalization' of gravity in this section. Nowadays, renormalization appears to be a necessary procedure for comparing the results of local quantum field theories, which involve ultraviolet divergences related to the idealized point particles, to experimental results.

Historically, the charged electron is assumed to have a finite radius in classical electrodynamics and the electron self-energy is identified with the electron mass. The advance of quantum electrodynamics (QED), however, has made this identification improbable.[13] In QED, the electron self-energy is a logarithmically divergent quantity. For QED to make sense and to agree with experiments, one must renormalize the theory. In this process, the electron self-energy is cancelled by a counter term so that the self-energy makes no contribution at all to the electron mass![b] Furthermore, if one looks at the photon self-energy in QED, it diverges quadratically by an 'honest' calculation of the Feynman diagrams.[13] This result is incompatible with the gauge invariance of QED and the observed photon mass. Finally, the graviton self-energy in general relativity and Yang–Mills gravity diverges quartically (at the one-loop level). What can one do with all these ultraviolet catastrophes? Thus far, the most effective way for dealing with them is

---

[b] This flip-flop in the interpretation of the electron mass from classical to quantum field theories suggests that such a renormalization is not the final story.

to renormalize the theory. Thus, a crucial property of a gauge theory is whether its ultraviolet divergences can be removed completely by a finite number of counter terms in the Lagrangian to make the S-matrix finite. In all gauge theories with internal gauge groups, this crucial property holds. In these renormalizable theories, all ultraviolet divergent quantities in higher order corrections can be canceled by a finite number of counter terms. Hence, the wonderful success of the electroweak theory and quantum chromodynamics.

In contrast, a theory of gravity with an external gauge symmetry involves a dimensional coupling constant and hence, cannot be renormalized by power counting. The reason is that such a theory has extra momentum in say, the graviton 3-vertex in comparison with the SU_2 gauge theory, as one can see from (11.26). This implies that each successive order has a worse divergence than the previous order. (Ordinarily, this would imply the need for an infinite number of counter terms in the Lagrangian to remove the infinities from a theory, if there were no special symmetry properties to alter the situation.) Thus, it is gratifying that the powerful dimensional regularization technique preserves the gauge invariance of a theory by putting all the different degrees of divergences in higher orders into the form of simple poles, which correspond to a logarithmical divergence.[14] Nevertheless, a new undesirable situation occurs in Einstein gravity with external space-time gauge groups related to the principle of general coordinate invariance. For example, in general relativity, when one calculates the one-loop self-energy of a graviton using Feynman diagrams, one obtains only simple pole terms. Nevertheless, the necessary counterterms to remove them are not part of a generally covariant term such as $\sqrt{-g}\,R^2$, $\sqrt{-g}R_{\mu\nu}R^{\mu\nu}$, etc.[15] Thus, in such a theory of gravity in curved space-time, the ultraviolet divergences cannot be removed completely by generally covariant counterterms in the Hilbert–Einstein Lagrangian to make the S-matrix finite.

Although Yang–Mills gravity is not renormalizable by power counting, the structure of its interactions has a close similarity to the usual renormalizable gauge theories. For example, the maximum number of graviton interactions in a vertex is 4, as shown in the Lagrangian L_ϕ in (10.2). In contrast, there exist vertices with an arbitrarily large number of gravitons in Einstein's theory of gravity and in other theories based on curved space-time.[16] It appears unlikely that the graviton coupling of Yang–Mills gravity can be further simplified because its gauge group is already the very simple Abelian group of 4-dimensional translation.[17] The problems of renormalizing Yang–Mills gravity is particularly interesting, because

(i) Yang–Mills gravity is formulated on the basis of flat space-time and has the maximum of 4-vertex graviton coupling, and (ii) the gravitational interaction is associated with a dimensional coupling constant. The combination of these two properties suggests a new and practical view of what it means to have a 'renormalizable' theory of quantum gravity, which may be simply stated as follows:

> In any theory of gravity involving a dimensional coupling constant, suppose that one can write down an action consisting solely of terms allowed by the gauge symmetry. Suppose the divergent terms in the finite-loop diagrams can be completely removed by counter terms in the action. Then, for all practical purposes, such a theory of gravity with finite-loop renormalization is essentially equivalent to a renormalized gauge theory.[18]

One may wonder: what is the physical meaning of (finite-loop and usual) renormalization? In particular, because of the experimental agreements of renormalization in quantum electrodynamics (QED) and electroweak theory, the success cannot be pure accidental and probably suggests some truth in the renormalization.[19] It is fair to say that the renormalization procedure to make a divergent field theory finite is, strictly speaking, not logical within the framework of local field theories.[20c] It may appear that physicists tried all kinds of tricks to avoid the divergent difficulties. Nevertheless, it seems reasonable to give the following explanation to the renormalization procedure in a local field theory: A truly satisfactory field theory must not have infinities and must agree with experiments. In such a finite field theory, the truly physical particle is probably not a point particle and, hence, a renormalized local field theory is some sort of 'non-local field theory.' For example, a physical particle may be more like a fuzzy point particle or associated with a fundamental length L_f or has a non-zero minimum position uncertainty, $(\Delta x_{min} \approx L_f)$.[21] However, so far we do not have the mathematics required to handle such a non-local quantum field theory and to preserve desirable symmetries, as one can see from the works of Yukawa, Katayama, and others.[22]

Under such a difficult situation, physicists avoid the insurmountable obstacles by constructing a local field theory and using renormalization (with dimensional regularization[16]) to remove unphysical infinities from the theory, so that *a field theory can employ simple mathematics with necessary symmetries to approximate the true physical theory*. This is indeed an ingenious idea because the renormalized QED has been a wonderful success experimentally. From this

---

[c] Local fields are continuous functions of space-time point x^μ and satisfy differential wave equations.

viewpoint, the un-renormalized QED with ultraviolet divergences is, logically, a truly local field theory, while the renormalized QED is some sort of effectively non-local field theory, which represents a better approximation to the physical world. Furthermore, whether a local field theory is renormalizable or finite-loop-renormalizable[23] does not seem to make much difference, as long as the theory provides a good description and can make predictions about nature. It is gratifying that gauge-field theorists circumscribe the difficulties encountered by Yukawa and others in a direct assault of the 'divergence dragons' in local field theory, and obtained splendid successes in QED and the unified electroweak theory. We hope that Yang–Mills gravity and its unification with other forces (to be discussed later) will have similar results.

References

1. However, if one chooses a non-linear gauge condition, the gauge condition cannot be imposed at all times, even in quantum electrodynamics. In this case, one needs to introduce scalar ghosts to restore gauge invariance and unitarity in electrodynamics. See, for example, J. P. Hsu, *Phys. Rev. D* **8**, 2609 (1973).

2. J. C. Ward, *Phys. Rev.* **77**, 2931 (1950); E. S. Fradkin, JETP **29**, 288 (1955); Y. Takahasi, *Nuovo Cimento* **6**, 370 (1957).

3. A. A. Slavnov, *Theor. Math. Phys.* **10**, 99 (1972); J. C. Taylor, *Nucl. Phys. B* **33**, 436 (1971).

4. R. P. Feynman, *Acta Phys. Polon.* **24**, 697 (1963).

5. B. S. DeWitt, *Phys. Rev.* **162**, 1195, 1239 (1967).

6. S. Mandelstam, *Phys. Rev.* **175**, 1604 (1968).

7. See, for example, B. W. Lee and J. Zinn-Justin, *Phys. Rev. D* **7**, 1049 (1973).

8. J. P. Hsu and J. A. Underwood, *Phys. Rev. D* **15**, 1668 (1977).

9. J. P. Hsu, *Eur. Phys. J. Plus* **127**: 35 (2012).

10. B. S. DeWitt, *Dynamical Theory of Groups and Fields* (Gordon and Breach, New York, 1965), pp. 102–105. DeWitt's discussion of the group of general coordinate transformations is in curved space-time, while our discussion of infinitesimal local translations and general coordinate transformations (13.40) is in flat space-time.

11. It appears that gauge identities (13.56) also contain some sort of 'coupling term' involving gravitons and FDM ghosts, which differ from the usual coupling vertex in the Feynman–Dyson rules. As a result, Yang–Mills gravity may not have gauge identities to all orders, which seems to be supported by explicit one-loop calculations with symbolic computing. The reason may be related to the gauge transformations (14.43) or (14.46),

which involve generators of infinite continuous (non-compact) group. The situation resembles that in Noether's Theorem II, which also involves generators of infinite continuous group. [See E. Noether, Goett. *Nachr.*, 235 (1918). English translation of Noether's paper by M. A. Tavel is online. Google search: M. A. Tavel, Noether's paper.]

12. D. M. Capper and M. Ramón Medrano, *Phys. Rev. D* **9**, 1641 (1974).

13. J. J. Sakurai, *Advanced Quantum Mechanics* (Addison-Wesley, 1984), p. 271. For a lucid discussion of the renormalization idea, see pp. 267–296.

14. In a sense, dimensional regularization effectively makes all couplings 'renormalizable,' just like those in theories that are renormalizable by power counting. The reason is that all ultraviolet divergences show up as simple poles $\Gamma(0)$ in dimensional regularization due to the relation $\Gamma(-n) = (-1)^n \Gamma(0)/n!$ for positive integers n. This property is quite different from that of Pauli–Villars regularization. (Cf. Ref. 13).

15. P. van Nieuwenhuizen, *Proc. Marcel Grossmann Meeting on General Relativity* (Ed. R. Ruffini, 1978 CERN Preprint TH2473); D. M. Capper and M. A. Namazie, *Nucl. Phys. B* **14**, 535 (1978); D. M. Capper, *J. Phys. A: Math. Gen.* **13**, 199 (1980).

16. R. Utiyama and T. Fukuyama, *Prog. Theor. Phys.* **45**, 612 (1971); Y. M. Cho, *Phys. Rev. D* **14**, 2521 (1976); K. Hayashi and T. Shirafuji, *Phys. Rev. D* **19**, 3524 (1979); A. A. Logunov and M. A. Mestvirishvili, *Progr. Theor. Phys.* **74**, 31 (1985); Ning Wu, *Commun. Theor. Phys. (Beijing, China)* **42**, 543 (2004).

17. Because of the translational gauge symmetry and the maximum 4-vertex coupling in Yang–Mills gravity, there is a possibility that the number of gauge invariant counter terms does not increase when the number of loops in Feynman diagrams keeps increasing.

18. Similar ideas have been used and discussed by some physicists. See B. L. Voronov and I. V. Tyutin, *Theore. Math. Phys.* **50**, 218 (1982); **52**, 628 (1982); M. Harada, T. Kugo and K. Yamawaki, *Prog. Theor. Phys.* **91**, 801 (1994); J. Gomis and S. Weinberg, see a discussion in *The Quantum Theory of Fields* (Cambridge Univ. Press, 1995; by S. Weinberg) Vol. 1, pp. 516–525; vol. 2, pp. 91–95. See also Ref. 23.

19. For example, the magnetic moment of the electron has been measured with a fantastic precision: $1.001\,159\,652\,180\,76 \pm 0.000\,000\,000\,000\,27$. The difference, i.e., experimental value minus theoretical value, is given by $-2.06(7.72) \times 10^{-12}$. See T. Kinoshita, talk at Nishina Hall, RIKEN (2010). (google search: Kinoshita, Magnetic moment.); Particle data group, *Particle Physics Booklet* (2012).

20. P. A. M. Dirac, *Lectures on Quantum Field Theory* (Academic Press, 1967) p. 2.

21. J. P. Hsu, *Nuovo Cimento B* **88**, 156 (1985); S. Y. Pei and J. P. Hsu, *Nuovo Cimento, B* **102**, 347, (1988). It appears that a departure from the concept of point particles

in field theories will lead to a departure from exact Lorentz and Poincaré invariance (or relativity theory) at very high energies.

22. See, for example, Y. Katayama, *Space-time Picture of Elementary Particles' in Proc. of the 1967 International Conference on Particles and Fields* (Ed. C. R. Hagen, G. Guralnik and V. A. Mathur, Interscience, 1967) p. 157.

23. The bias against needing an infinite number of counter terms to renormalize a Lagrangian seems analogous to earlier controversial views on infinite sets. Poincaré was dismayed by Cantor's infinite sets and said: "Actual infinity does not exist. What we call infinite is only the endless possibility of creating new objects no matter how many exist already." Gauss also had the view that the only access one had to the infinite is through the concept of limits and that one must not treat infinite sets as if they have an existence comparable to the existence of finite sets. However, one could interpret infinite numbers of counter terms as the endless 'steps' of approximation to the true physical theory. As long as the approximation gets closer and closer to the experimental value, why should one be dismayed by it? There seems to be no particular reason to believe that an experimentally consistent physical theory must be one that can be approximated by a local field theory with a finite number of 'steps'.

14.

A Unified Gravity–Electroweak Model

14-1. The gauge covariant derivative and gauge curvatures of a unified gravity–electroweak model

The electromagnetic and weak interactions have been 'satisfactorily' unified based on a gauge invariant action with the internal gauge group $SU_2 \times U_1$. The strong interaction can be described by quantum chromodynamics with color SU_3 groups. In previous chapters, we have shown that the gravitational interaction can be described by a gauge invariant action with an external space-time translational group. Because all the known interactions can be understood on the basis of gauge symmetries, they can, in principle, be unified. After all these years, the vision of Yang, Mills, Utiyama, Weyl and others may finally be realized.

Before considering a unification of all the fundamental interactions, in this chapter, let us first consider the gravity–electroweak sector of a unified model for simplicity. We would like to show that the gravitational interaction can be unified with the electroweak interaction of the Weinberg–Salam theory,[1,2] just as that theory unifies the electromagnetic and weak interactions. Furthermore, such a unification leads to new understandings and predictions, most notably that such a unified model predicts the violation of the electromagnetic U_1 and the weak SU_2 gauge symmetries by the gravitational interaction. This unification is based on the Yang–Mills–Utiyama-Weyl (YMUW) framework, which accommodates both external space-time gauge symmetry and internal gauge symmetries. Although this new unified model can be formulated in both inertial and non-inertial frames, the complications of a non-inertial frame may obscure the points we want to make and thus, our discussions of this new gravity–electroweak ('gravielecweak') model, its quantization and physical implications will be based on inertial frames with the metric tensor $\eta_{\mu\nu} = (+1, -1, -1, -1)$.

The basic natures of forces due to gauge fields are dictated by the mathematical properties of the generators associated with their gauge groups. Within our gravielecweak model, we use the gauge group $T_4 \times SU_2 \times U_1$, which allows us to introduce naturally a new symmetric tensor gauge field $\phi_{\mu\nu} = \phi_{\nu\mu}$

through replacement in the usual Yang–Mills framework,[3]

$$\partial_\mu \to \partial_\mu - ig\phi_\mu^v p_v + if W_\mu^a t^a + if' U_\mu t_o$$
$$= \partial_\mu + g\phi_\mu^v \partial_v + if W_\mu^a t^a + if' U_\mu t_o \equiv d_\mu,$$

where $p_v = i\partial/\partial x^v$ are the T_4 generators, t^a ($a = 1, 2, 3$) are the generators of SU_2 in general, and t_o (weak hypercharge) is the generator of U_1.[4] The operator d_μ may be called the unified gauge covariant derivative (UGCD). There are three independent gauge coupling constants: f and f' are dimensionless, and g has the dimension of length because the T_4 generator $p_\mu = i\partial/\partial x^\mu$ has the dimension of (1/length) in natural units.

The unified gravielecweak model leads to the following conceptual understanding of well-known forces in nature:

(A) Based on the T_4 gauge symmetry, the gravitational forces due to the exchange of virtual gravitons between two electrons, two positrons and between an electron and a positron, must be the same kind of force. The electron and the positron have the same attractive gravitational force characterized by one coupling constant g. The anti-particles are, roughly speaking, related to the complex conjugate of d_μ.

In contrast, the electromagnetic interaction can have both attractive and repulsive forces for matter–matter, matter–antimatter and antimatter–antimatter. These forces are characterized by two dimensionless coupling constants, $-e$ for electrons and $+e$ for positrons. This difference between the gravitational and the electromagnetic forces stems from the absence of an 'i' in the T_4 term involving $-ig\phi_\mu^v p_v = g\phi_\mu^v \partial_v$ in the UGCD d_μ.

(B) The UGCD d_μ can be expressed in terms of the gravitational potential $\phi_{\mu v}$ and the electromagnetic potential A_μ,

$$d_\mu = \partial_\mu + g\phi_\mu^v \partial_v + ieQA_\mu + \ldots.$$

where $Q = [f t_3 \sin\theta_W + f' t_o \cos\theta_W]/e = (t_3 + t_o)$.[4] The repulsive electric force between two electrons can be calculated from the exchange of a virtual photon using the formula,[5]

$$V = \frac{i}{(2\pi)^3} \int (-iM_{fi}) \exp[i\mathbf{k} \cdot \mathbf{r}] d^3k, \tag{14.1}$$

for the effective potential. The result is the well-known Coulomb potential in quantum electrodynamics. Similarly, the gravitational force between two electrons can be calculated from the exchange of a virtual graviton. The invariant amplitude $-iM_{fi}$ can be written using Feynman–Dyson rules (11.23) and the electron–graviton vertex (11.29). We have

$$-iM_{fi} = \left(\frac{ig}{2}\right)^2 \overline{u}(q)\gamma^\alpha u(p)(p^\beta + q^\beta)$$
$$\times G_{\alpha\beta\rho\sigma}\overline{u}'(q')\gamma^\rho u'(p')(p'^\sigma + q'^\sigma),$$

where the graviton propagator $G_{\alpha\beta\rho\sigma}$ in the DeWitt gauge is given by (11.23) with $\xi = 2$, $\zeta = 1$. The invariant amplitude $-iM_{fi}$ is for the scattering process $e^-(p)e^-(p') \to e^-(q)e^-(q')$ with the exchange of one graviton $\phi_{\mu\nu}(k)$, where we have $p + k = q$ and $p' = q' + k$ for two vertices of the Feynman diagram. In the classical and static limit, in which the electrons are at rest, we obtain the attractive potential $V = (-g^2m^2)/(8\pi r) = -Gm^2/r$ for $r > 0$, where we have used the relation in (9.1).[6] In quantum Yang–Mills gravity, the attractive gravitational force is interlocked with the property that gravitons are quanta of the symmetric tensor field $\phi_{\mu\nu}$.

To see the gauge curvatures associated with each group in the UGCD d_μ, let us calculate the commutators $[d_\mu, d_\nu]$. We obtain

$$[d_\mu, d_\nu] = -iC_{\mu\nu\alpha}p^\alpha + if W^a_{\mu\nu}t^a + if'U_{\mu\nu}t_o, \quad \left[t^a, t^b\right] = i\epsilon^{abc}t^c,$$
(14.2)

$$d_\mu = J^\nu_\mu \partial_\nu + if W^a_\mu t^a + if'U_\mu t_o, \quad J^\nu_\mu = \delta^\nu_\mu + g\phi^\nu_\mu,$$

where $J^\nu_\mu \partial_\nu \equiv \Delta_\mu$ is called the T_4 gauge covariant derivative. The T_4 gauge curvature $C_{\mu\nu\alpha}$ in flat space-time is given by (11.5), which is completely different from the Riemann-Christoffel curvature in general relativity. However, there exists a non-trivial gauge invariant action, which is quadratic in the new T_4 gauge curvatures $C_{\mu\nu\alpha}$, and is consistent with experiments, as discussed in chapters 8 and 9.

In the unified model, all gauge fields are assumed to appear in the UGCD d_μ on an equal footing just as in the Weinberg–Salam theory, as shown in (14.2). Consequently, the new SU_2 and U_1 gauge curvatures in the presence of the gravitational gauge potential are

$$W^a_{\mu\nu} = J^\sigma_\mu \partial_\sigma W^a_\nu - J^\sigma_\nu \partial_\sigma W^a_\mu - f\epsilon^{abc}W^b_\mu W^c_\nu, \quad \text{and} \qquad (14.3)$$
$$U_{\mu\nu} = J^\sigma_\mu \partial_\sigma U_\nu - J^\sigma_\nu \partial_\sigma U_\mu = \Delta_\mu U_\nu - \Delta_\nu U_\mu,$$

respectively. In the absence of gravity, i.e., $g = 0$, the gauge curvatures in (14.3) reduce to those in the Weinberg–Salam theory. We observe that the T_4 gauge potential appears in the SU_2 and U_1 gauge curvatures in (14.3), but that the SU_2 and U_1 gauge potentials do not affect the T_4 gauge curvature in (14.2). This shows the universal nature of the gravitational force, namely that it acts on all particles in the physical world.

14-2.   The Lagrangian in the gravielecweak model

Based on the gauge curvatures in (14.2)–(14.3), the new Lagrangian L_{gew} of the unified gravielecweak model (for leptons) can be assumed to take the form,

$$L_{gew} = L_\phi + L_{WS}, \tag{14.4}$$

$$L_\phi = \frac{1}{4g^2} \left(C_{\mu\nu\alpha} C^{\mu\nu\alpha} - 2C_{\mu\alpha}{}^\alpha C^{\mu\beta}{}_\beta \right),$$

$$L_{WS} = -\frac{1}{4} W^{a\mu\nu} W^a_{\mu\nu} - \frac{1}{4} U_{\mu\nu} U^{\mu\nu} + \overline{L} i\gamma^\mu d_\mu L + \overline{R} i\gamma^\mu d_\mu R$$
$$+ (d_\mu \phi)^\dagger (d^\mu \phi) - V(\phi^\dagger \phi) - \frac{m}{\rho_o} \left(\overline{L}\phi R + \overline{R}\phi^\dagger L \right),$$

using the usual notations,[4] i.e., R denotes the right-handed electron iso-singlet with only two independent spinor components, and L is the left-handed lepton iso-doublet, etc. It is possible to include all the quarks and leptons in the three families.[4] We observe that all terms in the original $SU_2 \times U_1$ Lagrangian are affected by the presence of the gravitational field $\phi_{\mu\nu}$ except the last two terms of L_{WS} in (14.4). This property suggests that the spontaneous symmetry breaking and the masses of particles in the model are independent of the gravitational interaction in the gauge invariant Lagrangian (14.4). Moreover, from the gravielecweak Lagrangian (14.4), we see that, similar to renormalizable Yang–Mills fields, the couplings of the gravitational gauge field and electroweak gauge fields are no more complex than a four-vertex in the Feynman diagrams.

14-3.   The equations of motion for quantum and classical particles

To explore the macroscopic implications of the gravielecweak model, it is more logically coherent to derive the equation of motion for a classical (or macroscopic) object from the wave equations for quantum particles in the model, rather than making a separate assumption about the motion of classical objects.

Thus, we investigate the classical limit (i.e., the geometric-optics limit) of the wave equations for gauge bosons, the Higgs boson, the electron, etc.

(A) The classical equation of motion for gauge bosons

For the bending of a light ray by the sun, we derived the eikonal equation for a light ray in the presence of gravity from the photon wave equation in Section 8–5, where the classical limit of the electron wave equation was also obtained. Now let us consider the classical limit of the wave equations for the gauge bosons. Similar to L_{em} in (8.34), we consider the following Lagrangian involving the massive gauge boson fields $W_\mu^a, a = 1, 2$, in the presence of gravity in the gravielecweak model:

$$L_W = -\frac{1}{4} W_{\mu\nu}^a W^{a\mu\nu} + \frac{1}{2} m_W^2 W_\mu^a W^{a\mu}, \quad a = 1, 2,$$

where the gauge curvature $W_{\mu\nu}^a$ is given by (14.3). This Lagrangian leads to a modified wave equation for the gauge boson with mass m_W in the presence of gravity,

$$\Delta_\mu W^{a\mu\lambda} + (\partial_\alpha J_\mu^\alpha) W^{a\mu\lambda} + m_W^2 W^{a\lambda} = 0, \quad \Delta_\mu = J_\mu^\nu \partial_\nu. \tag{14.5}$$

Using the gauge condition, $\partial_\mu W^{a\mu} = 0$, for vector gauge fields and the limiting expression[7] for the field

$$W^{a\lambda} = I^{a\lambda} \exp(i\Psi),$$

we can derive the 'eikonal equation' for massive vector bosons,

$$\left[\delta_\sigma^\lambda \left(G^{\mu\nu} \partial_\mu \Psi \partial_\nu \Psi - m_W^2 \right) - g\phi_\sigma^\mu J^{\lambda\nu} \partial_\mu \Psi \partial_\nu \Psi \right] I^{a\sigma} = 0,$$
$$G^{\mu\nu} = \eta_{\alpha\beta} J^{\alpha\mu} J^{\beta\nu},$$

in the geometric-optics limit (i.e., both the mass m_W and the wave 4-vector(or momentum) $\partial_\mu \Psi$ are very large).[7] Following the steps from (8.35) to (8.36), we obtain

$$\left[G_L^{\mu\nu} (\partial_\mu \Psi) \partial_\nu \Psi - m_W^2 \right] = 0, \tag{14.6}$$

where the effective Riemannian metric tensor for massive vector gauge bosons is

$$G_L^{\mu\nu} = G^{\mu\nu} - \frac{g}{4} \phi_\lambda^\mu J^{\lambda\nu} = \eta_{\alpha\beta}(\eta^{\mu\alpha} + g\phi^{\mu\alpha}) \left(\eta^{\nu\beta} + \frac{3g}{4} \phi^{\nu\beta} \right). \tag{14.7}$$

We can show that other gauge bosons such as Z^0 and the photon have the same eikonal Equations (14.6), except that m_W^2 is replaced by m_Z^2 and 0 respectively. (Cf. Equation (8.36).)

(B) The classical equations of motion for gravitons

It is interesting to ask whether there is an 'effective metric tensor' in the eikonal equation for gravitons in the geometric-optics limit. To answer this question, we use the free gravitational wave equation, i.e., (8.32) with $S^{\mu\nu} = 0$, and the following limiting expression in inertial frames,

$$\phi^{\mu\nu} = a^{\mu\nu} \exp(i\Phi_G), \quad a^{\mu\nu} = \epsilon^{\mu\nu}(\lambda)b(x),$$

for the tensor field, where the wave 4-vectors $\partial_\mu \Psi_G$ are very large. The polarization tensor $\epsilon^{\mu\nu}(\lambda)$ for gravitons satisfies $\sum_\lambda \epsilon_{\mu\nu}(\lambda)\epsilon_{\alpha\beta}(\lambda) \to -(\eta^{\mu\nu}\eta^{\alpha\beta} - \eta^{\mu\alpha}\eta^{\nu\beta} - \eta^{\mu\beta}\eta^{\nu\alpha})$ by taking the polarization sum, where the factor $-(\eta^{\mu\nu}\eta^{\alpha\beta} - \eta^{\mu\alpha}\eta^{\nu\beta} - \eta^{\mu\beta}\eta^{\nu\alpha})$ is the same as that in the graviton propagator (11.24), similar to the polarization vector for photons.[a] Following the steps (8.35)–(8.36), one can obtain the classical equations of motion for gravitons in flat space-time,

$$\eta^{\mu\nu}\partial_\mu \Psi_G \partial_\nu \Psi_G = 0, \tag{14.8}$$

which does not involve any effective Riemann metric tensor. This property is consistent with the fact that the gravitational field itself $\phi_{\mu\nu}$ is the physical origin of the effective Riemannian metric tensors in the classical equations of motion of all other particles.

(C) The classical equations of motion for leptons, quarks and Higgs particles

The wave equations for fermions (leptons and quarks) lead to the classical equation (i.e., the Einstein–Grossmann equation),

$$G^{\mu\nu}(\partial_\mu S)(\partial_\nu S) - m^2 = 0, \quad G^{\mu\nu} = \eta_{\alpha\beta}J^{\alpha\mu}J^{\beta\nu}, \tag{14.9}$$

---

[a]See discussions in (8.35)–(8.36).

as discussed in Section 8-5. We stress that only in the geometric-optics limit (i.e., the classical limit) of the modified Dirac wave Equation (8.37) with the space-time translational gauge symmetry does an 'effective Riemann metric tensor' $G^{\mu\nu}$ emerge, as shown in (14.9) or (8.40).

The Lagrangian for the Higgs field $\eta(x)$ in the presence of the gravitational gauge potential is given by

$$L_\eta = \frac{1}{2}\Delta_\mu\eta\Delta^\mu\eta - \frac{1}{2}m_H^2\eta^2. \tag{14.10}$$

The modified wave equation for the Higgs boson due to the presence of gravity can be derived from (14.10). We can show that it also reduces to the Einstein–Grossmann equation in the classical limit,

$$G^{\mu\nu}(\partial_\mu\Psi)(\partial_\nu\Psi) - m_H^2 = 0, \tag{14.11}$$

where we have used the following relations in the geometric-optics limit,

$$\eta = \eta_o\exp(i\Psi), \quad \partial_\lambda\partial_\alpha\Psi \ll (\partial_\alpha\Psi)\partial_\lambda\Psi. \tag{14.12}$$

The momentum $\partial_\mu\Psi$ and mass m_H in (14.12) are large quantities in the geometric-optics limit.

Thus, in the unified gravielecweak model, there are three different types of classical equations of motion for particles in the geometric-optics limit, in contrast to Einstein gravity:

(a) The eikonal (or Hamilton–Jacobi type) Equations (14.6) involving the metric tensor $G^{\mu\nu} = \eta_{\alpha\beta}J^{\alpha\mu}J^{\beta\nu} - (g/4)\phi_\lambda^\mu J^{\lambda\nu}$ for the massive gauge bosons $W^\pm$ and Z^0, and massless photons.
(b) The eikonal Equation (14.8) for gravitational rays involving the Minkowski metric tensors.
(c) The Einstein–Grossmann equation (14.9) involving the effective metric tensor $G^{\mu\nu} = \eta_{\alpha\beta}J^{\alpha\mu}J^{\beta\nu}$ for quarks, leptons and scalar particles with masses. All observable matter in the universe is composed of quarks and leptons and obeys the Einstein–Grossmann equation.

Equation (14.9) can describe both the motion of a classical object in free fall and the perihelion shift of the planet Mercury in an inertial frame. Thus, in the geometric-optics limit, Yang–Mills gravity is compatible with the equivalence principle in Einstein gravity.

14-4.   Violations of U_1 and SU_2 gauge symmetries by gravity

For gauge transformations in the gravielecweak model, we assume that the tensor gauge field $\phi_\mu^\nu(x)$ is an electrically neutral field and an iso-scalar, so that the T_4 gauge potential ϕ_μ^ν does not transform under SU_2 and U_1 gauge transformations. For other fields, the SU_2 and U_1 gauge transformations are the same as those in the Weinberg–Salam theory.[4] For example, in the absence of gravity (i.e., $g = 0$), local gauge transformations for $W_a^\mu \equiv W_a^\mu(x)$, $W_a^{\mu\nu} \equiv W_a^{\mu\nu}(x)$, $U_\mu \equiv U_\mu(x)$ and $U_{\mu\nu} \equiv U_{\mu\nu}(x)$ are given by

$$\delta W_a^\mu = \frac{1}{f}\partial^\mu \omega_a(x) + \epsilon_{abc}\omega_b(x)W_c^\mu, \quad [\delta W_a^{\mu\nu}]_{g=0}$$

$$= \epsilon_{abc}\omega_b(x)W_c^{\mu\nu}, \tag{14.13}$$

$$\delta U_\mu = \partial_\mu \omega(x), \quad \delta U_{\mu\nu} = 0, \tag{14.14}$$

for SU_2 and U_1, where $\omega_b(x)$ and $\omega(x)$ are infinitesimal functions.[4]

In the presence of gravity (i.e., $g \neq 0$), δU_μ is given by (14.14). However, the U_1 gauge transformations of the gauge curvature $U_{\mu\nu}$ differ from that in (14.14),

$$\delta U_{\mu\nu}(x) = J_m u^\sigma \partial_\sigma(\delta U_\nu) - J_\nu^\sigma \partial_\sigma(\delta U_\mu)$$

$$= g[\phi_\mu^\sigma \partial_\sigma \partial_\nu \omega(x) - \phi_\nu^\sigma \partial_\sigma \partial_\mu \omega(x)] \neq 0. \tag{14.15}$$

Similarly, the SU_2 gauge transformation of the gauge curvature $W_a^{\mu\nu}$ also differs from that in (14.13),

$$\delta W_a^{\mu\nu} = J^{\mu\sigma}\partial_\sigma(\delta W_a^\nu) - J^{\nu\sigma}\partial_\sigma(\delta W_a^\mu) + f\epsilon_{abc}(\delta W_b^\mu)(\delta W_c^\nu)$$

$$\neq [\delta W_a^{\mu\nu}]_{g=0}, \tag{14.16}$$

where δW_a^μ is given by (14.13). Thus, the Lagrangian L_{WS} in (14.4), which involves $W_{a\mu\nu}W_a^{\mu\nu}$ and $U_{\mu\nu}U^{\mu\nu}$, has only approximate $SU_2 \times U_1$ symmetry due to the presence of a very small quantity $g\phi_\mu^\sigma$ in (14.15) and (14.16).

As a result of this violation of $SU_2 \times U_1$ gauge symmetry by gravity, we have modified wave Equations (14.5) for the gauge bosons. This violation is interesting because in the classical limit we obtain the eikonal Equation (14.6) for the $W^\pm$ gauge boson, which involves an effective Riemann metric tensor $G_L^{\mu\nu}$ given by (14.7).

To see the effect of the violation of the U_1 gauge symmetry by gravity on the conservation of electric charge, consider the electromagnetic sector of the unified model. It suffices to concentrate on the action $\int d^4x L_{em}$ involving only the electromagnetic field A_μ and a charged fermion ψ, which are coupled to the gravitational field $\phi_{\mu\nu}$,[1,4]

$$L_{em} = -\frac{1}{4} F_{\mu\nu} F^{\mu\nu} + \frac{i}{2} \left[\overline{\psi} \gamma^\mu (\Delta_\mu - ieA_\mu) \psi \right.$$

$$\left. - \left[(\Delta_\mu + ieA_\mu) \overline{\psi} \right] \gamma^\mu \psi \right] - m \overline{\psi} \psi, \tag{14.17}$$

$$F_{\mu\nu} = \Delta_\mu A_\nu - \Delta_\nu A_\mu, \quad \Delta_\mu = (\delta_\mu^\nu + g\phi_\mu^\nu)\partial_\nu \equiv J_\mu^\nu \partial_\nu, \quad e > 0,$$

in inertial frames with the metric tensors $\eta_{\mu\nu} = (1, -1, -1, -1)$ and $c = \hbar = 1$, where A_μ is assumed to satisfy the usual gauge condition $\partial^\mu A_\mu = 0$. The $W^\pm$ and Z^o gauge bosons and others are not considered here. The unified gauge covariant derivative, $d_\mu = \Delta_\mu - ieA_\mu + \cdots = \partial_\mu + g\phi_\mu^\nu \partial_\nu - ieA_\mu + \cdots$, involves the generators of the space-time translational gauge symmetry, $p_\nu = i\partial_\nu$.

The generalized Maxwell's wave equations can be derived from (14.17). We have

$$\partial_\alpha (J_\mu^\alpha F^{\mu\nu}) = -e\overline{\psi} \gamma^\nu \psi, \quad J_\mu^\alpha = \delta_\mu^\alpha + g\phi_\mu^\alpha. \tag{14.18}$$

The wave Equation (14.18) implies a modified continuity equation for electric current in the presence of gravity,

$$\partial_\nu J_{total}^\nu = 0, \quad J_{tot}^\nu = -e\overline{\psi} \gamma^\nu \psi - g\partial_\alpha(\phi_\mu^\alpha F^{\mu\nu}), \tag{14.19}$$

where we have used the identity $\partial_\mu \partial_\nu F^{\mu\nu} = 0$.

Thus, the usual current $e\overline{\psi} \gamma^\nu \psi$ in the electroweak theory and quantum electrodynamics is no longer exactly conserved in the presence of gravity. Only a total current, composed of both the usual electromagnetic current and a new 'gravity-em current' [corresponding to the second terms in (14.19)], is conserved. The total conserved charge in the gravielecweak model is given by

$$Q_{tot} = \int J_{tot}^0 d^3x = \int \left[-e\overline{\psi} \gamma^0 \psi - g\partial_\alpha(\phi_\mu^\alpha F^{\mu 0}) \right] d^3x. \tag{14.20}$$

The new effect involves a factor $|g\phi_0^0| \approx |g\phi_1^1| \approx Gm/r$, which is roughly 10^{-9} on the surface of Earth. Such an effect could be tested experimentally if the electric field **E** and magnetic induction **B** (i.e., the electromagnetic field strength $F_{\mu\nu}$) are very large.[8]

Another important effect of the violation of U_1 and SU_2 gauge symmetries by gravity is that the classical equations of motion of fermions, bosons and scalar particles now involve effective Riemann metric tensors, as shown in Section 14-3. Thus the perihelion shift of the planet Mercury could be considered a demonstration of the violation of U_1 and SU_2 gauge symmetry in the unified gravielecweak model. As discussed in Section 9-3, a crucial test of the unified model and Yang–Mills gravity would come from a more accurate measurement of the deflection of light with a frequency in the optical range.

References

1. J. P. Hsu, *Intl. J. Mod Phys. A* **21**, 5119 (2006) and references therein; *Int. J. Mod Phys. A* **24**, 5217 (2009); J. P. Hsu and L. Hsu, *A Broader View of Relativity, General Implications of Lorentz and Poincaré Invariance* (World Scientific, 2006), Appendix D. Available online: 'google books, a broader view of relativity, Hsu.'

2. S. Weinberg, *The Quantum Theory of Fields*, (Cambridge Univ. Press, 1996), Vol. II, p. 305; R. E. Marshak, *Conceptual Foundations of Modern Particle Physics* (World Scientific, 1982) pp. 321–328. See also ref. 5.

3. J. P. Hsu, *Mod. Phys. Letts. A* **26**, 1707 (2011).

4. Kerson Huang, *Quarks, Leptons and Gauge Field* (World Scientific, 1982) pp. 108–117, 152–156, 241–242; R. E. Marshak, *Conceptual Foundations of Modern Particle Physics* (World Scientific, 1993) Ch. 2.

5. J. J. Sakurai, *Advanced Quantum Mechanics* (Addison-Wesley, 1967) p. 258–259.

6. The effective potential also involves divergent terms. For example, at $r = 0$, the effective potential due to the exchange of a graviton has a term involving $\delta^3(\mathbf{r})$. Moreover, the minus sign in the effective potential (14.1) occurs when the virtual quantum exchanged is a scalar particle (in Yukawa's attractive nuclear potential) or a tensor particle (in Yang–Mills gravity).

7. L. Landau and E. Lifshitz, *The Classical Theory of Fields* (Trans. by M. Hamermesh, Addison-Wesley, Cambridge, Mass. 1951), pp. 136–137.

8. The small new effect involves a term of the form $g\phi_\mu^\alpha(\partial_\alpha F^{\mu\nu})$, so if the field strength $F_{\mu\nu}$ has a very high frequency, one may be able to detect this new predicted effect.

15.

A Unified Gravity–Strong Force Model

15-1.   Unified gauge covariant derivatives and gauge curvatures

We now consider the unification of Yang–Mills gravity and quantum chromodynamics in the Yang–Mills–Utiyama–Weyl (YMUW) framework with two independent coupling constants. We call the resulting theory which involves the gauge group $T_4 \times [SU_3]_{color}$ 'quantum chromogravity' (QCG). Similar to the U_1 or SU_2 gauge symmetry in the electroweak theory, QCG involves T_4 gauge symmetry, so that the T_4 group generators appear in the Lagrangian through a gauge covariant derivative. In the YMUW framework, the T_4 gauge covariant derivatives are obtained through the following replacement,[1]

$$\partial_\mu \to \partial_\mu - ig\phi_\mu^\nu p_\nu = J_\mu^\nu \partial_\nu, \quad J_\mu^\nu = (\delta_\mu^\nu + g\phi_\mu^\nu), \tag{15.1}$$

in inertial frames with the metric tensor $\eta_{\mu\nu}$, where we use the representation $p_\mu = i\partial_\mu$. The operators p_ν denote the generators (in general) of the flat space-time translation group and satisfy $[p_\mu, p_\nu] = 0$. This is similar to the minimal substitution in quantum electrodynamics for introducing the electromagnetic coupling. In (15.1), a mixed tensor field ϕ_μ^ν is naturally associated with the T_4 generator $p_\mu = i\partial_\mu = i\partial/\partial x^\mu$ and the coupling constant g has the dimension of length.

In a general frame of reference (which may be non-inertial) with the Poincaré metric tensors $P_{\mu\nu}$, the form of the gauge covariant derivative (15.1) still holds, provided we replace ∂_μ by the covariant derivative D_μ associated with $P_{\mu\nu}$.[1] Therefore, the unified gauge covariant derivatives $\mathbf{d}_\mu$ in QCG for general frames are defined as

$$\mathbf{d}_\mu \equiv D_\mu + g\phi_\mu^\nu D_\nu + ig_s G_\mu^a \frac{\lambda^a}{2}, \tag{15.2}$$

where $\lambda^a (a = 1, 2, 3 \ldots 8)$ is the fundamental representation of the eight generators of color SU_3. The fields G_μ^a are the color gauge vector potentials, i.e., the gluon fields.[2-4]

In order to see the gauge curvatures associated with the unified gauge covariant derivative $\mathbf{d}_\mu$ in inertial and non-inertial frames, we calculate the commutators $[\mathbf{d}_\mu, \mathbf{d}_\nu]$ using the general expression (15.2). The purpose is to show that the T_4 gauge invariant quantum chromogravity can be formulated in both inertial and non-inertial frames. After some straightforward calculations, we have[5]

$$[\mathbf{d}_\mu, \mathbf{d}_\nu] = C_{\mu\nu\sigma} D^\sigma + ig_s G_{\mu\nu a} \frac{\lambda_a}{2}, \tag{15.3}$$

where the T_4 gauge curvature $C_{\mu\nu\sigma}$ is given by (11.5) with $\partial_\mu \to D_\mu$ and $\eta_{\mu\nu} \to P_{\mu\nu}$. The new $[SU_3]_{color}$ gauge curvature (i.e., the color gauge field strength tensor) $G_{a\mu\nu}$ in the presence of the gravitational gauge potential is now given by

$$G_{a\mu\nu} = J^\sigma_\mu D_\sigma G_{a\nu} - J^\sigma_\nu D_\sigma G_{a\mu} - g_s f_{abc} G_{b\mu} G_{c\nu} \tag{15.4}$$

$$= \Delta_\mu G_{a\nu} - \Delta_\nu G_{a\mu} - g_s f_{abc} G_{b\mu} G_{c\nu},$$

$$\Delta_\mu = J^\sigma_\mu D_\sigma, \quad J_{\mu\nu} = P_{\mu\nu} + g\phi_{\mu\nu}, \quad \phi_{\mu\nu} = \phi_{\nu\mu} = P_{\nu\lambda}\phi^\lambda_\mu, \tag{15.5}$$

where $a, b, c = 1, 2, \ldots 8$, the structure constants f_{abc} are anti-symmetric and Δ_μ is the T_4 gauge covariant derivative in arbitrary coordinates in flat space-time.

15-2. The action of the unified model and violations of local SU_3 gauge symmetry by gravity

Within the YMUW framework with the gauge curvatures (15.4) and (15.5), we can construct the Lagrangian L_G for gluon fields $G_{a\nu}(x)$ and the Lagrangian $L_{\phi q}$ for quark fields $q(x)$ and gravitational fields $\phi_{\mu\nu}$, similar to those in quantum chromodynamics and Yang–Mills gravity.[2,5] The action functional S_{QCG} and the Lagrangian L_{QCG} of the unified quantum chromogravity (QCG) model in general frames are naturally assumed to take the form,[5]

$$S_{QCG} = \int L_{QCG} d^4 x, \quad L_{QCG} = L_G + L_{\phi q}, \tag{15.6}$$

$$L_G = -\frac{1}{4} G^{a\mu\nu} G^a_{\mu\nu} \sqrt{-P}, \quad L_{\phi q} = \left[L_\phi + \bar{q} \left(i\Gamma^\mu \mathbf{d}_\mu - M \right) q \right] \sqrt{-P}, \tag{15.7}$$

$$L_\phi = \frac{1}{4g^2} \left(C_{\mu\nu\alpha} C^{\mu\nu\alpha} - 2 C^\alpha_{\mu\alpha} C^{\mu\beta}_\beta \right), \quad \{\Gamma_\mu, \Gamma_\nu\} = 2P_{\mu\nu},$$

$$\Gamma_\mu = \gamma_a e^a_\mu, \quad \{\gamma_a, \gamma_b\} = 2\eta_{ab}, \quad \eta_{ab} e^a_\mu e^b_\nu = P_{\mu\nu},$$

where $P = \det P_{\mu\nu}$ and $q's$ denote collectively spinor quarks with three colors and six flavors.[2,3] There are two independent quadratic forms that one can construct with the T_4 gauge curvature $C_{\mu\nu\alpha}$. Their linear combination, as shown in L_ϕ, is consistent with gravitational experiments, as discussed in Chapter 9. Although there is a small difference in the field equations for quarks and anti-quarks due to the presence of the gravitational gauge field $\phi_{\mu\nu}$ in (15.7), this difference can be eliminated by symmetrizing the fermion Lagrangian.[1] (See Equation (15.13) below.) We ignore the difference here since it does not affect the observable physical implications of the theory.

The $[SU_3]_{color}$ gauge transformations in the unified model are the same as those in the usual quantum chromodynamics, provided the gravitational gauge fields $\phi_{\mu\nu}(x)$ are assumed to transform trivially under $[SU_3]_{color}$ because they are not affected by the strong force. Under the $[SU_3]_{color}$ gauge transformations, the gauge curvature $G_{a\mu\nu}$ in (15.4) transforms according to the usual relation in quantum chromodynamics,

$$\delta G_{a\mu\nu}(x) = f_{abc}\omega_b(x)G_{c\mu\nu}(x),$$

only when $g = 0$, i.e., in the absence of the gravitational field or when the gauge function $\omega_b(x)$ is a constant. Thus, the local color SU_3 gauge invariant Lagrangian L_G in (15.7) is violated by the presence of gravity due to the presence of $\Delta_\mu = J_\mu^\sigma D_\sigma$ rather than D_μ in (15.4).

Although the unified model does not have the exact local color SU_3 gauge symmetry, the action S_{QCG} in (15.6) still has T_4 gauge invariance. The T_4 gauge transformations are more involved because the infinitesimal and local space-time translations are given by

$$x^\mu \to x'^\mu = x^\mu + \Lambda^\mu(x) \tag{15.8}$$

with an arbitrarily infinitesimal vector function $\Lambda^\mu(x)$. Because these local translations (15.8) with an arbitrary $\Lambda^\mu(x)$ are also general infinitesimal coordinate transformations, the T_4 gauge transformations for vector or tensor fields (in a general frame of reference with arbitrary coordinates) are formally the same as the Lie variation of tensors.[1] For a space-time-dependent tensor $Q_{\alpha_1...\alpha_n}^{\mu_1...\mu_m}(x)$ in the action (15.6) including $P_{\mu\nu}$, the T_4 gauge transformations are assumed to be

$$Q_{\alpha_1...\alpha_n}^{\mu_1...\mu_m}(x) \to (Q_{\alpha_1...\alpha_n}^{\mu_1...\mu_m}(x))^\$$$

$$= \frac{\partial x'^{\mu_1}}{\partial x^{\nu_1}} \cdots \frac{\partial x'^{\mu_m}}{\partial x^{\nu_m}} \frac{\partial x^{\beta_1}}{\partial x'^{\alpha_1}} \cdots \frac{\partial x^{\beta_n}}{\partial x'^{\alpha_n}} (1 - \Lambda^\lambda(x)\partial_\lambda) Q_{\beta_1...\beta_n}^{\nu_1...\nu_m}(x), \tag{15.9}$$

where $\mu_1, \nu_1, \alpha_1, \beta_1$, etc. are flat space-time indices. Here, we have suppressed indices of internal groups because they do not change the external gauge transformations (15.9). As usual, both the (Lorentz) spinor field ψ and the (Lorentz) scalar field Φ are treated as 'coordinate scalars' and have the same translational gauge transformation: $\psi \to \psi^\$ = \psi - \Lambda^\lambda \partial_\lambda \psi$, and $\Phi \to \Phi^\$ = \Phi - \Lambda^\lambda \partial_\lambda \Phi$. Moreover, the covariant derivative of a vector V_ν, i.e., $D_\mu V_\nu$, transforms like a covariant tensor $T_{\mu\nu}$.

In the QCG model, the symmetry related to the transformation (15.8) is particularly interesting because symmetries usually imply a conservation law. In this case, the space-time translational symmetry implies the conservation of the energy-momentum tensor. This is important because otherwise, the QCG model would not be satisfactory as it would not be consistent with an established conservation law. As we have mentioned previously, the Yang–Mills–Utiyama–Weyl framework is just general enough to realize this symmetry-conservation-law connection in all reference frames[6] (see Appendix A). In a more general framework such as curved space-time, this important symmetry-conservation-law connection is, in general, lost according to Noether's Theorem II. (Cf. Section 0-1 in overview).

Based on the translational gauge transformations (15.9) or (8.4)–(8.6), we can show that the scalar Lagrangian $L_Q = L_{QCG}/\sqrt{-P}$ transforms as follows,

$$L_Q \to (L_Q)^\$ = L_Q - \Lambda^\lambda (\partial_\lambda L_Q). \tag{15.10}$$

Since Λ^μ is an infinitesimal gauge vector function, the gauge transformation of $P_{\mu\nu}$ leads to

$$\sqrt{-P} \to \sqrt{-P^\$} = [(1 - \Lambda^\sigma \partial_\sigma)\sqrt{-P}](1 - \partial_\lambda \Lambda^\lambda), \quad P = \det P_{\mu\nu}. \tag{15.11}$$

As a result, the Lagrangian $\sqrt{-P}\, L_Q$ changes only by a divergence under the gauge transformation (15.9),

$$\int L_{QCG} d^4 x = \int \sqrt{-P}\, L_Q d^4 x \to \int [\sqrt{-P}\, L_Q - \partial_\lambda (\Lambda^\lambda L_Q \sqrt{-P})] d^4 x$$

$$= \int \sqrt{-P}\, L_Q d^4 x = \int L_{QCG} d^4 x. \tag{15.12}$$

The divergence term in (15.12) does not contribute to the field equations because one can transform an integral over a 4-dimensional volume into the integral of a vector over a hypersurface on the boundaries of the volume of integration where fields and their variations vanish. Thus, we have shown that, although the QCG Lagrangian is not explicitly gauge invariant, the action S_{QCG} given by (15.6) is invariant under the T_4 gauge transformations (15.9).

In order to see the big picture of gauge symmetries in all fundamental interactions in physics, consider both the unified gravielecweak and chromogravity models. In the gravielecweak model, the action $S_{gew} = \int d^4 x L_{gew}$, where L_{gew} is given in (14.4), can be generalized to arbitrary coordinates, just like S_{QCG} in (15.6). One can follow the steps from (15.8) to (15.12) to show that the gravielecweak action S_{gew} is invariant under the T_4 gauge transformations in a general frame of reference. Thus, the only exact local gauge symmetry in the actions of (14.4) and (15.6) is the local space-time translation gauge symmetry. All other internal gauge symmetries such as $SU_2 \times U_1$ in (14.4) and color SU_3 in (15.6) are violated by the gravitational interaction, as discussed in Sections 14–4 and 15–2. These are only approximate symmetries in the physical world. The violation of all internal gauge symmetries by gravity is an important result in the unified models because it leads to an effective curved space-time for the motion of all classical objects. (Cf. Sections 14–5, ch. 14 and 15–3 below.) Such violations of fundamental gauge symmetries are very difficult to detect in the high-energy laboratory however, unless a macroscopic object is involved.

15-3.   Effective curved space-time for motions of quarks and gluons in the classical limit

As usual, for simplicity, we discuss classical equations of motion in chromogravity in inertial frames. We have, for example,

$$\Delta_\mu = J_\mu^\sigma \partial_\sigma, \quad J_{\mu\nu} = \eta_{\mu\nu} + g\phi_{\mu\nu}, \quad \phi_{\mu\nu} = \phi_{\nu\mu} = \eta_{\nu\lambda}\phi_\mu^\lambda,$$

instead of (15.5). As mentioned previously, the presence of the translation gauge symmetry of gravity in the QCG model leads to an effective curved space-time for the motion of all classical objects. The reason for this is that T_4 gauge symmetry dictates the presence of T_4 gauge covariant derivatives $\Delta_\mu = J_{\mu\nu}\partial^\nu$ in the Lagrangians of all quark and gluon fields in the unified model. Since we are interested in the motion of classical objects, we set the strong coupling constant $g_s = 0$ in (15.4), (15.6) and (15.7) for simplicity. As shown in Equation (15.7) for the presence of gravitational interaction, the ordinary partial derivative ∂_μ in the free Lagrangian for scalar and fermion fields will be replaced by the T_4 gauge covariant derivative Δ_μ:

$$\overline{\psi}i\gamma^\mu\partial_\mu\psi \rightarrow \overline{\psi}i\gamma^\mu\Delta_\mu\psi, \quad (\partial_\mu\Phi)\partial^\mu\Phi \rightarrow (\Delta_\mu\Phi)\Delta^\mu\Phi,$$

where we consider inertial frames with the metric tensor $\eta_{\mu\nu}$ and ignore mass terms, which have little to do with the gravitational interaction. The effects of this replacement can be seen in Equations (15.2)–(15.7), where the gauge curvatures of gauge fields are also modified to be consistent with (15.1). This replacement implies that there is a universal coupling of gravity to all other fields through the unified gauge covariant derivatives (15.2), where $D_\mu = \partial_\mu$ for inertial frames, and the first two terms in (15.2) are the T_4 gauge covariant derivatives. Furthermore, all wave equations of quarks and gluons reduce to classical equations with 'effective Riemann metric tensors' in the geometric-optics (or classical) limit. This limiting property is key for the unified chromogravity theory to be consistent with all classical tests of gravity.

We now examine the classical limit of the quark and gluon wave equations. First, let us consider only quarks and a gravitational field for simplicity. In inertial frames, the Lagrangian for a quark q in the presence of the gravitational gauge potential takes the usual form

$$L_q = \frac{i}{2}\left[\bar{q}\gamma_\mu\Delta^\mu q - (\Delta^\mu\bar{q})\gamma_\mu q\right] - m\bar{q}q, \tag{15.13}$$

$$\gamma_\mu\gamma_\nu + \gamma_\nu\gamma_\mu = 2\eta_{\mu\nu}, \quad \Delta_\mu = J_{\mu\nu}\partial^\nu, \quad J_{\mu\nu} = J_{\nu\mu},$$

where we have symmetrized the quark Lagrangian so that both quarks and anti-quarks have the same coupling to gravity. Although as far as we know, free quarks do not exist, the wave equation for such a quark can be derived from (15.13)

$$i\gamma_\mu\Delta^\mu q - mq + \frac{i}{2}\gamma_\mu\left[\partial_\nu(J^{\mu\nu})\right]q = 0. \tag{15.14}$$

Using the limiting expression[7] for the fermion field $q = q_0\exp(iS)$, where q_0 is a constant spinor, and the properties that $g\phi_{\mu\nu}$ and $g\partial_\alpha\phi_{\mu\nu}$ are extremely small for gravity, we can write the quark Equation (15.14) in the form

$$\left[\gamma_\mu J^{\mu\sigma}\partial_\sigma S + m\right]q_0 = 0, \quad J^{\mu\sigma} = (\eta^{\mu\sigma} + g\phi^{\mu\sigma}). \tag{15.15}$$

In the classical limit, the momentum $\partial_\mu S$ and mass m are large quantities.[5,7] To eliminate the spin variables, we multiply (15.15) by a factor of $(\gamma_\sigma J^{\sigma\mu}\partial_\mu S - m)$. With the help of the anti-commutation relation for γ_σ in (15.13), Equation (15.15) leads to the Einstein–Grossmann equation with an effective Riemann metric $G_{\mu\nu}$,

$$G^{\mu\nu}(\partial_\mu S)(\partial_\nu S) - m^2 = 0, \quad G^{\mu\nu} = \eta_{\alpha\beta}J^{\alpha\mu}J^{\beta\nu}. \tag{15.16}$$

This is the equation of motion of a classical particle in the presence of the gravitational tensor field $\phi^{\mu\nu}$.

For the classical limit of the gluon wave equation in the presence of a gravitational field, let us consider the wave equation of a gluon field coupled only to the gravitational tensor field $\phi_{\mu\nu}$. In inertial frames with $P_{\mu\nu} = \eta_{\mu\nu}$, the wave equation of the gluon can be derived from the Lagrangian L_G in (15.7),

$$\partial_\mu(J_\sigma^\mu G_a^{\sigma\nu}) - g_s f_{abc} G_{b\mu} G_c^{\mu\nu} = 0. \tag{15.17}$$

Using the limiting expression $G_a^\mu = b_a^\mu \exp(iS')$ and choosing a 'Lorentz gauge' for the gluon field $\partial_\mu G_a^\mu = 0$, the wave Equation (15.17) for the massless gluon reduces to

$$J_\sigma^\mu J^{\sigma\lambda} \partial_\mu \partial_\lambda \left(b_a^\mu e^{iS'} \right) - g \phi_\sigma^\mu J^{\nu\lambda} \partial_\lambda \partial_\mu \left(b_a^\sigma e^{iS'} \right) = 0. \tag{15.18}$$

In the geometric-optics limit, the wave 4-vector $\partial_\mu S'$ is large so that (15.18) can be written as

$$\overline{G}^{\mu\nu}(\partial_\mu S')(\partial_\nu S') = 0, \quad \overline{G}^{\mu\nu} = G^{\mu\nu} - \frac{g}{4}\phi_\sigma^\mu J^{\sigma\nu}, \tag{15.19}$$

where $G^{\mu\nu} = \eta_{\alpha\beta} J^{\alpha\mu} J^{\beta\nu}$. Since macroscopic objects are not made of gluons and, as far as we know, free gluons do not exist, the additional term $(g/4)\phi_\sigma^\mu J^{\sigma\nu}$ in (15.19) probably cannot be tested experimentally. The result (15.19) for gluon field is the same as the eikonal Equation (8.36) for a light ray. The calculations in (15.13)–(15.19) can also be carried out in general frames of reference with arbitrary coordinates.

15-4. Discussion

Just as in the usual QCD theory, the Faddeev–Popov method is used to quantize the gauge fields in the unified QCG model. The quantization of the strong-interaction sector (i.e., the chromodynamics sector) is well-known,[2] while the quantization of the gravitational gauge field $\phi_{\mu\nu}$ has been discussed in Chapter 11, where the Feynman–Dyson rules for Yang–Mils gravity were also obtained.

The emergence of the effective metric tensor $G^{\mu\nu}$ in the geometric-optics limit is a general property of any field equation because all fields are coupled to gravity through the T_4 gauge covariant derivative $\Delta_\mu = J_{\mu\nu}\partial^\nu$ in Equation (15.1). Thus, any classical object moves as if in a 'curved space-time' with the metric tensor $G^{\mu\nu} = \eta_{\alpha\beta} J^{\alpha\mu} J^{\beta\nu}$, and behaves in a way consistent with the equivalence principle. However, the true underlying space-time of quantum chromogravity is flat. Consequently, one still has the true conservation of the energy-momentum tensor by Noether's theorem or by the Fock–Hilbert approach (see Appendix A). Furthermore, the usual quantization procedure for the gravitational field[5] can be carried out within QCG in inertial frames. The unified model suggests that the curvature of space-time revealed by the classical tests of gravity is a manifestation of the translational gauge symmetry in the classical limit within flat space-time.

Although (15.14) or (15.5) is, strictly speaking, only a formal wave equation for a single quark, one can consider the classical Einstein–Grossmann Equation (15.16) as the equation for a bound system of a large number of quarks with a large mass. Physically, the large mass and large eikonal in the classical limit are actually due to the summation of the individual masses and eikonals for each particle in the system under consideration, as discussed in Chapter 8.

In light of previous discussions of the gravielecweak and QCG models, it seems fair to say that the gravitational, strong, and electroweak interactions are all dictated by gauge symmetries (in the Yang–Mills–Utiyama–Weyl framework), even though all internal gauge symmetries are subtly violated by the extremely weak gravitational coupling. That gauge symmetries dictate interactions was advocated by Utiyama,[8] *et al.* and particularly stressed by Yang.[9] Dynamical symmetries such as U_1, SU_2, SU_3 and T_4 appear to be more profound (in the sense that they dictate fundamental interactions of nature) than the geometrical symmetries of the rotational group and Poincaré group, which are valid for all physical laws.[10] Furthermore, the unified model of chromogravity suggests that the fundamental role for our understanding of nature is played by the gauge curvatures of symmetry groups associated with conservation laws rather than by the space-time curvature of Riemannian geometry.

References

1. J. P. Hsu, *Int. J. Mod. Phys. A* **21**, 5119 (2006); J. P. Hsu and L. Hsu, *A Broader View of Relativity, General Implications of Lorentz and Poincaré Invariance* (World Scientific, 2006), Appendix D.
2. Kerson Huang, *Quarks, Leptons and Gauge Field* (World Scientific, 1982), pp. 241–242.
3. R. E. Marshak, *Conceptual Foundations of Modern Particle Physics* (World Scientific, 1993) Ch. 5.

4. S. Weinberg, *The Quantum Theory of Fields, vol. II* (Cambridge Univ. Press, 1996), pp. 152–157.

5. J. P. Hsu, *Chin. Phys. C*, 36(5), 403 (2012).

6. V. Fock, *The Theory of Space Time and Gravitation* (Pergamon Press, 1958) pp. 163–166.

7. L. Landau and E. Lifshitz, *The Classical Theory of Fields* (Trans. by M. Hamermesh, Addison-Wesley, Cambridge, Mass. 1951), pp. 136–137.

8. R. Utiyama, in *100 Years of Gravity and Accelerated Frames, The Deepest Insights of Einstein and Yang–Mills* (Ed. J. P. Hsu and D. Fine, World Scientific, 2005), p. 157. See also T. W. B. Kibble, *ibid*, p. 168.

9. C. N. Yang, in ref. 8, pp. 527–538.

10. E. Wigner, *Proc. Nat. Acad. Sci.* **51**, 5 (1964).

16.

Outlook

In this final chapter, we explore the possible implications of taiji space-time symmetry and the YMUW framework for some of the current problems in physics, including the accelerated expansion of the universe, the existence of the Higgs boson and the physical origin of particle masses, finite field theory and possible violation of Lorentz invariance at extremely high energies, and a complete unification of all known interactions.

16-1.   Taiji space-time — a basic framework for all physics

Thus far in physics, there are gauge theories for the electroweak and strong interactions based on a flat space-time, and a separate theory for the gravitational interaction based on a curved space-time. All of these theories have been impressively successful in explaining and predicting experimental results and all enjoy enthusiastic support. Nevertheless, physicists still search for a single comprehensive theory for all interactions based on a single space-time framework that can also accommodate both inertial and non-inertial reference frames. In view of the recently established accelerated expansion of the universe, it is clear that our understanding of space-time is far from complete.

The taiji symmetry framework we have discussed in this book is character-ized by the following features:

(a) It is a 4-dimensional flat space-time with arbitrary coordinates. These space-time coordinates have operational meanings.

(b) It can accommodate both inertial and non-inertial frames.

(c) All known conservation laws are natural consequences of the inherent symmetries of this framework.

(d) It has the most general symmetry under infinitesimal local space-time translations (15.8), which is mathematically the same as arbitrary infinitesimal coordinate transformations. However, the generators (and their associated covariant derivatives) of the local space-time translations and the arbitrary coordinate transformations groups are completely different.[a] Taiji symmetry (with the generators of the space-time translation group) is intimately related to Yang–Mills gravity and dictates a universal attractive force for all physical objects, including all gauge bosons and gravitons.

(e) It allows the quantization of all fields and the unification of all known interactions based on the Yang–Mills–Utiyama–Weyl framework, which is a generalization of the Yang–Mills framework.

Thus, taiji space-time symmetry provides a framework for a physically coherent model in which all interactions are unified. It avoids the problems that mar the existing conventional theory based on curved space-time with general coordinate invariance. In particular, one does not have to worry about losing either the operational meaning of space-time coordinates or the well-defined frames of reference in the presence of gravity.[1]

16-2. The cosmic Lee–Yang force and a linear potential for the accelerated expansion of the universe

Concerning the apparent accelerating expansion of the universe, what is 'written in the stars'[2] is no longer so clear. What contributions can the YMUW framework make towards an understanding of this phenomenon? This great enigma challenges the power of Yang–Mills' idea of gauge symmetry and of physicists' imaginations.

In 1955, Lee and Yang proposed a new gauge field to understand the experimentally established conservation of baryon charge (or number).[3,4] Their theory is formally identical to electromagnetism, except that the electric charge is replaced by a baryonic charge. They discussed a long-range repulsive force ($\propto 1/r^2$) between baryons. Using the Eötvös experiment, they estimated that the strength of such a repulsive force between nucleons (or baryons) was at least one million times smaller than that of the gravitational force.

Can such an extremely weak long-range repulsive force have cosmological significance for baryon matter? As we shall see, the answer is 'YES,' provided that this extremely weak baryonic gauge field B_μ has a linear potential. This requirement

---

[a]For the group generators, see (13.46) and (8.1). For their covariant derivatives, see (7.2), (7.3), and (7.34).

of a linear potential can be satisfied if the U_{1b} gauge field B_μ obeys fourth order differential equations rather than the usual second order differential equations.[5]

The gauge invariant Lagrangian for the massless U_{1b} gauge field is not unique because besides the usual Lagrangian, which is quadratic in the gauge invariant field strength $B_{\mu\nu} = \partial_\mu B_\nu - \partial_\nu B_\mu$, there is another simple gauge invariant Lagrangian, which is quadratic in $\partial_\lambda B_{\mu\nu}$.[5] Let us consider such a gauge invariant Lagrangian involving up and down quarks and the baryonic gauge field B_μ for an accelerated cosmic expansion (ACE):

$$L_{ace} = -\frac{L_s^2}{4} \partial_\lambda B_{\mu\nu} \partial^\lambda B^{\mu\nu} + L_{ud}, \tag{16.1}$$

$$L_{ud} = i\bar{u}_n \gamma_\mu \left(\partial^\mu - \frac{ig_b}{3} B_\mu \right) u_n - m_u \bar{u}_n u_n + (u \to d), \tag{16.2}$$

in the absence of gravity, where the color index n is summed from 1 to 3. One can easily include the gauge field associated with the conserved lepton numbers in (16.1), but we shall not discuss it here.

The new gauge-invariant field equation derived from (16.1) is a fourth-order partial differential equation of B_μ,

$$\partial^2 \partial_\nu B^{\nu\mu} - g_b' J_q^\mu = 0, \quad g_b' = g_b/(3L_s^2), \tag{16.3}$$

where the source of the gauge field is $J_q^\mu = \bar{u}_n \gamma^\mu u_n + \bar{d}_n \gamma^\mu d_n$.

In the static case, the scalar potential field B_0 satisfies the fourth order differential equation

$$\nabla^2 \nabla^2 B^0 = -\frac{g_b}{3L_s^2} J^0 \equiv \rho_B. \tag{16.4}$$

The general solution to B^0 can be expressed in terms of a modified version of the usual Green's function for a second order partial differential equation,

$$B^0(\mathbf{r}) = \int F(\mathbf{r}, \mathbf{r}') \rho_B(\mathbf{r}') d^3 r',$$

$$F(\mathbf{r}, \mathbf{r}') = \frac{1}{\nabla^2} \frac{-1}{4\pi|\mathbf{r} - \mathbf{r}'|} = \int \frac{e^{-i\mathbf{k}\cdot(\mathbf{r}-\mathbf{r}'')}}{k^2} \frac{d^3 k}{(2\pi)^3} \frac{1}{4\pi|\mathbf{r}'' - \mathbf{r}'|} d^3 r'', \tag{16.5}$$

where $1/\nabla^2$ is an integral operator.

Suppose we impose a 'Coulomb-like gauge' $\partial_k B^k = 0$, the static exterior potential then satisfies the equation $\nabla^2 \nabla^2 B^0 = 0$. The solution for such a potential is not unique: B^o may take the form A'/r, $B'r$, or $C'r^2$. However, it turns out that only the solution $B'r$ can be understood as a potential due to the exchange of virtual quanta described by the fourth order field equation (16.3).

The reason for this is that from the viewpoint of quantum field theory, the fermion (baryon) source ρ_B in (16.4) is represented by the usual delta-function because these fermions satisfy the Dirac equations. Based on this property, the solution for the potential B_0 should be proportional to r rather than to r^2 or $1/r$. This result can be seen explicitly by substituting $\rho_B(\mathbf{r}) = g'_b \delta^3(\mathbf{r})$ in (16.4), where $g' = g_b/(3L_s^2)$ for quarks (and $g' = g_b/(L_s^2)$ for protons and neutrons). We obtain a linear potential in the static limit,

$$B^0(\mathbf{r}) = g'_b \int_{-\infty}^{\infty} \frac{1}{(\mathbf{k}^2)^2} e^{i\mathbf{k}\cdot\mathbf{r}} d^3 k = -\frac{g'_b}{8\pi} |\mathbf{r}|, \tag{16.6}$$

which can be understood as a generalized function.[6]

This linear potential (16.6) produces a constant repulsive Lee–Yang force between baryonic galaxies. The equation of motion of a freely moving test particle in the Newtonian limit then becomes[7]

$$\frac{d^2\mathbf{r}}{dt^2} = \mathbf{g} + \mathbf{g}_{LY}, \tag{16.7}$$

where $\mathbf{g}$ is the gravitational acceleration produced by the distribution of ordinary matter, while the constant acceleration $\mathbf{g}_{LY}$ is due to the modified Lee–Yang baryon gauge field.[7]

Because of the extreme weakness of the coupling strength of the baryonic gauge field B_μ, its existence cannot be detected in our solar system or even within our galaxy. However, the source of this field is ubiquitous baryonic matter rather than some mysterious 'dark energy' in the universe. This model suggests that the accelerated cosmic expansion could be understood in terms of the cosmic Lee–Yang force produced by the baryonic gauge field. One advantage of the Yang–Mills idea of gauge symmetry is that the Lagrangian (16.1) for this field can be easily included in a model that unifies all the known interactions.

In contrast, if one were to assume a suitable cosmological constant λ in Einstein's gravitational field equation, one would obtain a solution for a static potential involving the term $+\lambda r^2/6$ in the Newtonian limit.[7] Instead of the constant acceleration in (16.7), a cosmological constant leads to an r-dependent acceleration for the cosmic expansion: $d^2\mathbf{r}/dt^2 = \mathbf{g} + C\mathbf{r}$. Thus, this difference between the modified Lee–Yang baryon gauge theory and Einstein gravity with a cosmological constant can be tested experimentally. We note, however, that such a cosmological constant is neither in harmony with local field theory nor with the conservation law of energy-momentu. In the Newtonian limit, the cosmological constant λ appears as a constant source in the universe and produces the quadratic potential field, $+\lambda r^2/6$ for the accelerated cosmic expansion. In this sense, the cosmological constant appears to be some sort of 'new aether' with a constant density everywhere in the universe,[7] resembling the old aether of 19th century electromagnetism.

In connection with linear potentials, the phenomenon of quark confinement could also be understood in terms of a linear potential associated with an attractive force at low energies. For a linear potential to produce only attractive forces between quarks and quarks, quarks and anti-quarks, and anti-quarks and anti-quarks, the gauge fields must be tensors of the second rank or a scalar field. This property also suggests that the relevant gauge symmetry is likely to be an external rather than an internal symmetry.[b] It would be very interesting if the potentials at the smallest known distance scales responsible for the confinement of quarks and at the largest known scales responsible for the accelerated expansion between galaxies were both due to linear potentials associated with the fourth-order differential field equations in the low energy approximation.[8] If this were to be true, then 'particle cosmology' would be a very interesting facet of the physics of the future.

16-3. Possible origins of mass in a unified model: constant vacuum field or Higgs field?

The origin of mass is one of the greatest enigmas in physics of all time. In the electroweak theory, the masses are generated by the Higgs boson. However, we would like to explore a new concept of symmetry breaking in which, instead of having the Higgs boson with spontaneous symmetry breaking, the correct gauge boson masses M_W and M_Z are generated by the coupling of a 'constant vacuum field' (isodoublet) to gauge fields. Such a mechanism of generating mass may be called a 'vacuum symmetry breaking,' in which an isodoublet constant vacuum field is assumed to break the gauge symmetry of a massless Lagrangian.[9]

---

[b]It appears that gauge fields associated with internal groups produce both attractive and repulsive forces, just like electromagnetic forces.

Suppose we have a massless Lagrangian L_I with the $SU_2 \times U_1$ gauge symmetry

$$L_I = -\frac{1}{4}(\partial_\mu W_\nu - \partial_\nu W_\mu - f W_\mu \times W_\nu)^2 + \bar{L} i \gamma^\mu D_\mu L$$

$$- \frac{1}{4}(\partial_\mu U_\nu - \partial_\nu U_\mu)^2 + \bar{e}_R i \gamma^\mu D_\mu e_R, \tag{16.8}$$

$$D_\mu = \partial_\mu + if W_\mu \cdot \mathbf{t} + if' U_\mu t_o.$$

$$L = \begin{pmatrix} \nu_L \\ e_L \end{pmatrix} \quad e_R = \frac{1}{2}(1 + \gamma^5)e(x), \tag{16.9}$$

where $\mathbf{t}$ and t_o are generators of SU_2 and U_1 respectively.

According to the idea of vacuum symmetry breaking, there exist constant vacuum isodoublets, which are assumed to have the form[9]

$$V_+ = \begin{pmatrix} \sqrt{F} \\ 0 \end{pmatrix}, \quad V_- = \begin{pmatrix} 0 \\ \sqrt{G} \end{pmatrix}. \tag{16.10}$$

These can couple to the dynamical fields W_μ and U_μ in $SU_2 \times U_1$ through the gauge covariant derivative

$$L_M = (D_\mu V_+)^\dagger (D^\mu V_-) - (\bar{L} V_- e_R + \bar{e}_R V_-^\dagger L)$$

$$= M_W^2 W_\mu W^{+\mu} + \frac{1}{2} M_Z^2 Z_\mu Z^\mu - m_e \bar{e}(x)e(x), \tag{16.11}$$

where we have used the usual definitions[10]

$$M_W^2 = f^2 F/2, \quad M_Z^2 = F(f^2 + f'^2)/2, \quad m_e = G, \tag{16.12}$$

$$W_\mu = (W_\mu^1 + iW_\mu^2)/\sqrt{2}, \quad Z_\mu = (fW_\mu^3 - f'U_\mu)/\sqrt{f^2 + f'^2}. \tag{16.13}$$

The photon field A_μ and the electric charge e are identified as follows,

$$A_\mu = (f'W_\mu^3 + fU_\mu)/\sqrt{f^2 + f'^2}, \quad ff'/\sqrt{f^2 + f'^2} = e, \tag{16.14}$$

and the U_1 gauge symmetry is not broken due to the absence of gravity.

In this model, the total lepton Lagrangian is assumed to be

$$L_{lepton} = L_I + L_M. \tag{16.15}$$

Since the model involves the 'vacuum isodoublets,' which cause the symmetry breaking that generates mass, it may be termed the 'VI-unified model.'

It follows from (16.12) that the ratio M_Z/M_W is given by

$$\frac{M_Z}{M_W} = \frac{\sqrt{f^2 + f'^2}}{f} = \frac{1}{\cos\theta_W}, \tag{16.16}$$

which is the same as that in Weinberg's lepton model and is consistent with high-energy experiments. Thus, in this framework, the Higgs boson is not necessary for producing the correct masses of gauge bosons, $W^\pm$ and Z^0. Instead, the necessary element is the $SU_2 \times U_1$ gauge symmetry.

This VI-unified model with an isodoublet vacuum field is not renormalizable however, because the mass term L_M in (16.15) is not generated through the spontaneous symmetry breaking of the gauge invariant massless Lagrangian L_I, in contrast to Weinberg-Salam unified model. Since L_I has $SU_2 \times U_1$ symmetry, the VI-unified model based on (16.15) is expected only to be finite-loop renormalizable, similar to Yang–Mills gravity.

The VI-unified model suggests an important experimental test of the key roles played by the two concepts of spontaneous symmetry breaking and renormalizability in a unified model. If the Higgs boson is detected, then these two concepts prove to be necessary and indispensable in the electroweak theory.[c] On the other hand, if the Higgs boson is not detected, then these two concepts are not indispensable and may be replaced by the simpler concepts of vacuum symmetry breaking and finite-loop renormalizability in the VI-unified model.

16-4. Finite quantum gravity and a possible departure from exact Lorentz invariance at high energies

Finite quantum gravity and gauge field theories may not have any experimental implications in the foreseeable future, but they have stirred physicists' curiosity and engrossed their minds for decades.

A physical field can be pictured as a set of oscillators in which each oscillator corresponds to a state for a boson,[10] similar to the radiation field of a blackbody or in a cavity. The state of an oscillator is specified by its momentum. Local field theory assumes that all states with large or small momenta are equally probable for an oscillator. For a blackbody, this assumption leads to a divergent result

---

[c] Recent experiments in July 2012 suggest the detection of a new particle which may be the Higgs boson. If this turns out to be the case, the electroweak theory with spontaneous symmetry breaking would be supported and the idea of vacuum symmetry breaking would be excluded.

for, say, the total energy per unit volume. We know that Planck's distribution law dictates the number of photons with a given momentum (or energy) and suppresses the contribution of states with large momenta to the total energy per unit volume in the cavity so that we have a finite result consistent with experiments. This appears to suggest that in a quantum system there is a natural probability distribution for oscillators that suppresses the contribution of large momentum states to physical amplitudes, so that all observable results are finite. However, any non-trivial inherent probability distribution for oscillators modifies the point particle picture and causes a departure from the Lorentz and Poincaré invariance at (extremely) high energies, if one insists on the unitarity (i.e., the conservation of probability) of the theory.

To see these properties, let us consider a simple finite model, in which the concept of local field is changed. This finite model assumes an inherent and universal probability distribution $P(p, m)$ for oscillators of all fields:

$$P(p, m) \to 1, \quad \mathbf{p} \to 0,$$
$$\to 0, \quad \mathbf{p} \to \infty. \tag{16.17}$$

These limiting properties assure that at low energies, the physical implications of the model are essentially identical to the standard results and that contributions from high energy (or large momentum) oscillators are suppressed. For a suitable function $P(p, m)$, all results will be finite no matter how complicated the interactions of local gauge fields and matter fields. In other words, all ultraviolet divergences in quantum Yang–Mills gravity, QED and quantum chromodynamics will disappear.

To see the physical implications of the inherent probability distribution (16.17), we expand the photon field $A(x^\mu), x^\mu = (w, x, y, z)$ in the radiation gauge as follows,

$$A(x^\mu) = \int \frac{d^3k}{\sqrt{2k_0(2\pi)^3}} \sum_{\lambda=1}^{2} \epsilon(k, \lambda) P^{1/2}(k, 0)[a(k, \lambda)e^{-ik\cdot x} + a^\dagger(k, \lambda)e^{ik\cdot x}],$$

$$\tag{16.18}$$

$$k_o = |\mathbf{k}|, \quad \mathbf{k} \cdot \vec{\epsilon}(k, \lambda) = 0, \quad \epsilon(k, \lambda) \cdot \epsilon(k, \lambda') = \delta_{\lambda\lambda'}, \tag{16.19}$$

$$[a(k, \lambda), a^\dagger(k', \lambda')] = \delta^3(k - k')\delta_{\lambda\lambda'},$$

$$[a(k, \lambda), a(k', \lambda')] = [a^\dagger(k, \lambda), a^\dagger(k', \lambda')] = 0. \tag{16.20}$$

The gauge invariant Lagrangian for the photon field has the usual form in equation (8.34) with $P_{\mu\nu} = \eta_{\mu\nu}$ in inertial frames. As expected, the Hamiltonian H_A and the momentum $\mathbf{k}_A$ of the photon field are modified by the presence of the inherent probability distribution $P(k,0)$ as follows

$$H_A = \int d^3k \, |\mathbf{k}| P(k,0) \sum_{\lambda=1}^{2} a^\dagger(k,\lambda) a(k,\lambda), \tag{16.21}$$

$$\mathbf{k}_A = \int d^3k \, \mathbf{k} P(k,0) \sum_{\lambda=1}^{2} a^\dagger(k,\lambda) a(k,\lambda), \tag{16.22}$$

where the contributions from large momenta are suppressed by the probability distribution $P(k,0)$. The modified photon propagator is

$$iD_{\mu\nu}^{tr}(x,P) = \langle 0|A_\mu(x)A_\nu(0)\theta(w)|0\rangle + \langle 0|A_\nu(0)A_\mu(x)\theta(-w)|0\rangle$$

$$= \int \frac{d^4k}{(2\pi)^4} \frac{P(k,0)\exp(-ik\cdot x)}{k^2 + i\epsilon} \sum_{\lambda=1}^{2} \epsilon_\mu(k,\lambda)\epsilon_\nu(k,\lambda),$$

$$\epsilon^\mu(k,\lambda) = (0, \epsilon(k,1), \epsilon(k,2), \mathbf{k}/|\mathbf{k}|) = (0, \epsilon(k,\lambda)).$$

In momentum space, the modified Feynman propagator for photons takes the form

$$\frac{-i\eta_{\mu\nu}P(k,0)}{k^2 + i\epsilon}. \tag{16.23}$$

The modified propagator (16.23) reduces to the Feynman propagator in the limit $P(k,0) \to 1$. The modified Coulomb potential is given by[10]

$$V(r) = e \int \frac{d^3k}{(2\pi)^3} \frac{P(k,0)}{\mathbf{k}^2} e^{i\mathbf{k}\cdot\mathbf{r}}. \tag{16.24}$$

To be specific, suppose $P(k,0) = \exp[-Rk/2]$ where $k = |\mathbf{k}|$, we have

$$V(r) = \frac{e}{2\pi^2 r} tg^{-1}\frac{r}{R}, \tag{16.25}$$

which is finite everywhere, including at the origin $r = 0$. It is interesting to note that the force at short distance becomes weaker and weaker as $r \to 0$:

$$F = -e\frac{dV}{dr} = \frac{e^2 r}{3\pi^2 R^3}, \quad r \ll R, \tag{16.26}$$

which resembles the situation in the asymptotically free Yang–Mills theory.

The length R plays the role of the fundamental length in the sense that physics at short distances ($r \ll R$) or large momenta ($p \gg \hbar/R$) is modified. If the inherent probability distribution $P(k, 0)$ is not a Lorentz scalar, the Lorentz and Poincaré invariance is violated.[11,12] Nevertheless, as long as R is small enough, these modifications cannot be detected by known experiments.

16-5.   Toward a total unification of all interactions

Let us take a step back and examine the big picture of all possible interaction forces in nature and speculate on the most general implications of the Yang–Mills–Utiyama–Weyl framework.

Apart from the well-known forces in quantum chromodynamics, the electroweak theory, and gravity, there are other possible forces related to quark confinement and the accelerated cosmic expansion. The quark confinement force should be attractive for all quarks and anti-quarks, and such a force may be related to a linear potential.[13] The force responsible for the accelerated cosmic expansion should be extremely weak and repulsive between baryon matter, as discussed in section 16-2.

Collecting these together, a total-unified group G_{tu}, consistent with established conservation laws, might be

$$G_{tu} = T_4 \times [SU_3]_{color} \times \{SU_2 \times U_1\} \times U_{1b}, \tag{16.27}$$

where a possible quark-confinement force and the U_{1l} group associated with the lepton number have not been included explicitly. In a total-unified model, the total gauge covariant derivative δ_μ in a general frame (with the metric tensor $P_{\mu\nu}$ and the associated covariant derivative D_μ) takes the form

$$\delta_\mu \equiv D_\mu + g\phi_\mu^\nu D_\nu + ig_s G_\mu^a \frac{\lambda^a}{2} + if W_\mu^k t^k + if' U_\mu t_o + \frac{ig_b}{3} B_\mu, \tag{16.28}$$

where λ_a ($a = 1, 2, 3, \ldots, 8$) are the eight color SU_3 generators, $t_k (k = 1, 2, 3)$ are SU_2 generators, and t_o (weak hypercharge) is the U_1 generator.[14] Such a total unified model suggests that the coupling constants in (16.28) are fundamental and inherent constants of nature.

The gauge curvature associated with each individual group and the Lagrangian for such a total-unified model can be obtained by calculating the commutator of the covariant derivative defined in (16.28). Let us calculate the commutator of the total-unified gauge covariant derivative (16.28),

$$[\delta_\mu, \delta_\nu] = C_{\mu\nu\sigma} D^\sigma + ig_s G^a_{\mu\nu} \frac{\lambda^a}{2} + if W^k_{\mu\nu} t^k + if' U_{\mu\nu} t_o + ig_b B_{\mu\nu}, \qquad (16.29)$$

where $G_{a\mu\nu}$ and $C_{\mu\nu\sigma}$ are given by (15.4) and (11.5) with $\partial_\mu \to D_\mu$ and $\eta_{\mu\nu} \to P_{\mu\nu}$, respectively. In general frames, the modified $SU_2 \times U_1$ gauge curvatures $W_{\mu\nu i}$ and $U_{\mu\nu}$, in the presence of the gravitational gauge potential, are given by

$$W_{\mu\nu i} = J^\sigma_\mu D_\sigma W_{\nu i} - J^\sigma_\nu D_\sigma W_{\mu i} - f \epsilon_{ijk} W_{\mu j} W_{\nu k}, \qquad (16.30)$$

$$U_{\mu\nu} = J^\sigma_\mu D_\sigma U_\nu - J^\sigma_\nu D_\sigma U_\mu, \quad J^\sigma_\mu = \delta^\sigma_\mu + g\phi^\sigma_\mu, \qquad (16.31)$$

where $i, j, k = 1, 2, 3$, for the SU_2 group.[14] The U_{1b} gauge curvature $B^{\mu\nu}$ associated with the baryon charge is

$$B_{\mu\nu} = J^\sigma_\mu D_\sigma B_\nu - J^\sigma_\nu D_\sigma B_\mu, \qquad (16.32)$$

in the presence of gravity.

In inertial frames with the metric tensor $P_{\mu\nu} = \eta_{\mu\nu}$ and $D_\mu = \partial_\mu$, the total-unified Lagrangian is assumed to be

$$L_{tu} = L_{QCG} - \frac{1}{4}(W^{\mu\nu a} W^a_{\mu\nu} + U^{\mu\nu} U_{\mu\nu}) - \frac{L_s^2}{4} \partial_\lambda B_{\mu\nu} \partial^\lambda B^{\mu\nu} + L_{matter}, \qquad (16.33)$$

where L_{QCG} is given in equations (15.6) and (15.7), and the matter Lagrangian L_{matter} includes all fermions and Higgs bosons.[14,15]

We make three observations regarding this model.

First, it is interesting to observe that, for example, the U_1 gauge curvature (or field strength) $U_{\mu\nu}$ in (16.31) [or $B_{\mu\nu}$ in (16.32)] involves a very small violation of the U_1 [or U_{1b}] gauge invariance due to the presence of gravity.[16] For example, one can see that the U_{1b} gauge curvature $B_{\mu\nu}$ in (16.32) is invariant under U_1 gauge transformations, i.e., $B'_\mu(x) = B_\mu(x) + \partial_\mu \Lambda(x)$, only when $g = 0$ (i.e., in the absence of a gravitational field $\phi_{\mu\nu}$). Such an extremely small violation of internal gauge symmetries due to gravity is intimately related to the property that all gauge fields and generators associated with the groups T_4, SU_3, SU_2 and U_1 are on equal footing in the total gauge covariant derivative δ_μ in (16.28), just like that in the $SU_2 \times U_1$ electroweak theory. This 'equal footing' of all gauge fields in (16.28) leads to the simplest coupling between gauge fields and should be considered as a basic principle for the total-unification of external and internal gauge symmetries in the model.

Second, one may wonder whether the total-unified model, with its extremely small violation of U_{1b} 'local' baryon gauge symmetry, has something to do with the imbalance between baryons and anti-baryons in the observable universe. The violation of baryon number conservation is one of the three necessary conditions for the observable imbalance between baryons and anti-baryons discussed by Sakharov.[17] This problem deserves further investigation.[18]

Third, at first glance, one may get the impression that in the total covariant derivative (16.28), the various interactions are simply juxtaposed, and not truly integrated or unified. However, such a model for complete unification may well be the best that one can do in the physical world as we know it. The reasons are as follows:

(i) Mathematically, it appears unlikely that one can further unite these external and internal gauge symmetries into a single Lie group.

(ii) The usual asymptotic freedom and running coupling constants in non-Abelian gauge theories are no longer true in the total-unified model, which is not renormalizable by power counting because the gravitational interaction has the dimensional coupling constant g. Therefore, it is unlikely that the different interactions can be unified at high energies and described by one single constant.

The usual argument based on the running coupling constant may not be airtight, because the gravitational coupling has not been included. Thus, the usual properties and conclusions in gauge field theories cannot be taken for granted in the total-unified model.

One interesting consequence of such a total-unified model is that it has implications that can be tested experimentally. Specifically, there is an extremely small violation of all internal gauge symmetries, $[SU_3]_{color}$, $SU_2 \times U_1$ and U_{1b}, due to gravity. In other words, all these dynamic symmetries for strong, electroweak and cosmic Lee–Yang forces are not exact in the presence of gravity, and only the space-time translational gauge symmetry of gravity is exact in the total-unified model. Such a violation in the unified model leads to the 'effective Riemann metric tensors' in the geometric-optics limit of wave equations of photons, quarks, etc. As a result, all classical objects move as if they were in a 'curved space-time, even though the real physical space-time at the quantum level is flat. As discussed in Chapter 9, the experimental results of the perihelion shift of the Mercury could be considered as a support of the violations of internal gauge symmetries. Furthermore, the violation of, say, the electromagnetic U_1 gauge symmetry leads to a modification of the continuity equation for the electric current given by (14.19) and predicts a 12% difference in (9.27) for the deflection of light by gravity,[19] in comparison with Einstein's theory of gravity.

16-6. Conclusion

The taiji symmetry framework provides a broader view of the four-dimensional symmetry of space-time and preserves an operational meaning for space-time coordinates in inertial and non-inertial frames. Within this framework, Yang–Mills gravity reveals the flat space-time translational origin of the gravitational field $\phi_{\mu\nu}$ while leading to an effective curved space-time for the motion of classical objects. It is a logically and experimentally consistent gauge theory, which represents an alternative to general relativity with its general coordinate invariance in an inherently curved space-time and the accompanying difficulties of quantization, etc. The importance of experimental tests of Yang–Mills gravity versus general relativity by the deflection of light in chapter 9 cannot be over-emphasized.[19] Furthermore, Yang–Mills gravity with space-time translational gauge symmetry paves the way to the unification of all interactions based on Yang–Mills–Utiyama–Weyl's framework. We believe that the taiji symmetry framework is a satisfactory framework that can accommodate not only all conservation laws in both inertial and non-inertial frames and all fundamental interactions in a unified manner,[20] but also novel phenomena such as the accelerated expansion of the universe.

References

1. E. P. Wigner, *Symmetries and Reflections, Scientific Essays* (The MIT Press, 1967), pp. 52–53.
2. The Royal Swedish Academy of Sciences, Poster: *Nobel Prize 2011 in Physics*.
3. T. D. Lee and C. N. Yang, in *100 Years of Gravity and Accelerated Frames, The Deepest Insights of Einstein and Yang–Mills* (Ed. J. P. Hsu and D. Fine, World Scientific, 2005), p. 155. The baryon conservation law was first postulated by E. C. G. Stueckelberg, *Helv. Phys. Acta*, 11, 299 (1938), see also ref. 1, p. 25.
4. T. D. Lee, *Particle Physics and Introduction to Field Theory* (Harwood, London, 1981) pp. 210–213.

5. J. P. Hsu, *Mod. Phys. Lett. A* **20**, 2855 (2005).

6. I. M. Gel'fand and G. E. Shilov, *Generalized Functions* vol. 1 (Academic Press, New York, 1964), pp. 359–363.

7. J. P. Hsu, *Int. J. Mod. Phys. A* **24**, 5217 (2009).

8. J. P. Hsu, *Nuovo Cimento B* **91**, 100 (1986).

9. J. P. Hsu, *Nuovo Cimento B* **89**, 30 (1985); *Nuovo Cimento B* **78**, 85 (1983).

10. J. P. Hsu and S. Y. Pei, *Phys. Rev. A* **37**, 1406 (1988); J. P. Hsu and Chagarn Whan, *Phys. Rev. A* **38**, 2248 (1988); J. P. Hsu, *Nuovo Cimento B* **89**, 30 (1985); *B* **78**, 85 (1983).

11. Suppose one assumes a suitable function for $P(k, 0)$ so that QED is finite. It is not known whether the agreement between such a new finite QED and experiment could be as good as that of the usual renormalized QED, which has a wonderful agreement with high-precision experiments such as that measuring the magnetic moment of the electron, as shown in ref. 19 in ch. 13.

12. P. A. M. Dirac, *Lecture on Quantum Field Theory* (Belfer Graduate School of Science, Academic Press, 1966), p. 23. Mathematically, one may assume $P(k, 0)$ to be a function of $k_\mu k^\mu$, which is Lorentz invariant. However, a Lorentz invariant distribution destroys the unitarity of the S matrix, e.g., the probability for certain physical process may be negative. Thus far, there is no satisfactory solution.

13. J. P. Hsu, *Nuovo Cimento B* **88**, 140 (1985).

14. Kerson Huang, *Quarks, Leptons and Gauge Field* (World Scientific, 1982), p. 108 and p. 241. R. E. Marshak, *Conceptual Foundations of Modern Particle Physics* (World Scientific, 1993) Ch. 2. For standard model based on $SU_3 \times SU_2 \times U_1$ groups, see M. Kaku, *Quantum Field Theory, A Modern Introduction* (Oxford Univ. Press, 1993) pp. 363–403.

15. S. Weinberg, *The Quantum Theory of Fields* (Cambridge Univ. Press, 1996), Vol. II, pp. 305–318.

16. It is possible to restore the internal gauge symmetries in the unified model by giving up the principle of 'equal footing' for all gauge fields in (16.28). However, the resultant Lagrangian will no longer be unique and will also become more complicated in the number of interaction terms and the structure of the interactions. For example, a 6-vertex for the photon–graviton coupling would then appear in the Feynman–Dyson rules in the model. In the total-unified model based on the covariant derivative (16.28), one has a maximum 4-vertex for the photon–graviton coupling, just like that in usual gauge theories.

17. A. P. Sakharov, *JETP* **5**, 24 (1967).

18. In the total unified model, one also has a modified continuity equation, which involves a new term in (14.19). One may speculate that the new term may have some cosmological implications, even though its effect is extremely small.

19. J. P. Hsu, *Eur. Phys. J. Plus*, (2013) **128**: 31, DOI 10.1140/epjp/i2013-1303-3.

20. All conservation laws of electric charges, baryon numbers, lepton numbers, etc. in the total-unified model appear to suggest a picture of a 'dumbbell universe': Our portion of the universe is dominated by matter and the other not-yet-observed portion of the universe is dominated by anti-matter. Furthermore, these two gigantic clusters of matter and anti-matter probably rotate about a common center of mass due to the cosmic 'Lee–Yang forces.' Such forces are produced by baryon and lepton gauge fields, which satisfy fourth order differential equations. These forces are attractive between matter and anti-matter. Their strengths are independent of the distance and, hence, stronger than the gravitational force for such a huge distance between the 'two dumbbells.'

Appendix A

The Fock–Hilbert Approach to Local Symmetry and Conservation Laws in General Frames of Reference

In chapter 6, we discussed the conserved and symmetric energy-momentum tensor $t_{\mu\nu}$ of a physical system in the context of a flat space-time framework with external and local translational symmetry, involving arbitrary vector functions. This approach for deriving the symmetric energy-momentum tensor based on the variation of metric tensors was first discussed by Hilbert in 1915, before Noether's theorem was published in 1918.[a] Hilbert obtained the symmetric energy-momentum tensor in a paper discussing Einstein's theory of gravity based on a new action (the Hilbert–Einstein action) with general coordinate invariance in curved space-time.

Somewhat later, Fock proved that the covariant continuity equation in curved space-time, which is satisfied by Hilbert's energy-momentum tensor, implies the genuine conservation of energy-momentum if and only if space-time is endowed with a constant curvature (i.e., a maximally symmetric space-time).[b] Since taiji space-time has a vanishing space-time curvature and can accommodate both inertial and non-inertial frames of reference, Fock's proof implies that within the taiji framework, the conservation laws for energy-momentum hold in all physical reference frames. The purpose of Appendix A is to show how the 'local' gauge symmetries of internal and external (space-time translation) groups lead to conservation laws based on the Fock–Hilbert approach.

A-1. Conserved Charges in Inertial and Non-inertial Frames

The usual derivation of the conservation law for electric charge in inertial frames is based on the global U(1) symmetry. However, in order to see the

---

[a] See Ref. 1. in ch. 6.
[b] See Refs. 2, 3 and 4 in ch. 6

implication of the local symmetry on conservation laws in general frames of reference, let us consider the gauge invariant actions of QED and gauge theories.[1] Using arbitrary coordinates with the metric tensor $P_{\mu\nu}$ in the taiji space-time framework, we have

$$S_{QED} = \int \left[-\frac{1}{4}F_{\mu\nu}F^{\mu\nu} + i\{\overline{\psi}\Gamma^{\mu}(D_{\mu} - ieA_{\mu})\psi\} - m\overline{\psi}\psi \right] \sqrt{-P}d^4x, \quad \text{and}$$

$$\tag{A.1}$$

$$S_{YMU} = \int \left[-\frac{1}{4}f_{k\mu\nu}f_k^{\mu\nu} + \overline{\Psi}[i\Gamma^{\mu}(D_{\mu} - igA_{k\mu}G_k) - m]\Psi \right] \sqrt{-P}d^4x,$$

for the QED action and the Yang–Mills–Utiyama action, respectively. The Yang–Mills–Utiyama action is taken to have a Lie group (or an internal gauge group) with N generators G_k ($k = 1, 2, \ldots N$) that have constant matrix representations in non-inertial frames. We have used the following relations in the action (A.1),

$$F_{\mu\nu} = D_{\mu}A_{\nu} - D_{\nu}A_{\mu} = \partial_{\mu}A_{\nu} - \partial_{\nu}A_{\mu},$$

$$f_{i\mu\nu} = \partial_{\mu}A_{i\nu} - \partial_{\nu}A_{i\mu} + gC_{ijk}A_{j\mu}A_{k\nu}, \tag{A.2}$$

$$\Gamma^{\mu} = \gamma^a e_a^{\mu}, \quad P_{\mu\nu}e_a^{\mu}e_b^{\nu} = \eta_{ab}, \quad \eta^{ab}e_{\mu a}e_b^{\nu} = \delta_{\mu}^{\nu},$$

where γ^a are the usual constant Dirac matrices and e_a^{μ} are tetrads.[1] The symbol D_{μ} is the covariant derivative associated with the Poincaré metric tensor $P_{\mu\nu}$ for non-inertial frames. The gauge variations of the electromagnetic potential fields A_{μ} and the gauge fields $A_{i\mu}$ are given by

$$\delta A_{\mu} = \partial_{\mu}\Lambda, \quad \delta A_{\mu i} = -(1/g)\partial_{\mu}\omega_i + C_{ijk}\omega_j A_{\mu k}, \tag{A.3}$$

where Λ and ω_i are arbitrary infinitesimal functions, and C_{ijk} are the completely anti-symmetric structure constants of a Lie group.[1] These structure constants, which are associated with the N generators G_k and the Jacobi identity, are necessary for the formulation of the Yang-Mills theory and have been discussed by Utiyama.[2] Since we are interested in the continuity equations involving fermion currents, it suffices to consider the variation of the actions in (A.1) with respect to gauge fields A_{μ} and $A_{i\mu}$. We have

$$\delta S_{QED} = \int \left[\partial_{\mu}\left(\sqrt{-P}F^{\mu\nu}\right) + j^{\nu}\sqrt{-P} \right] \delta A_{\nu}d^4x, \tag{A.4}$$

$$\delta S_{YMU} = \int \left[\left(\delta_{ik}\partial_{\mu} + gC_{ijk}A_{j\mu}\right)\left(\sqrt{-P}f_k^{\mu\nu}\right) + J_i^{\nu}\sqrt{-P} \right] \delta A_{i\nu}d^4x, \tag{A.5}$$

$$j^{\nu} = e\overline{\psi}\Gamma^{\nu}\psi, \quad J_i^{\nu} = g\overline{\Psi}\Gamma^{\nu}G_i\Psi.$$

From (A.3), (A.4) and (A.5), the variations of the gauge invariant actions lead to the following covariant relations involving the electric and Utiyama currents, j^ν and J_i^ν, in non-inertial frames,

$$\partial_\nu \left[\partial_\mu (\sqrt{-P} F^{\mu\nu}) + j^\nu \sqrt{-P} \right] = 0, \tag{A.6}$$

$$\left(\delta_{ij} \partial_\nu + g C_{inj} A_{n\nu} \right) \left[\left(\delta_{ik} \partial_\mu + g C_{imk} A_{m\mu} \right) \left(\sqrt{-P} f_k^{\mu\nu} \right) + J_i^\nu \sqrt{-P} \right] = 0, \tag{A.7}$$

where Λ and ω_i are arbitrary functions. Note that the variations of the fermion fields $\delta\psi$ and $\delta\overline{\psi}$ do not affect the results (A.6) and (A.7). With the help of $F^{\mu\nu} = -F^{\nu\mu}$ and gauge field equations

$$(\delta_{ik} \partial_\mu + g C_{imk} A_{m\mu})(\sqrt{-P} f_k^{\mu\nu}) + J_i^\nu \sqrt{-P} = 0, \tag{A.8}$$

(A.6) and (A.7) can take the form of covariant continuity equations in non-inertial and inertial frames

$$D_\mu j^\mu = \frac{1}{\sqrt{-P}} \partial_\mu (\sqrt{-P}\, j^\mu) = 0, \tag{A.9}$$

$$\frac{1}{\sqrt{-P}} \partial_\nu (\sqrt{-P}\, [J_i^\nu + g C_{ijk} A_{j\mu} f_k^{\mu\nu}]) = 0. \tag{A.10}$$

As shown in Section 6-5, these continuity equations lead to the conservation of charge in general frames of reference in flat space-time,

$$Q_e = \int \sqrt{-P}\, j^\mu \, dS_\mu, \tag{A.11}$$

$$Q_{Ui} = \int \sqrt{-P}\, [J_i^\nu + g C_{ijk} A_{j\mu} f_k^{\mu\nu}] \, dS_\nu. \tag{A.12}$$

The vector dS_μ is dual to the anti-symmetric tensor $dS^{\nu\gamma\sigma}$ (which is an infinitesimal element of a hypersurface) and $dS_0 = dxdydz$ is the projection of the hypersurface on the hyperplane $w = x^0 = constant$. The electric charge Q_e in (A.11) characterizes the electromagnetic interaction and Utiyama's charges Q_{Uj} in (A.12) characterize the interaction of the gauge fields $A_{j\mu}$.[2] The values of these constant charges are the same in both inertial and non-inertial frames and are truly universal constants. Such results indicate that the coupling constants of gauge fields in quantum chromodynamics and electroweak theory are truly universal.

A-2.   Energy-Momentum Tensor of Fermions

Since there is no GL(4) representation for spinors under the Lorentz group, the Hilbert energy-momentum tensor (6.2) can be defined only for tensor objects under the Lorentz transformations. For fermions, we must introduce tetrads or vierbeins e_μ^a or e_b^ν, i.e., four covariant or contravariant vector functions. By definition, $e_b^\nu \equiv \eta_{ab} P^{\mu\nu} e_\mu^a$, and they satisfy the identities $e_\mu^a e_a^\nu = \delta_\mu^\nu$, etc. Under the T(4) gauge transformations, these quantities transform as four-vectors in (7.4),

$$\delta e_\mu^a = -\Lambda^\lambda \partial_\lambda e_\mu^a - e_\lambda^a \partial_\mu \Lambda^\lambda, \quad \delta e_a^\mu = -\Lambda^\lambda \partial_\lambda e_a^\mu + e_a^\lambda \partial_\lambda \Lambda^\mu. \tag{A.13}$$

For the energy-momentum tensor of fermions, the variation δe_μ^a (or δe_μ^a) in the invariant action S_ψ involving pure fermions[3] (i.e., the Lagrangian in S_{YMU} with $f_{k\mu\nu} = 0$ in (A.5)) leads to

$$\delta S_\psi = \int d^4 x \sqrt{-P} T_\mu^a \delta e_a^\mu = 0,$$

$$\delta S_\psi = \int d^4 x \sqrt{-P} T_a^\mu \delta e_\mu^a = 0. \tag{A.14}$$

Because Λ^λ are arbitrary functions, from Equations (A.13) and (A.14) we have

$$\frac{1}{\sqrt{-P}} \partial_\sigma (\sqrt{-P} T_\lambda^a e_a^\sigma) + T_\mu^a \partial_\lambda e_a^\mu = 0,$$

$$\frac{1}{\sqrt{-P}} \partial_\sigma (\sqrt{-P} T_a^\sigma e_\lambda^a) - T_a^\nu \partial_\lambda e_\nu^a = 0, \tag{A.15}$$

where we have used integration by parts.

We define a new symmetric energy-momentum tensor by the relation in taiji space-time with arbitrary coordinates,

$$T_{.\nu}^\mu = \frac{1}{2}(e_a^\mu T_\nu^a + e_\nu^a T_a^\mu) = T_{\nu.}^{.\mu}, \tag{A.16}$$

which is a mixed tensor.[4] We have used a dot to mark the position of the indices for clarity. For symmetric tensors, the dots will eventually be unnecessary. From (A.16), we have

$$Q_\nu \equiv \frac{1}{\sqrt{-P}} \partial_\mu (\sqrt{-P} T_\nu^\mu) = (-e_\sigma^a e_b^\sigma T_\mu^b \partial_\nu e_a^\mu + e_\sigma^b e_a^\sigma T_b^\mu \partial_\nu e_\mu^a), \tag{A.17}$$

where we have used (A.15) and the identity $e_\nu^a e_b^\nu = \delta_b^a$. Differentiation of this identity gives

$$e_\sigma^a \partial_\nu e_a^\mu = -(\partial_\nu e_\sigma^a) e_a^\mu. \tag{A.18}$$

We also have the relations

$$\partial_\nu P_{\alpha\mu} = -P_{\alpha\rho} P_{\mu\sigma} \partial_\nu P^{\rho\sigma} = -P_{\alpha\rho} P_{\mu\sigma} \left(-P^{\gamma\sigma} \Gamma_{\gamma\nu}^\rho - P^{\gamma\rho} \Gamma_{\gamma\nu}^\sigma \right)$$

$$= P_{\alpha\rho} \Gamma_{\mu\nu}^\rho + P_{\mu\sigma} \Gamma_{\alpha\nu}^\sigma, \tag{A.19}$$

$$\Gamma_{\mu\nu}^\lambda = \frac{1}{2} P^{\lambda\alpha} (\partial_\nu P_{\mu\alpha} + \partial_\mu P_{\nu\alpha} - \partial_\alpha P_{\mu\nu}).$$

It follows from (A.17), (A.18) and (A.19) that

$$0 = Q_\nu - T_\rho^\alpha e_a^\rho \partial_\nu e_\rho^a = Q_\nu - T^{\alpha\sigma} e_{\sigma a} \partial_\nu e_\alpha^a = Q_\nu - \frac{1}{2} T_\rho^\alpha P^{\rho\mu} \partial_\nu P_{\alpha\mu}$$

$$= \frac{1}{\sqrt{-P}} \partial_\mu (\sqrt{-P} T_\nu^\mu) - T_\rho^\alpha \Gamma_{\alpha\nu}^\rho = D_\mu T_\nu^\mu. \tag{A.20}$$

Thus, in non-inertial (and inertial) frames, we have the conservation law of the fermion energy-momentum tensor $D_\mu T^{\mu\nu} = 0$, where $T^{\mu\nu} = T_\sigma^\mu P^{\sigma\nu}$. In taiji space-time, Equation (A.20) implies the true conservation of the energy-momentum tensor.[4]

References

1. See, for example, K. Huang, *Quarks Leptons & Gauge Fields* (World Scientific, 1982) Ch. IV.

2. C. N. Yang and R. L. Mills, *Phys. Rev.* **96**, 191 (1954); R. Utiyama, *Phys. Rev.* **101**, 1597 (1956).

3. For fermion Lagrangians with external space-time gauge groups, see J. P. Hsu, *Phys. Rev. Lett.* **42**, 934 (1979); E **42**, 1720 (1979); *Phys. Lett. B* **119**, 328 (1982).

4. If a theory is invariant under a local Lorentz transformation, the quantity $e_a^\mu T^{a\nu}$ can be shown to be symmetric in μ and ν. See S. Weinberg, *Gravitation and Cosmology* (John Wiley and Sons, 1972), pp. 370–371.

Appendix B

Calculations of $H^{\mu\nu}$ in the Gravitational Field Equation

B-1. Evaluations of the component H^{00}

To see and appreciate the wonderful physical implications of the new T_4 gauge curvature in (8.14) and the invariant action (8.16), one must explore the gravitational Equation (8.26) in detail. The new gravitational field Equation (8.26) and the 'effective' metric tensors embedded in the eikonal Equation (8.36) and the classical Hamilton–Jacobi Equation (8.40) are key properties of Yang–Mills gravity in flat space-time.

In this section, we carry out explicitly the detailed calculations for the perihelion shift of Mercury and the deflection of light. For this purpose, it suffices to solve for the static and spherically symmetric solutions for $J_0^0(r)$, $J_1^1(r)$, and $J_2^2(r) = J_3^3(r)$ in the tensor field Equation (8.26). It is convenient to use the spherical coordinates $x^\mu = (w, r, \theta, \phi)$ with the metric tensor $P_{\mu\nu} = (1, -1, -r^2, -r^2 \sin^2\theta)$. The tensor $H^{\mu\nu}$ is given in (8.26) and contains 6 terms:

$$H_1^{\mu\nu} = D_\lambda(J_\rho^\lambda C^{\rho\mu\nu}), \quad H_2^{\mu\nu} = -D_\lambda(J_\alpha^\lambda C^{\alpha\beta}{}_\beta P^{\mu\nu}), \quad H_3^{\mu\nu} = D_\lambda(C^{\mu\beta}{}_\beta J^{\nu\lambda}),$$

$$H_4^{\mu\nu} = -C^{\mu\alpha\beta} D^\nu J_{\alpha\beta}, \quad H_5^{\mu\nu} = C^{\mu\beta}{}_\beta D^\nu J_\alpha^\alpha, \quad H_6^{\mu\nu} = -C^{\lambda\beta}{}_\beta D^\nu J_\lambda^\mu. \tag{B.1}$$

With the help of the following relations of covariant derivatives,

$$D_\lambda C^{\rho\mu\nu} = \partial_\lambda C^{\rho\mu\nu} + \Gamma_{\lambda\sigma}^\sigma C^{\sigma\mu\nu} + \Gamma_{\lambda\sigma}^\mu C^{\rho\sigma\nu} + \Gamma_{\lambda\sigma}^\nu C^{\rho\mu\sigma}, \tag{B.2}$$

$$D_\lambda J_\beta^\alpha = \partial_\lambda J_\beta^\alpha + \Gamma_{\lambda\sigma}^\alpha J_\beta^\sigma - \Gamma_{\lambda\beta}^\sigma J_\sigma^\alpha, \tag{B.3}$$

$$D_\nu J_{\alpha\beta} = \partial_\nu J_{\alpha\beta} - \Gamma_{\nu\alpha}^\sigma J_{\sigma\beta} - \Gamma_{\nu\beta}^\sigma J_{\alpha\sigma}, \tag{B.4}$$

and $D_\lambda P_{\alpha\beta} = 0$, we can express $J^{\mu\nu}$ and $H^{\mu\nu}$ in terms of three functions, S, R and T,

$$J^0_0 = S(r), \quad J^1_1 = R(r), \quad J^2_2 = J^3_3 = T(r), \quad J^{00} = J_{00} = S(r),$$
(B.5)

$$J^{11} = J_{11} = -R(r), \quad -r^2 J^{22} = \frac{J_{22}}{-r^2} = -r^2 \sin^2\theta J^{33} = \frac{J_{33}}{-r^2 \sin^2\theta} = T(r).$$

All other components of $J^{\mu\nu}$ vanish.

Let us calculate the first term H^{00}:

$$
\begin{aligned}
H^{00}_1 &= D_\lambda \left(J^\lambda_\rho C^{\rho 00} \right) \\
&= \left(\partial_\lambda J^\lambda_\rho - \Gamma^\sigma_{\lambda\rho} J^\lambda_\sigma + \Gamma^\lambda_{\sigma\lambda} J^\sigma_\rho \right) C^{\rho 00} + J^\lambda_\rho \left(\partial_\lambda C^{\rho 00} + \Gamma^\rho_{\lambda\sigma} C^{\sigma 00} \right. \\
&\quad \left. + \Gamma^0_{\lambda\sigma} C^{\rho\sigma 0} + \Gamma^0_{\lambda\sigma} C^{\rho 0\sigma} \right) \\
&= (\partial_1 J^1_1) C^{100} - \left(\Gamma^2_{21} J^2_2 + \Gamma^3_{31} J^3_3 \right) C^{100} \\
&\quad + \left(\Gamma^2_{12} J^1_1 + \Gamma^3_{13} J^1_1 \right) C^{100} + J^1_1 \partial_1 \left(J^{11} \partial_1 J^{00} \right) \\
&\quad + J^2_2 \Gamma^2_{21} C^{100} + J^3_3 \Gamma^3_{31} C^{100} \\
&= \left[\left(\partial_1 R + \frac{2}{r} R \right) \right] C^{100} - R \partial_1 (R \partial_1 S),
\end{aligned}
$$
(B.6)

$$\partial_1 R = \frac{dR(r)}{dr}, \quad \text{etc.,}$$
(B.7)

where we have used (B.2)–(B.5) and

$$C^{100} = J^{11} \partial_1 J^{00} = -R \partial_1 S, \quad C^{000} = C^{200} = C^{300} = 0,$$
(B.8)

$$C^{\alpha\beta\lambda} = -C^{\beta\alpha\lambda}, \quad P_{\mu\nu} = (1, -1, -r^2, -r^2 \sin^2\theta), \quad P_{\mu\nu} P^{\nu\alpha} = \delta^\alpha_\mu,$$
(B.9)

$$\Gamma^\mu_{\alpha\beta} = \frac{1}{2} P^{\mu\sigma} (\partial_\alpha P_{\sigma\beta} + \partial_\beta P_{\sigma\alpha} - \partial_\sigma P_{\alpha\beta}),$$
(B.10)

$$\Gamma^1_{22} = -r, \quad \Gamma^1_{33} = -r \sin^2\theta, \quad \Gamma^2_{12} = \frac{1}{r},$$

$$\Gamma^2_{33} = -\sin\theta\cos\theta, \quad \Gamma^3_{13} = \frac{1}{r}, \quad \Gamma^3_{23} = \cot\theta,$$
(B.11)

where all other components of the Christoffel symbols $\Gamma^\mu_{\alpha\beta}$ vanish.

For the second term H_2^{00}, we find

$$
\begin{aligned}
H_2^{00} &= -\left(\partial_\lambda J_\rho^\lambda - \Gamma_{\lambda\rho}^\sigma J_\sigma^\lambda + \Gamma_{\sigma\lambda}^\lambda J_\rho^\sigma\right) C^{\rho\beta}{}_\beta - J_\alpha^\lambda \left(\partial_\lambda C^{\alpha\beta}{}_\beta + \Gamma_{\lambda\sigma}^\alpha C^{\sigma\beta}{}_\beta\right.\\
&\quad \left. + \Gamma_{\lambda\sigma}^\beta C^{\alpha\sigma}{}_b - \Gamma_{\lambda\beta}^\sigma C^{\alpha\beta}{}_\sigma\right)\\
&= -\left(\partial_1 J_1^1\right) C^{1\beta}{}_\beta + J_\sigma^\lambda \Gamma_{\lambda\alpha}^\sigma C^{\alpha\beta}{}_\beta - J_\alpha^\sigma \Gamma^l d_{\sigma\lambda} C^{\alpha\beta}{}_\beta - J_1^1 \partial_1 C^{1\beta}{}_\beta\\
&\quad - J_2^2 \Gamma_{2\sigma}^2 C^{\sigma\beta}{}_\beta - J_3^3 \Gamma_{3\sigma}^3 C^{\sigma\beta}{}_\beta\\
&= -\left[\frac{dR}{dr} + \frac{2R}{r} + R\frac{d}{dr}\right] C^{1\beta}{}_\beta,
\end{aligned}
\tag{B.12}
$$

where

$$
C^{1\beta}{}_\beta = -R\left(\frac{dS}{dr} + 2\frac{dT}{dr}\right) - \frac{2T^2}{r} + \frac{2TR}{r},
\tag{B.13}
$$

$$
C^{0\beta}{}_\beta = C^{2\beta}{}_\beta = C^{3\beta}{}_\beta, \quad P^{00} = 1.
\tag{B.14}
$$

For other components of H^{00}, we obtain

$$
H_3^{00} = D_\lambda \left(C^{0\beta}{}_\beta J^{0\lambda}\right) = 0, \quad H_4^{00} = -C^{0\alpha\beta} D^0 J_{\alpha\beta} = 0,
\tag{B.15}
$$

$$
H_5^{00} = C^{0\beta}{}_\beta D^0 J_\beta^\beta = 0, \quad H_6^{00} = -C^{\lambda\beta}{}_\beta D^0 J_\lambda^0 = 0,
\tag{B.16}
$$

where we have used $D^\lambda = P^{\lambda\sigma} D_\sigma$. From previous results, we obtain H^{00} as follows,

$$
\begin{aligned}
-H^{00} &= -\left(\frac{dR}{dr} + \frac{2R}{r}\right) C^{100} + R\frac{d}{dr}\left(R\frac{dS}{dr}\right)\\
&\quad + \left[\frac{dR}{dr} + \frac{2R}{r} + R\frac{d}{dr}\right]\left(-R\frac{dS}{dr} - 2R\frac{dT}{dr} - \frac{2T^2}{r} + \frac{2TR}{r}\right)\\
&= \frac{d}{dr}\left(R^2\frac{dS}{dr}\right) + \frac{2R^2}{r}\frac{dS}{dr} + \left(\frac{dR}{dr} + \frac{2R}{r} + R\frac{d}{dr}\right)\\
&\quad \times \left(-R\frac{dS}{dr} - 2R\frac{dT}{dr} - \frac{2T^2}{r} + \frac{2TR}{r}\right),
\end{aligned}
\tag{B.17}
$$

where C^{100} is given by (B.8). This result is consistent with Equation (9.4).

B-2. Evaluations of the component H^{11}

Next, we calculate H^{11}. Its first term is

$$H_1^{11} = \left(D_\lambda J_\rho^\lambda\right) C^{\rho 11} + J_\rho^\lambda \left(D_\lambda C^{\rho 11}\right) = J_2^2 \Gamma_{22}^1 C^{212} + J_3^3 \Gamma_{33}^1 C^{313}.$$
(B.18)

The T_4 gauge curvature $C^{\mu\nu\sigma}$ in H_1^{11} can be evaluated:

$$
\begin{aligned}
C^{212} &= J^{22} \left(\partial_\sigma J^{12} + \Gamma_{2\sigma}^1 J^{\sigma 2} + \Gamma_{2\sigma}^2 J^{1\sigma}\right) \\
&\quad - J^{11} \left(\partial_1 J^{22} + \Gamma_{1\sigma}^2 J^{\sigma 2} - \Gamma_{2\sigma}^1 J^2 \sigma\right) \\
&= -r J^{22} J^{22} + J^{22} \frac{1}{r} J^{11} - J^{11} \partial_1 J^{22} + \frac{2}{r} J^{22} \\
&= -\frac{T^2}{r^3} + \frac{1}{r^3} RT - R\partial_1 \left(\frac{T}{r^2}\right) - \frac{2}{r^3} TR \\
&= -\left[\frac{R}{r}\frac{d}{dr}\left(\frac{T}{r}\right) + \frac{T^2}{r^3}\right] = -C^{122},
\end{aligned}
$$
(B.19)

$$C^{313} = \frac{C^{212}}{\sin^2 \theta} = -C^{133}.$$
(B.20)

Thus, we have

$$H_1^{11} = 2rTC^{122} = 2rT\left[\frac{R}{r}\frac{d}{dr}\left(\frac{T}{r}\right) + \frac{T^2}{r^3}\right].$$
(B.21)

Similarly, we obtain

$$H_2^{11} = \left[S' + \frac{2}{r}R + R\frac{d}{dr}\right] C^1{}^\beta{}_\beta, \quad S' = \frac{dS}{dr},$$
(B.22)

$$
\begin{aligned}
C^1{}^\beta{}_\beta &= J^{1\sigma} D_\sigma J_\beta^\beta - J^{\beta\sigma} D_\sigma J_\beta^1 \\
&= J^{11}\partial_1 \left(J_\beta^\beta - J_1^1\right) - J^{\beta\sigma}\Gamma_{\rho\sigma}^1 J_\beta^\rho + J^{\beta\sigma}\Gamma_{\alpha\beta}^\rho J_\rho^1 \\
&= -R(S' + 2T') - \frac{2}{r}T^2 + \frac{2}{r}TR, \quad T' = \frac{dT}{dr},
\end{aligned}
$$
(B.23)

$$H_3^{11} = \left[\frac{dJ^{11}}{dr} + \frac{2T}{r} - \frac{2R}{dr} + J^{11}\frac{d}{dr}\right] C^1{}^\beta{}_\beta,$$
(B.24)

$$H_4^{11} = -C^{1\alpha\beta}p^{1\sigma}(\partial_\sigma J_{\alpha\beta} - \Gamma^\rho_{\sigma\alpha}J^{\rho\beta} - \Gamma^\rho_{\sigma\beta}J^{\alpha\rho})$$

$$= -C^{100}p^{11}\partial_1 J_{00} - C^{122}p^{11}\partial_1 J_{22} - C^{133}p^{11}\partial_1 J_{33}$$

$$+ 2[C^{122}p^{11}\Gamma^\rho_{12} + C^{133}p^{11}\Gamma^\rho_{13}]J_{\rho 3}$$

$$= -R\left(\frac{dS}{dr}\right)^2 - 2C^{122}(Tr^2)' + 4rTC^{122}, \tag{B.25}$$

$$H_5^{11} = C_\beta^{1\beta}p^{1\sigma}D_\sigma J_\alpha^\alpha$$

$$= -C_{\beta\beta}^1\partial_1 J_\alpha^\alpha - \left[-R(S' + 2T') - \frac{2}{r}T^2 + \frac{2}{r}TR\right]$$

$$\times (S' + R' + 2T'), \tag{B.26}$$

$$H_6^{11} = -C_\beta^{\lambda\beta}p^{1\sigma}\left(\partial_\sigma J_\lambda^1 - \Gamma^\rho_{\sigma\alpha\lambda}J_\rho^1 + \Gamma^1_{\rho\sigma}J_\lambda^\rho\right)$$

$$= \left[-RS' - 2RT' - \frac{2}{r}T^2 + \frac{2}{r}TR\right]R', \quad R' = \frac{dR}{dr}, \quad etc. \tag{B.27}$$

The first two terms in (9.5) can be identified with

$$-(H_1^{11} + H_4^{11}) = -2rTC^{122} + R\left(\frac{dS}{dr}\right)^2 + 2C^{122}\frac{d(Tr^2)}{dr} - 4rTC^{122}$$

$$= R\left(\frac{dS}{dr}\right)^2 + 2r^3\frac{d(T/r)}{dr}\left[\frac{R}{r}\frac{d}{dr}\left(\frac{T}{r}\right) + \frac{T^2}{r^3}\right]. \tag{B.28}$$

The third term in (9.5) is

$$-(H_2^{11} + H_3^{11} + H_5^{11} + H_6^{11}) = \left(S' + 2T' - \frac{2T}{r}\right)C^{1\beta}_{\ \beta}, \tag{B.29}$$

where $C^{1\beta}_{\ \beta}$ is given by (B.13).

B-3.   Evaluations of the component H^{22}

Now let us consider H^{22}. The first term H_1^{22} is

$$H_1^{22} = D_\lambda\left(J_\rho^\lambda C^{\rho 22}\right)$$

$$= \left(D_\lambda J_1^\lambda\right)C^{122} + \left(D_\lambda J_3^\lambda\right)C^{322} + J_\rho^\lambda\left(\partial_\lambda C^{\rho 22} + \Gamma^\rho_{\sigma\lambda}C^{\sigma 22}\right.$$

$$\left. + \Gamma^\rho_{\sigma\lambda}C^{\sigma 22} + \Gamma^\rho_{\sigma\lambda}C^{\sigma 22}\right)$$

$$= \left(\partial_1 J_1^1 - \Gamma_{21}^2 J_2^2 - \Gamma_{31}^3 J_3^3 + \Gamma_{12}^2 J_1^1 + \Gamma_{13}^3 J_1^1 \right) C^{122} + J_1^1 \partial_1 C^{122}$$

$$+ J_2^2 \Gamma_{12}^2 C^{122} + J_3^3 \Gamma_{13}^3 C^{122} + J_1^1 \Gamma_{21}^2 C^{122}$$

$$+ J_2^2 \Gamma_{12}^2 C^{212} + J_1^1 \Gamma_{21}^2 C^{122}$$

$$= \left(\frac{dR}{dr} - \frac{T}{r} + \frac{4RT}{r} + R\frac{d}{dr} \right) C^{122}, \tag{B.30}$$

where

$$C^{122} = \frac{R}{r^2} \frac{dT}{dr} - \frac{RT}{r^3} + \frac{T^2}{r^3}. \tag{B.31}$$

Similarly, we also have

$$H_2^{22} = -P^{22} D_\lambda \left(J_\alpha^\lambda C^{\alpha\beta}{}_\beta \right)$$

$$= \frac{1}{r^2} \left[\left(D_\lambda J_0^\lambda \right) C^{0\beta}{}_\beta + \left(D_\lambda J_1^\lambda \right) C^{1\beta}{}_\beta + \left(D_\lambda J_2^\lambda \right) C^{2\beta}{}_\beta + \left(D_\lambda J_3^\lambda \right) C^{3\beta}{}_\beta \right.$$

$$\left. + J_0^\lambda D_\lambda C^{0\beta}{}_\beta + J_1^\lambda D_\lambda C^{1\beta}{}_\beta + J_2^\lambda D_\lambda C^{2\beta}{}_\beta + J_3^\lambda D_\lambda C^{3\beta}{}_\beta \right]$$

$$= \frac{1}{r^2} \left[\left(\partial_\lambda J_1^\lambda - \Gamma_{\lambda 1}^\rho J_\rho^\lambda + \Gamma_{\lambda \rho}^\rho J_1^\lambda \right) C^{1\beta}{}_\beta + J_1^1 \partial_1 C^{1\beta}{}_\beta \right.$$

$$\left. + J_2^2 \Gamma_{21}^2 C^{1\beta}{}_\beta + J_3^3 \Gamma_{31}^3 C^{1\beta}{}_\beta \right]$$

$$= \frac{1}{r^2} \left[\frac{dR}{dr} + \frac{2R}{r} + R\frac{d}{dr} \right] \left[-R\frac{dS}{dr} - 2R\frac{dT}{dr} - \frac{2T^2}{r} + \frac{2RT}{r} \right], \tag{B.32}$$

where we have used $C^{0\beta}{}_\beta = C^{2\beta}{}_\beta = C^{3\beta}{}_\beta = 0$.

The rest of terms in H^{22} can also be obtained in the same way:

$$H_3^{22} = (D_\lambda J^{\lambda 2}) C^{2\beta}{}_\beta + J^{\lambda 2} D_\lambda C^{2\beta}{}_\beta = J^{\lambda 2} \Gamma_{\lambda \rho}^2 C^{\rho\beta}{}_\beta = \frac{1}{r} J^{22} C^{1\beta}{}_\beta$$

$$= -\frac{T}{r^3} \left[-R\frac{dS}{dr} - 2R\frac{dT}{dr} - \frac{2T^2}{r} + \frac{2TR}{r} \right], \tag{B.33}$$

$$H_4^{22} = C^{2\alpha\beta} P^{2\sigma} D_\sigma J_{\alpha\beta} = C^{2\alpha\beta} P^{22} \left(\partial_2 J_{\alpha\beta} - \Gamma_{2\alpha}^\rho J_{\rho\beta} - \Gamma_{2\beta}^\rho J_{\alpha\rho} \right)$$

$$= -C^{2\alpha\beta} \left(-\Gamma_{2\alpha}^1 J_{1\beta} - \Gamma_{2\alpha}^2 J_{2\beta} - \Gamma_{2\alpha}^3 J_{3\beta} - \Gamma_{2\beta}^1 J_{\alpha 1} - \Gamma_{2\beta}^2 J_{\alpha 2} - \Gamma_{2\beta}^3 J_{\alpha 3} \right)$$

$$= \frac{1}{r^3} C^{212} J_{22} - \frac{1}{r} C^{212} J_{11} = -\frac{1}{r^3} C^{122} \left(J_{22} - r^2 J_{11} \right)$$

$$= \frac{1}{r} (R - T) \left(\frac{R}{r^2} \frac{dT}{dr} - \frac{RT}{r^3} + \frac{T^2}{r^3} \right), \tag{B.34}$$

$$H_5^{22} = C^{2\beta}{}_\beta D^2 J_\alpha^\alpha = 0, \quad C^{2\beta}{}_\beta = 0, \tag{B.35}$$

$$H_6^{22} = -C^{\lambda\beta}{}_\beta P^{2\sigma} D_\sigma J_\lambda^2 = C^{\lambda\beta}{}_\beta P^{22} \Gamma^\rho_{2\lambda} J_\rho^2 + C^{\lambda\beta}{}_\beta P^{22} \Gamma^2_{2\rho} J_\lambda^\rho$$

$$= \left(-\frac{T}{r^3} + \frac{R}{r^3}\right)\left[-R\frac{dS}{dr} - 2R\frac{dT}{dr} - \frac{2T^2}{r} + \frac{2TR}{r}\right], \tag{B.36}$$

where we have used $C^{233} = -C^{323} = 0$.

Collecting all 6 terms H_1^{22} through H_6^{22} together, we have

$$H^{22} = \left(R\frac{d}{dr} + \frac{dR}{dr} + \frac{5R}{r} - \frac{2T}{r}\right)\left(\frac{R}{r^2}\frac{dT}{dr} - \frac{RT}{r^3} + \frac{T^2}{r^3}\right)$$

$$+ \left(\frac{R}{r^3} - \frac{2T}{r^3}\right)C^{1\beta}{}_\beta + \frac{1}{r^2}\left(\frac{dR}{dr} + \frac{2R}{r}\right)C^{1\beta}{}_\beta$$

$$= \left(R\frac{d}{dr} + \frac{dR}{dr} + \frac{5R}{r} - \frac{2T}{r}\right)\left(\frac{R}{r^2}\frac{dT}{dr} - \frac{RT}{r^3} + \frac{T^2}{r^3}\right)$$

$$+ \left(\frac{1}{r^2}\frac{dR}{dr} + \frac{R}{r^2}\frac{d}{dr} - \frac{2T}{r^3} + \frac{3R}{r^3}\right)$$

$$\times \left(-R\frac{dS}{dr} - 2R\frac{dT}{dr} - \frac{2T^2}{r} + \frac{2RT}{r}\right). \tag{B.37}$$

This result is consistent with Equation (9.6), whose first two terms are just H^{22}. Note that there is no minus sign associated with H^{22} in (B.37), in contrast to the previous terms in H^{00} and H^{11} in (B.17), (B.28) and (B.29).

The tensor $A^{\mu\nu}$ in the field Equation (8.28) can also be calculated by the same method discussed above.

Appendix C

Tensor Properties of Physical Quantities in Taiji Space-Time

The mathematical structure of taiji space-time is neither that of the Minkowski space-time of special relativity nor the Riemannian space-time of general relativity, but something in between the two. Its metric tensors are more general than the Minkowski metric tensors so that it can describe physics and the absolute space-time properties in non-inertial frames, and are simpler than the metric tensors of curved space-time so that all conservation laws (including the conservation of energy-momentum) and quantum gravity and its unifications with all other forces can be formulated. Similar to the concept of the golden mean, taiji space-time appears to be the most suitable framework for describing physics that is supposed to be covered by special relativity and general relativity.

Taiji space-time or the taiji symmetry framework is a mathematical framework with arbitrary coordinates and Poincaré metric tensors $P_{\mu\nu}$ that are consistent with vanishing Riemann–Christoffel curvature tensors. The differential coordinate $dx^\mu = (dw, dx, dy, dz)$ in general frames is, by definition, a contravariant vector. The coordinate x^μ is also a contravariant vector in inertial frames with the Minkowski metric tensor $\eta_{\mu\nu} = (1, -1, -1, -1)$. However, x^μ is not a contravariant vector in, say, non-inertial frames (or in spherical coordinates).

In taiji space-time, the coordinate transformations are obtained on the basis of the principle of limiting continuation of physical laws, which plays the same fundamental role as the principle of relativity in special relativity. To discuss experiments involving non-inertial frames, we often use coordinate transformations between an inertial frame $F_I(x_I^\mu)$ and a non-inertial frame $F(x^\nu)$,

$$x_I^\mu = x_I^\mu(x^\lambda), \quad \text{or} \quad dx_I^\mu = T^\mu{}_\nu dx^\nu, \quad T^\mu{}_\nu = \frac{\partial x_I^\mu}{\partial x^\nu}. \tag{C.1}$$

One can obtain the covariant metric tensor $P_{\mu\nu}$ from (C.1) using

$$ds^2 = \eta_{\mu\nu} dx_I^\mu dx_I^\nu = P_{\mu\nu} dx^\mu dx^\nu. \tag{C.2}$$

Based on the known covariant metric $P_{\mu\nu}$ in (C.2), one can derive the contravariant metric tensor $P^{\mu\nu}$ using the relation $P^{\alpha\lambda}P_{\lambda\mu} = \delta^\alpha_\mu$. The principle of limiting continuation of physical laws requires the accelerated transformations (C.1) and the associated Poincaré metric $P_{\mu\nu}$ to reduce to the Lorentz transformations and the Minkowski metric tensor $\eta_{\mu\nu}$, respectively, in the limit of zero acceleration.

The Poincaré metric tensors $P_{\mu\nu}$ and $P^{\mu\nu}$ can be used to raise or to lower the indices of vectors (e.g., V^μ) and tensors (e.g., $\phi_{\mu\nu}$, $C_{\mu\nu\alpha}$),

$$V_\mu = P_{\mu\nu}V^\nu, \quad V^\nu = P^{\nu\lambda}V_\lambda, \quad \phi^\lambda_\mu = P^{\alpha\lambda}\phi_{\mu\alpha}, \text{ etc.,} \tag{C.3}$$

where V_μ may be the covariant coordinate differential dx_μ, the covariant momentum p_μ, the covariant wave vector k_μ, etc. All these covariant vectors have the same transformation properties, which can be derived from the transformations of the corresponding contravariant vector. For example, any contravariant vector V^ν or k^ν transforms like dx^μ in (C.1),

$$V^\nu_I = T^\nu_\lambda V^\lambda, \quad k^\nu_I = T^\nu_\lambda k^\lambda. \tag{C.4}$$

The transformation of a covariant wave vector k_μ can be obtained by using $V^\nu_I = \eta^{\nu\mu}V_{I\mu}$, (C.3), and (C.4),

$$\eta^{\nu\alpha}k_{I\alpha} = T^\nu_\lambda P^{\lambda\mu}k_\mu, \tag{C.5}$$

where (C.4) is the transformation of a contravariant vector and $\eta^{\nu\alpha}k_{I\alpha} = (k_{I0}, -k_{I1}, -k_{I2}, -k_{I3})$. The transformations (C.5) may be useful if the inverse transformations are difficult to obtain.

The components of a contravariant (covariant) vector or tensor are denoted by a superscript (subscript). A tensor such as ϕ^ν_μ is a mixed tensor. This distinction between contravariant, covariant, and mixed tensors is necessary to maintain, especially in discussing experiments involving non-inertial frames, as we shall see below.

The physical momentum of a particle is a covariant vector p_μ because it is originally introduced in, say, the Lagrangian formulation of mechanics through the relation,

$$p_\mu = \frac{\partial S}{\partial x^\mu} \equiv \partial_\mu S, \quad \left[\frac{\partial S}{\partial x^\mu_I} = \frac{\partial S}{\partial x^\nu}\frac{\partial x^\nu}{\partial x^\mu_I}\right], \tag{C.6}$$

which holds in inertial and non-inertial frames because S is a scalar action for the motion of a particle.

The contravariant momentum vector is defined by $p^\mu = P^{\mu\nu}p_\nu$ and, hence, depends on the Poincaré metric tensors. Other physical functions such as the covariant electromagnetic potential $A_\mu(x)$ or the mixed gravitational potential ϕ_μ^ν were originally introduced through the replacement $\partial_\mu \rightarrow \partial_\mu + ieA_\mu$ in electrodynamics and $\partial_\mu \rightarrow \partial_\mu + g\phi_\mu^\nu\partial_\nu$ in Yang–Mills gravity. These functions A_μ and ϕ_μ^ν are, by definition, independent of the Poincaré metric tensors. Thus, the contravariant electromagnetic potential $A^\mu = P^{\mu\nu}A_\nu$ and the covariant tensor $\phi_{\mu\alpha} = P_{\mu\nu}\phi_\alpha^\nu$ depend on the metric tensors. These properties are useful when one performs variation of a Lagrangian with respect to the metric tensors in the Fock–Hilbert approach to local symmetry and conservation laws. One must know which functions are dependent on the metric tensor.

To illustrate the experimental difference between a covariant and a contravariant tensor, let us consider the rotational transformations of wave vectors between an inertial frame $F_I(x_I^\lambda) \equiv F_I(x_I)$ and a rotating frame $F(x)$, which can be derived from the principle of limiting Lorentz–Poincaré invariance, as discussed in chapter 5. The frequency shift of electromagnetic waves emitted from an orbiting radiation source have two relations,

$$k_{I0} = \gamma^{-1}\left[k_0 + \Omega y k_1 - \Omega x k_2\right], \ldots \tag{C.7}$$

for the covariant wave vector $k_\mu = (k_0, k_1, k_2, k_3)$, and

$$k_I^0 = \gamma^{-1}\left[k^0 + \gamma^2\Omega^2 wxk^1 + \gamma^2\Omega^2 wyk^2\right], \ldots \tag{C.8}$$

for the contravariant wave vector $k^\mu = (k^0, k^1, k^2, k^3)$. These two relations predict different frequency shifts, as measured in an inertial frame. Such an ambiguity can be resolved by a proper understanding of the mathematical nature of physical quantities. Since the physical wave vector must have the same transformation properties as the covariant momentum, $p_\mu = \hbar k_\mu$, one should use the relation (C.7) to predict the observable frequency shift of an orbiting radiation source, and not (C.8).

Author Index

Subject Index